Chemical Principles in the Laboratory

Twelfth Edition

Emil J. Slowinski

Macalester College
St. Paul, Minnesota

Wayne C. Wolsey

Macalester College
St. Paul, Minnesota

Robert C. Rossi

Macalester College
St. Paul, Minnesota

Cengage

Australia • Brazil • Canada • Mexico • Singapore • United Kingdom • United States

Chemical Principles in the Laboratory,
Twelfth Edition
**Emil J. Slowinski, Wayne C. Wolsey,
and Robert C. Rossi**

Product Director: Thais Alencar

Product Manager: Katherine Caudill-Rios

Learning Designer: Peter McGahey

Subject Matter Expert: Theresa Dearborn

Associate Content Manager: Alexander Sham

Product Assistant: Nellie Mitchell

Marketing Manager: Timothy Cali

Senior Digital Delivery Lead: Beth McCracken

IP Analyst: Ann Hoffman

IP Project Manager: Carly Belcher

Text and Image Permissions Researcher: Lumina
Datamatics

Production Service: MPS Limited

Compositor: MPS Limited

Art Director: Lizz Anderson

Manufacturing Planner: Rebecca Cross

Cover Designer: Lizz Anderson

Cover Image: Alfred Pasieka/Science Source

About the cover: The cover image is a micrograph (a very close-up picture taken through a microscope) of a substance called a ferrofluid, by scientific photographer and artist Alfred Pasieka. Ferrofluids are liquids loaded with tiny solid magnetic particles, such that the mixture itself (an example of a colloid) reacts strongly to magnetic fields. When near a strong permanent magnet, ferrofluids take on a distinct, symmetric shape consisting of closely spaced circular cones (spikes). This micrograph is of a single such spike, near the edge of the region influenced by the magnetic field, taken under specialized lighting. Alfred passed away in 2018, but he enjoyed a long and productive career creating stunning science-related images like this one.

For product information and technology assistance, contact us at
Cengage Customer & Sales Support, 1-800-354-9706 or
support.cengage.com.

For permission to use material from this text or product, submit all
requests online at **www.copyright.com**.

Library of Congress Control Number: 2019954762

ISBN: 978-0-357-36453-6

Cengage
200 Pier 4 Boulevard
Boston, MA 02210
USA

Cengage is a leading provider of customized learning solutions
with employees residing in nearly 40 different countries and sales in more
than 125 countries around the world. Find your local representative at:
www.cengage.com.

To learn more about Cengage platforms and services, register or access
your online learning solution, or purchase materials for your course,
visit **www.cengage.com.**

Printed in Mexico
Printed Number: 4   Print Year: 2023

Contents

Preface

A Few Words to the Students

In spite of its many successful theories, chemistry remains an experimental science. As our world has grown increasingly digital, many initiatives have emerged based on the hope that students might learn chemistry by working with software and observing reactions on a computer screen rather than actually carrying them out in a laboratory. While we are proponents of innovation and follow these efforts closely, even dabbling in them ourselves, we remain convinced that chemistry is, at its core, a hands-on laboratory science, in the vast majority of cases best learned by carrying out first-hand experiments with real chemicals. Your instructors must agree, because teaching "wet" laboratory sections is a difficult and expensive undertaking, not the path of least resistance. Please keep in mind that a lot of resources are going into giving you the opportunity to learn chemistry in this way; we urge you to make the most of it!

It is not easy to do good experimental work. It requires experience, thought, and care. As beginning students, you have not had much opportunity to do experiments. We feel that the effort you put into your laboratory sessions can pay off in many ways. You can gain a better understanding of how the chemical world works, manual dexterity in manipulating apparatus, an ability to apply mathematics to chemical systems, and, perhaps most importantly, a way of thinking that allows you to better analyze many problems in and out of science. Who knows, perhaps you will find you enjoy doing chemistry and go on to a career as a chemist, as many of our students have. We should make clear, however, that while professional chemists rely on many of the skills that you will learn from the experiments in this manual, the experiments they conduct differ in several important respects. First, the experiments in this manual have all been carried out many times, and their design vetted and refined, such that they offer reasonably reproducible results—at least as much as is possible with an experimental science. In research chemistry, a great deal of thought and effort (and some measure of luck!) must go into the design of experiments if they are to have any hope of producing a reproducible result; we have done that work for you. All of the reagents that go into an experiment must also be correctly prepared, and that work will largely be done for you by your instructors or their colleagues. Second, the "correct" outcome of these experiments is known: If something else is obtained (because we can be confident in the experimental design and the reagents used), it is almost always the result of an error made in the course of carrying out the experiment. When doing research-level chemistry, we must always consider whether the experimental design or some other unknown factor is responsible for an interesting result, and the experiment must be repeated multiple times to ensure that it is reproducible. So even if you find great success in carrying out the experiments in this manual, don't get too cocky. You will emerge well-prepared to do what professional chemists do . . . but you will not have done it yet!

In writing this manual, we have attempted to illustrate many established principles of chemistry with experiments that are as interesting and challenging as possible. These principles are basic to the science but are usually not intuitively obvious. With each experiment we introduce the theory involved, state in detail the procedures that are used, describe how to draw conclusions from your observations, and, in an Advance Study Assignment (ASA), ask you to answer questions similar to those you will encounter in the experiment. Each ASA is designed to help prepare you for and assist you in the analysis of the experiment it is associated with. Where applicable, we furnish you with sample data and show (in some detail) how that data can be used to obtain the desired results. The Advance Study Assignments generally include the guiding principles as well as the specific relationships to be employed. If you work through the steps in each calculation by yourself, you should understand how to proceed when you are called upon to analyze the data you obtain in the laboratory. *Before coming to lab, you should read over the experiment for that week and do the Advance Study Assignment.* If you prepare for lab as you should, you will get more out of it. To give an experiment a bit of a challenge, we occasionally ask you to work with chemical "unknowns," whose identity is unknown to you (but not to your instructors).

This edition includes an appendix entitled "Statistical Treatment of Laboratory Data." When professional scientists collect experimental data, especially in doing chemical analyses to determine compositions, they usually summarize their results by means of an average (the arithmetic mean) and a measure of how consistent the data are (through a statistical parameter called the standard deviation). In several experiments of this type you will be asked to calculate these quantities from your results. Your instructor may decide to make these calculations an optional part of the experiments, but these values are easily obtained on many standard calculators or via any spreadsheet (such as Excel).

A Few More Words, This Time to the Instructors

While a great deal of this manual remains his handiwork, dating back as far as to his first days teaching labs at Macalester College, Professor Emeritus Emil Slowinski is, unfortunately, no longer with us save in spirit and in our hearts. We miss both his experimental perspicacity and his knack for clear, direct prose aimed at students, but we have done our humble best to emulate both in putting forth this new edition.

If this is the first time that you are using this manual, we have done what we can to make the transition to a new manual as easy as possible. The *Instructor's Manual*, available online, contains a list of required equipment and chemicals, directions for preparing solutions, and suggestions for dealing with the disposal of chemical waste for each experiment. It details the time required to do the experiments and the approximate cost per student. It also offers comments and suggestions for each experiment that may be helpful, sample data and calculations, and answers to the Advance Study Assignment questions (see next paragraph).

Each experiment in this manual includes an "Advance Study Assignment," or "ASA," designed to assist students in preparing for the experiment and particularly (where applicable) in making the calculations required for it. In such cases, the ASAs offer sample data and (in some detail) step the students through how that data can be used to obtain the desired results. If students work through the ASA for an experiment before coming to lab, they will be well prepared to make the necessary calculations on the basis of the data they obtain in the laboratory, and we encourage you to employ the ASAs as pre-lab exercises.

As with any endeavor, there are many people who contribute to the effort. We value input and queries from users of the manual, and these have guided many changes in this edition. We would like to thank in particular Nicole Gill of Ramapo College in Mahwah, New Jersey; Kevin Milani of Hibbing Community College, up north in our own state of Minnesota; Mary Stroud of Xavier University in Cincinnati, Ohio; and Ryan Hayes of Andrews University in Berrien Springs, Michigan. We appreciate the assistance of many members of the Cengage staff, in the form of editorial and technical support in the preparation of this edition, and would like to thank in particular Alex Sham for his diligence and patience in getting this edition past the finish line, as well as his superb editing. The careful, thorough, and thoughtful copyediting work of Patricia Daly, of the Paley Company, improved this edition significantly and is most appreciated. The cooperation of our chemistry colleagues at Macalester College over the years has also been essential.

It has been a great experience being involved with this laboratory manual over its many editions. We appreciate the support of our users. We encourage any comments, questions, or suggestions you may have. Please send them to wolsey@macalester.edu and rossi@rrts.us.

Finally, we acknowledge the patient support of our families in the preparation of this edition.

Wayne C. Wolsey
Robert C. Rossi

New in This Edition

This twelfth edition of *Chemical Principles in the Laboratory* features a new experiment, entitled "Fundamentals of Quantum Mechanics," which provides an accessible and hands-on introduction to the concept and general principles of quantized behavior. It was developed to address the steady increase in quantum chemistry coverage in chemistry courses, as this topic has become less exotic and more mainstream, all the while remaining something that few students can readily wrap their head around, and for which little accessible lab work has been generally available.

As in all previous editions, we have also changed the Advance Study Assignment questions; however, for this edition we have changed our approach to doing so. Instructors are urged to read the Preface of the Instructor's Manual for more details.

We continuously work to improve the clarity and usability of this manual, and this edition certainly reflects that. An effort has been made to more rigorously systematize the numbering of sections, subsections, and questions, such that these are consistent throughout the manual. Significant clarity improvements have been made in the qualitative analysis experiments, and particularly in Experiment 40, the Ten Test Tube Mystery. The structure and content of Appendix II have also been improved as part of that process. Acting on helpful suggestions from users of this manual, as well as in response to questions from them, the procedural instructions for many other experiments have been improved, including: in Experiment 10, mention of safe agitation of the test tube; in 13, a nod to the possible use of other model kits; in 16, mention of the Pythagorean Theorem; in 17, clarification of the procedure for estimating melting points; in 23, optional instructions for student preparation of a calibration curve; in 25, extra help on the calculations around the preparation of a buffer; in 26, some rearranging to improve clarity; in 28, additional detail concerning the blank; and in 36, a tweak aimed at reducing the risk of botching the experiment by adding chromate to the wrong test tube. Also based on user feedback, the text of Experiment 15 has undergone a major overhaul and a figure depicting the "device" has been added, in an effort to make the experimental approach and calculations more readily understood. Similarly, the procedure in Experiment 19 has been extensively revised to streamline and clarify the procedure.

The terminology in Experiment 14 has been adjusted to properly treat "heat" as exclusively an energy transfer, not a property of matter per the debunked caloric theory. The terms "specific heat" and "calorimeter" have also been made more precise therein, changed to "specific heat capacity" and "adiabatic calorimeter."

Throughout this new edition, the use, treatment, and rigor of significant figures have been improved. As a key part of this process, the final appendix has been revised to explain the relevance of significant figures when attempting to follow lab instructions.

Finally, we work to keep this manual up to date, and in that pursuit the RDI of vitamin C has been updated in Experiment 43 and Appendix VII has been updated to accommodate the newest versions of Excel. Appendix VIII has been added, introducing Google Sheets, which is a free alternative to Excel and is seeing ever-wider use on college campuses.

Safety in the Laboratory

Read This Section Before Performing Any of the Experiments in This Manual

A chemistry laboratory can be, and should be, a safe place in which to work. Yet each year in academic and industrial laboratories accidents occur that in some cases injure seriously, or even kill, chemists. Most of these accidents could have been foreseen and prevented had the chemists involved used proper judgment and taken proper precautions.

The experiments you will be performing have been selected at least in part because they can be done safely. Instructions in the procedures should be followed carefully and in the order given. Sometimes even a change in the concentration of one reagent is sufficient to change the conditions of a chemical reaction so as to make it occur in a different way, perhaps at a highly accelerated rate. So, do not deviate from the procedure specified in the manual when performing experiments, unless specifically told to do so by your instructor.

Eye Protection: One of the simplest, and most important, things you can do in the laboratory to avoid injury is to protect your eyes by routinely wearing safety glasses. Your instructor will tell you what eye protection to use, and you should use it. Glasses worn up in your hair may be fashionable, but they will not protect your eyes. If you use contact lenses, it is even more critical that you wear safety glasses as well.

Chemical Reagents: Chemicals in general are toxic materials. This means that they can act as poisons or carcinogens (causes of cancer) if they get into your digestive or respiratory system. Never taste a chemical substance, and avoid getting any chemical on your skin. If contact should occur, wash off the affected area promptly with plenty of water. Also, wash your face and hands when you are through working in the laboratory. Never pipet by mouth; when pipetting, use a rubber bulb or other device to suck up the liquid. Avoid breathing vapors given off by reagents or reactions. If directed to smell a vapor, do so cautiously. Use a fume hood when the directions call for it.

Some reagents, such as concentrated acids or bases, or gases like chlorine and bromine, are corrosive, which means that they can cause chemical burns on your skin and eat through your clothing. Where such reagents are being used, we note the potential danger with a **CAUTION:** box at that point in the procedure. Be particularly careful when carrying out that step. Always read the label on a reagent bottle before using it; there is a great difference between the properties of 1 M H_2SO_4 and those of concentrated (18 M) H_2SO_4.

A few of the chemicals we use are flammable. These include heptane, ethanol, and acetone. Keep ignition sources, like hotplates and Bunsen burners, well away from any open beakers containing such chemicals, and be careful not to spill them on the laboratory bench where they might be easily ignited.

When disposing of the chemical products from an experiment, use good judgment. Some dilute, nontoxic solutions can be poured down the sink, flushing with plenty of water. Insoluble or toxic materials should be put in the waste containers provided for that purpose. Your lab instructor may give you instructions for treatment and disposal of the products from specific experiments.

Safety Equipment: In the laboratory there are various pieces of safety equipment, which may include a safety shower, an emergency eye wash, a fire extinguisher, and a fire blanket. Learn where these items are, so that you will not have to waste precious moments looking for them if you ever need them in a hurry.

Laboratory Attire: Come to the laboratory in sensible clothing. Long, flowing fabrics are out, as are bare feet. Sandals and open-toed shoes offer less protection than regular shoes. Keep long hair tied back, out of the way of flames and reagents. Tight-fitting jewelry (especially rings) can exacerbate the harm caused by chemical exposure and should not be worn in lab.

If an Accident Occurs: During the laboratory course a few accidents will probably occur. For the most part these will not be serious and might involve a spilled reagent, a beaker of hot water that gets tipped over, a dropped test tube, or a small fire.

A common response in such a situation is panic. A student may respond to an otherwise minor accident by doing something irrational, like running from the laboratory, when the remedy for the accident is close at hand. If a mishap befalls a fellow student, watch for signs of panic and tell the student what to do; if it seems necessary, help him or her do it. Call the instructor for assistance.

Most chemical spills are best handled by quickly absorbing wet material with a paper towel and then washing the area with water from the nearest sink. Use the eye wash fountain if you get something in your eye. In case of a severe chemical spill on your clothing or shoes, use the emergency shower and *immediately* take off the affected clothing. In case of a fire in a beaker, on a bench, or on your clothing or that of another student, do not panic and run. Smother the fire with an extinguisher, with a blanket, or with water, as seems most appropriate at the time. If the fire is in a piece of equipment or on the lab bench and does not appear to require instant action, have your instructor put the fire out. If you cut yourself on a piece of broken glass, tell your instructor, who will assist you in treating it.

A Message to English as a Second Language (ESL) Students: Many students for whom English is not their first language take courses in chemistry before they are completely fluent. If you are such a student, it may be that in some experiments you will be given directions that you do not completely understand. If that happens, do not try to do that part of the experiment by simply doing what the student next to you seems to be doing. Ask that student, or the instructor, what the confusing word or phrase means, and when you understand what you should do, go ahead. You will soon learn the language well enough, but until you feel comfortable with it, do not hesitate to ask others to help you with unfamiliar phrases and expressions.

Although we have spent considerable time here describing some of the things you should be concerned with in the laboratory from a safety point of view, this does not mean you should work in the laboratory in fear and trepidation. Chemistry is not a dangerous activity when practiced properly. Chemists as a group live longer than other professionals, in spite of their exposure to potentially dangerous chemicals. In this manual we have attempted to describe safe procedures and to employ chemicals that are safe when used properly. Many thousands of students have performed these experiments without having accidents, and you can too. However, we authors cannot be in the laboratory when you carry out the experiments to be sure that you observe the necessary precautions. You and your laboratory instructor must, therefore, see to it that the experiments are done properly and assume responsibility for any accidents or injuries that may occur.

Disclaimer: Chemistry experiments employing the use of chemicals and laboratory equipment can be dangerous, and misuse may cause serious bodily injury. Cengage Learning encourages you to speak with your instructors and become acquainted with your school's laboratory safety regulations before attempting any experiments. Cengage Learning and the authors have provided, for your convenience only, safety information intended to serve as a starting point for good practices. Cengage Learning and the authors make no guarantee or representations as to the accuracy or sufficiency of such information and/or instructions.

Experiment 1

The Densities of Liquids and Solids

Given a sample of a pure liquid, we can measure many of its characteristics. Its temperature, mass, color, and volume are among the many properties we can determine. We find that if we measure the mass and volume of different samples of the liquid, the mass and volume of each sample are related in a simple way: if we divide the mass by the volume, the result we obtain is the same for each sample, independent of its mass. That is, for samples A, B, and C of the liquid at constant temperature and pressure,

$$\text{Mass}_A / \text{Volume}_A = \text{Mass}_B / \text{Volume}_B = \text{Mass}_C / \text{Volume}_C = \text{a constant}$$

That constant, which is clearly independent of the size of the sample, is called its density and is one of the fundamental properties of the liquid. The density of water is exactly $1.00000 \text{ g}/\text{cm}^3$ at 4°C and is slightly less than that at room temperature ($0.9970 \text{ g}/\text{cm}^3$ at 25°C). Densities of liquids and solids range from values that are less than that of water to values that are much greater. Osmium metal has a density of $22.5 \text{ g}/\text{cm}^3$ and is probably the densest material known at ordinary pressures.

In any density determination, two quantities must be determined—the mass and the volume of a given quantity of matter. The mass can easily be determined by weighing a sample of the substance on a balance. The quantity we usually think of as "weight" is really the mass of a substance. In the process of "weighing" we find the mass, taken from a standard set of masses, that experiences the same gravitational force as that experienced by the given quantity of matter we are weighing. The mass of a sample of liquid in a container can be found by taking the difference between the mass of the container plus the liquid and the mass of the empty container.

The volume of a liquid can easily be determined by means of a calibrated container. In the laboratory a graduated cylinder is often used for routine measurements of volume. Accurate measurement of liquid volume is made using a pycnometer, which is simply a container having a precisely definable volume. The volume of a solid can be determined by direct measurement if the solid has a regular geometric shape. Such is not usually the case, however, with ordinary solid samples. A convenient way to determine the volume of a solid is to accurately measure the volume of liquid displaced when an amount of the solid is immersed in the liquid. The volume of the solid will equal the volume of the liquid that it displaces.

In this experiment we will determine the density of a liquid and a solid by the procedure we have outlined. First we weigh an empty flask and its stopper. We then fill the flask completely with water, measuring the mass of the filled stoppered flask. From the difference in these two masses we find the mass of water, and then, from the known density of water, we determine the volume of the flask. We empty and dry the flask, fill it with an unknown liquid, and weigh again. From the mass of the liquid and the volume of the flask we find the density of the liquid. To determine the density of an unknown solid metal, we add the metal to the dry empty flask and weigh. This allows us to find the mass of the metal. We then fill the flask with water, leaving the metal in the flask, and weigh again. The increase in mass is that of the added water; from that increase, and the density of water, we calculate the volume of water we added. The volume of the metal must equal the volume of the flask minus the volume of water. From the mass and volume of the metal we calculate its density. The calculations involved are outlined in detail in the Advance Study Assignment (ASA).

Experimental Procedure

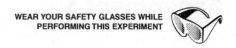
WEAR YOUR SAFETY GLASSES WHILE
PERFORMING THIS EXPERIMENT

A. Mass of a Coin

After you have been shown how to operate the analytical balances in your laboratory, read the section on balances in Appendix IV. Take a coin and measure its mass to ± 0.0001 g. Record the mass on the report page. If your balance has a TARE bar, use it to re-zero the balance. Take another coin and weigh it, recording its mass. Remove both coins, zero the balance, and weigh both coins together, recording the total mass. If you have no TARE bar on your balance, add the second coin and measure and record the mass of the two coins. Then remove both coins and find the mass of the second one by itself. When you are satisfied that your results are those you would expect, obtain a glass-stoppered flask, which will serve as a pycnometer, and samples of an unknown liquid and an unknown metal.

B. Density of a Liquid

If your flask is not clean and dry, clean it with detergent solution and water, rinse it with a few milliliters of acetone, and dry it by letting it stand for a few minutes in the air or by *gently* blowing compressed air into it for a few moments.

Weigh the dry flask with its stopper on an analytical balance, or a top-loading balance if so directed, to the nearest milligram. Fill the flask with deionized water until the liquid level is nearly to the *top* of the ground surface in the neck. Put the stopper in the flask in order to drive out *all* the air and any excess water. Work the stopper gently into the flask, so that it is firmly seated in position. Wipe any water from the outside of the flask with a towel and soak up all excess water from around the top of the stopper.

Again weigh the flask, which should be completely dry on the outside and full of water, to the nearest milligram. Given the density of water at the temperature of the laboratory (see Appendix I) and the mass of water in the flask, you should be able to determine the volume of the flask very precisely. Empty the flask, dry it, and fill it with your unknown liquid. Stopper and dry the flask as you did when working with the water, and then weigh the stoppered flask full of the unknown liquid, making sure its surface is dry. This measurement, used in conjunction with those you made previously, will allow you to accurately determine the density of your unknown liquid.

C. Density of a Solid

Pour your sample of liquid from the flask into its container. Rinse the flask with a small amount of acetone and dry it thoroughly. Add small chunks of the metal sample to the flask until the flask is at least half full. Weigh the flask, with its stopper and the metal, to the nearest milligram. You should have at least 50 g of metal in the flask.

Leaving the metal in the flask, fill the flask with water and then replace the stopper. Roll the metal around in the flask to make sure that no air remains between the metal pieces. Refill the flask if necessary, and then weigh the dry, stoppered flask full of water plus the metal sample. Properly done, the measurements you have made in this experiment will allow a calculation of the density of your metal sample that will be accurate to about 0.1%.

> **DISPOSAL OF REACTION PRODUCTS.**   Pour the water from the flask. Put the metal in its container. Dry the flask and return it with its stopper and your metal sample, along with the sample of unknown liquid.

Name _____ Section _____

Experiment 1

Data and Calculations: The Densities of Liquids and Solids

A. **Mass of a Coin**

Mass of coin 1 _____ g Mass of coin 2 _____ g

Mass of coins 1 and 2 weighed together _____ g
What general law is illustrated by the results of this part of the experiment?

B. **Density of a Liquid**

Mass of empty flask plus stopper   _____ g

Mass of stoppered flask plus water   _____ g

Mass of stoppered flask plus liquid   _____ g

Mass of water

                                                                         _____ g

Temperature in the laboratory   _____ °C

Volume of flask (Obtain the density of water from Appendix I; note that a density in g/mL is identical to a density in
g/cm^3, because $\underline{1}$ mL = $\underline{1}$ cm^3.)

                                                                         _____ cm^3

Mass of liquid

                                                                         _____ g

Density of liquid   _____ g/cm^3

To how many significant figures can the liquid density be
properly reported? (See Appendix V.)   _____

(continued on following page)

C. Density of a Solid

Mass of stoppered flask plus metal   _____ g

Mass of stoppered flask plus metal plus water   _____ g

Mass of metal   _____ g

Mass of water   _____ g

Volume of water

                                                                                               _____ cm^3

Volume of metal   _____ cm^3

Density of metal

                                                                                               _____ g/cm^3

To how many significant figures can the density of the metal be
properly reported?   _____

Explain why the value obtained for the density of the metal is likely to have a larger percentage error
than that found for the liquid.

Unknown liquid # _____   Unknown solid # _____

Experiment 1

Advance Study Assignment: Densities of Solids and Liquids

1. **Finding the volume of a flask**

   A student obtained a clean, dry glass-stoppered flask. She weighed the flask and stopper on an analytical balance and found the total mass to be 34.166 g. She then filled the flask with water and obtained a mass for the full stoppered flask of 68.090 g. From these data, and the fact that at the temperature of the laboratory the density of water was 0.997 g/cm^3, find the volume of the stoppered flask.

   a. First we need to obtain the mass of the water in the flask. This is found by recognizing that the mass of a sample is equal to the sum of the masses of its parts. For the filled and stoppered flask:

   Mass of filled stoppered flask = (mass of empty stoppered flask) + (mass of water)

   So

   mass of water = (mass of filled flask) − (mass of empty flask)

   Mass of water = _____ g − _____ g = _____ g

   Many mass and volume measurements in chemistry are made by the method used in 1(a). This method is called measuring by difference, and it is a very useful one.

   b. The density of a pure substance is equal to its mass divided by its volume:

   $$\text{density} = \frac{\text{mass}}{\text{volume}} \quad \text{or} \quad \text{volume} = \frac{\text{mass}}{\text{density}}$$

   The volume of the flask is equal to the volume of the water it contains. Since we know the mass and density of the water, we can find its volume and that of the flask. Calculate the volume of the flask.

   Volume of flask = volume of water = $\dfrac{\text{mass of water}}{\text{density of water}}$ = $\dfrac{\underline{} \text{ g}}{\underline{} \text{ g/cm}^3}$ = _____ cm^3

2. **Finding the density of an unknown liquid**

   Having obtained the volume of the flask, the student emptied the flask, dried it, and filled it with an unknown liquid whose density she wished to determine. The mass of the stoppered flask when completely filled with liquid was 57.418 g. Find the density of the liquid.

   a. First we need to find the mass of the liquid by measuring the difference:

   Mass of liquid = _____ g − _____ g = _____ g

(continued on following page)

b. Since the volume of the liquid equals that of the flask, we know both the mass and volume of the liquid and can easily find its density using the equation in 1(b). Calculate the density of the unknown liquid.

Density of liquid = _____ g/cm^3

3. Finding the density of a solid

The student then emptied the flask and dried it once again. To the empty flask she added pieces of a metal until the flask was about three-fourths full. She weighed the stoppered flask and its metal contents and found that the mass was 306.150 g. She then filled the flask with water, stoppered it, and obtained a total mass of 309.827 g for the flask, stopper, metal, and water. Find the density of the metal.

a. To find the density of the metal we need to know its mass and volume. We can easily obtain its mass by the method of differences:

Mass of metal = _____ g − _____ g = _____ g

b. To determine the volume of metal, we note that the volume of the flask must equal the volume of the metal plus the volume of water in the filled flask containing both metal and water. If we can find the volume of water, we can obtain the volume of metal by the method of differences. To obtain the volume of the water we first calculate its mass:

Mass of water = mass of (flask + stopper + metal + water) − mass of (flask + stopper + metal)

Mass of water = _____ g − _____ g = _____ g

The volume of water is found from its density, as in 1(b).

$$\text{Volume of water} = \frac{\text{mass of water}}{\text{density of water}} = \frac{\text{_____ g}}{\text{_____ g/cm}^3} = \text{_____ cm}^3$$

c. From the volume of the water, we calculate the volume of metal:

Volume of metal = (volume of flask) − (volume of water)

Volume of metal = _____ cm^3 − _____ cm^3 = _____ cm^3

From the mass and volume of metal, find the density, using the equation in 1(b):

Density of metal = _____ g/cm^3

Now go back to Question 1 and check to see that you have reported the proper number of significant figures in each of the results you calculated in this assignment. Use the rules on significant figures as given in your chemistry text or Appendix V.

Experiment 2

Resolution of Matter into Pure Substances, I. Paper Chromatography

The fact that different substances have different solubilities in a given solvent can be used in several ways to effect a separation of substances from mixtures in which they are present. We will see in an upcoming experiment how fractional crystallization allows us to obtain pure substances by relatively simple procedures based on solubility properties. Another widely used resolution technique, which also depends on solubility differences, is chromatography.

In a chromatographic experiment a mixture is deposited on some solid adsorbing substance, which might consist of a strip of filter paper, a thin layer of silica gel on a piece of glass, some finely divided charcoal packed loosely in a glass tube, or even a resin coating the walls of an incredibly small-diameter tube.

The components of a mixture are adsorbed on a solid to varying degrees, depending on the nature of the component, the nature of the adsorbent, and the temperature. A solvent is then made to flow through the adsorbent solid under applied or gravitational pressure, or by the capillary effect. As the solvent passes the deposited sample, the various components tend, to varying extents, to be dissolved and swept along the solid. The rate at which a component will move along the solid depends on its relative tendency to be dissolved in the solvent as opposed to adsorbed on the solid. The net effect is that, as the solvent passes slowly through the solid, the components separate from each other and move along as rather diffuse zones. With the proper choice of solvent and adsorbent, it is possible to resolve many complex mixtures by this procedure. If necessary, we can usually recover a given component by identifying the position of the zone containing the component, removing that part of the solid from the system, and eluting the desired component with a suitable (good) solvent.

The name given to a particular kind of chromatography depends upon the manner in which the experiment is conducted. Thus, we have column, thin-layer, paper, and gas chromatography, all in very common use (Fig. 2.1). Chromatography in its many possible variations offers the chemist one of the best methods, if not the best method, for resolving a mixture into pure substances, regardless of whether that mixture

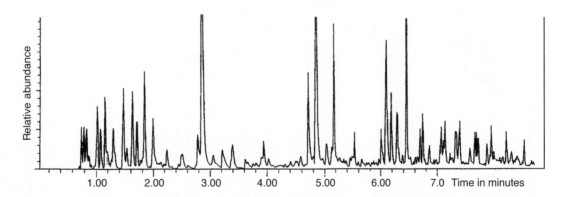

Figure 2.1 This is a gas chromatogram of a sample of unleaded gasoline. Each peak corresponds to a different molecule, so gasoline has many components: at least fifty, each of which can be identified. The molar masses vary from about fifty to about one-hundred fifty, with the largest peak, at about three minutes, being due to toluene, $C_6H_5CH_3$. The sample size for this chromatogram was less than 10^{-6} g (less than 1 μg)! Gas chromatography offers the best method for resolution of complex volatile mixtures.

Chromatogram courtesy of Prof. Becky Hoye at Macalester College.

consists of a gas, a volatile liquid, or a group of nonvolatile, relatively unstable, complex organic compounds.

In this experiment we will use paper chromatography to separate a mixture of metallic ions in solution. A sample containing a few micrograms of ions is applied as a spot near one edge of a piece of filter paper. That edge is immersed in a solvent, with the paper held vertically. As the solvent rises up the paper by capillary action, it will carry the metallic ions along with it to a degree that depends upon the relative tendency of each ion to dissolve in the solvent rather than adsorb on the paper. Because the ions differ in their properties, they move at different rates and become separated on the paper. The position of each ion during the experiment can be recognized if the ion is colored, as some of them are. At the end of the experiment their positions are established more clearly by treating the paper with a staining reagent that reacts with each ion to produce a colored product. By observing the position and color of the spot produced by each ion, and the positions of the spots produced by an unknown mixture containing some of those ions, you can readily determine the ions present in the unknown.

It is possible to describe the position of spots such as those you will be observing in terms of a quantity called the R_f ("retention factor") value. In the experiment the solvent rises a certain distance, say L centimeters. At the same time, a given component will usually rise a smaller distance, say D centimeters. The ratio of D/L is called the R_f value (or "the retention factor") for that component:

$$R_f = \frac{D}{L} = \frac{\text{distance component moves}}{\text{distance solvent moves}} \tag{1}$$

The R_f value is a characteristic property of a given component in a chromatographic experiment conducted under particular conditions. Within some limits, it does not depend upon concentration or upon the other components present. Hence, it can be reported in the literature and used by other researchers doing similar analyses. In the experiment you will be doing, you will be asked to calculate the R_f values for each of the cations studied.

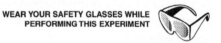

WEAR YOUR SAFETY GLASSES WHILE PERFORMING THIS EXPERIMENT

Experimental Procedure

Obtain an unknown mixture and a piece of filter paper 19 cm long and 11 cm wide. Along the 19-cm edge, draw a pencil line about one centimeter from that edge. Starting 1.5 cm from the end of the line, mark the line at 2-cm intervals. Label the segments of the line as shown in Figure 2.2, with the formulas of the ions to be studied and the known and unknown mixtures. (This must all be done in pencil, not pen.)

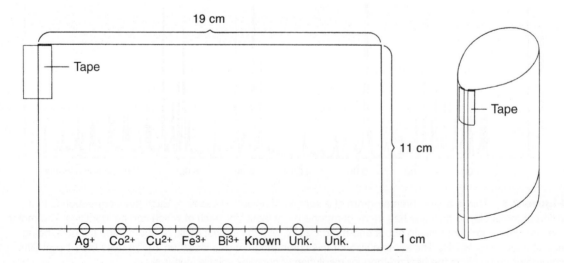

Figure 2.2   This figure shows how the filter paper should be set up for this experiment.

Put two or three drops of 0.10 M solutions of the following compounds in separate micro test tubes, one solution to a tube:

$$AgNO_3 \quad Co(NO_3)_2 \quad Cu(NO_3)_2 \quad Fe(NO_3)_3 \quad Bi(NO_3)_3$$

In solution these substances exist as ions. The metallic cations are Ag^+, Co^{2+}, Cu^{2+}, Fe^{3+}, and Bi^{3+}, respectively, and each drop of solution contains about fifty micrograms (micrograms is abbreviated as μg, and $\underline{1}$ μg = 10^{-6} g) of cation. Into a sixth micro test tube put 2 drops of each of the five solutions; swirl until the solutions are well mixed. This mixture will be our known, since we know it contains all of the cations.

Your instructor will furnish you with a fine capillary tube, which will serve as an applicator. Test the application procedure by dipping the applicator into one of the colored solutions and touching it momentarily to a round piece of filter paper. The liquid from the applicator should form a spot no larger than 8 mm in diameter. Practice making spots until you can reproduce the spot size each time.

Clean the applicator by dipping the tip about one centimeter into a beaker of deionized water and then touching the round filter paper to remove the liquid. In this case, continue to hold the applicator against the paper until all the liquid in the tube is gone. Repeat the cleaning procedure one more time. Dip the applicator into one of the cation solutions and put a spot on the line on the rectangular filter paper in the region labeled for that cation. Clean the applicator twice, and repeat the procedure with another solution. Continue this approach until you have put a spot for each of the five cations and the known and unknown mixtures on the paper, cleaning the applicator between solutions. Dry the paper by moving it in the air or holding it briefly in front of a hair dryer or heat lamp (low setting). Apply the known and unknown mixtures three more times to the same spots; the known and unknown mixtures are less concentrated than the cation solutions, so this procedure will increase the amount of each ion present in the spots. Be sure to dry the spots between applications, since otherwise they will get larger. Don't heat the paper more than necessary, just enough to dry the spots.

Draw 15 mL of eluting solution from the supply on the reagent shelf. This solution is made by mixing a solution of hydrochloric acid (HCl) in water with ethanol and butanol, which are organic solvents. Pour the eluting solution into a 600-mL beaker and cover with a watch glass.

Check to make sure that the spots on the filter paper are all dry. Place a 4- to 5-cm length of tape along the upper end of the left edge of the paper, as shown in Figure 2.2, so that about half of the tape is on the paper. Form the paper into a cylinder by attaching the tape to the other edge in such a way that the edges are parallel but do not overlap. When you are finished, the pencil line at the bottom of the cylinder should form a circle (approximately, anyway) and the two edges of the paper should not quite touch. Stand the cylinder up on the lab bench to check that such is the case and readjust the tape if necessary. *Do not* tape the lower edges of the paper together.

Place the cylinder in the eluting solution in the 600-mL beaker, with the sample spots down near the liquid surface. The paper should not touch the walls of the beaker. Cover the beaker with the watch glass. The solvent will gradually rise by capillary action up the filter paper, carrying along the cations at different rates. After the process has gone on for a few minutes, you should be able to see colored spots on the paper, showing the positions of some of the cations.

While the experiment is proceeding, you can test the effect of the staining reagent on the different cations. Put an 8-mm spot of each of the cation solutions on a clean piece of round filter paper, labeling each spot in pencil and cleaning the applicator between solutions. Dry the spots as before. Some of them will have a little color; record those colors on the report page. Put the filter paper on a paper towel, and, using the spray bottle on the lab bench, spray the paper evenly with the staining reagent, getting the paper moist but not really wet. The staining reagent is a solution containing potassium ferrocyanide and potassium iodide. This reagent forms colored precipitates or reaction products with many cations, including all of those used in this experiment. Note the colors obtained with each of the cations. Considering that each spot contains less than 50 μg of cation, the tests are surprisingly definitive.

When the eluting solution has risen as high as time permits, but not higher than within about two centimeters of the top of the filter paper (which typically requires between seventy-five minutes and two hours), remove the filter paper cylinder from the beaker and take off the tape. Draw a pencil line along the solvent front. Dry the paper with gentle heat until it is dry. Note any cations that must be in your unknown by virtue of your being able to see their colors. Then, with the filter paper set out flat on a paper towel, spray it as before with the staining reagent. Any cations you identified in your unknown before staining should be observed, as well as any that require staining for detection.

Measure the distance from the straight line on which you applied the spots to the solvent front, which is distance L in Equation 1. Then measure the distance from the pencil line to the center of the spot made by each of the cations, when pure and in the known; this is distance D. Calculate the R_f value for each cation. Then calculate R_f values for the cations in the unknown. How do the R_f values compare?

DISPOSAL OF REACTION PRODUCTS.   When you are finished with the experiment, pour the eluting solution into the waste container, not down the sink. Wash your hands before leaving the laboratory.

Experiment 2

Data and Calculations: Resolution of Matter into Pure Substances,
I. Paper Chromatography

	Ag^+	Co^{2+}	Cu^{2+}	Fe^{3+}	Bi^{3+}
Colors (if observed)					
Dry	_____	_____	_____	_____	_____
After staining	_____	_____	_____	_____	_____
Pure Cations					
Distance solvent moved (*L*)	_____	_____	_____	_____	_____
Distance cation moved (*D*)	_____	_____	_____	_____	_____
R_f	_____	_____	_____	_____	_____
Known Mixture					
Distance solvent moved	_____	_____	_____	_____	_____
Distance cation moved	_____	_____	_____	_____	_____
R_f	_____	_____	_____	_____	_____

(continued on following page)

Unknown Mixture

Cations identified

   Dry   _____ _____ _____ _____ _____

   After staining   _____ _____ _____ _____ _____

Distance solvent moved   _____ _____ _____ _____ _____

Distance cation moved   _____ _____ _____ _____ _____

R_f _____ _____ _____ _____ _____

Composition of unknown   _____ _____ _____ _____ _____

Unknown # _____

Experiment 2

Advance Study Assignment: Resolution of Matter into Pure Substances,
I. Paper Chromatography

1. A student chromatographs a mixture, and after developing the spots with a suitable reagent he observes the following:

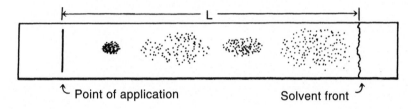

L

↳ Point of application   Solvent front ↗

What are the R_f values?

2. Explain, in your own words, why samples can often be separated into their components by chromatography.

(continued on following page)

3. The solvent moves 3 cm in about ten minutes. Why shouldn't the experiment be stopped at that time, instead of waiting much longer for the solvent to move 10. cm?

4. In this experiment it takes about ten microliters (microliters is abbreviated as μL, and $\underline{1\ \mu L}$ = $\underline{10^{-6}\ L}$) of solution to produce a spot 1 cm in diameter. If the $AgNO_3$ solution contains 10.8 g Ag^+ per liter, how many micrograms of Ag^+ ion are there in one spot?

_____ micrograms

Experiment 3

Resolution of Matter into Pure Substances, II. Fractional Crystallization

One of the important problems faced by chemists is that of determining the nature and state of purity of the substances with which they work. In order to perform meaningful experiments, chemists must ordinarily use essentially pure substances, which are often prepared by separation from complex mixtures.

In principle the separation of a mixture into its component substances can be accomplished by carrying the mixture through one or more physical changes, experimental operations in which the nature of the components remains unchanged. Because the physical properties of various pure substances are different, physical changes frequently allow an enrichment of one or more substances in one of the fractions that is obtained during the change. Many physical changes can be used to accomplish the resolution of a mixture, but in this experiment we will restrict our attention to one of the simpler ones in common use, namely, fractional crystallization.

The solubilities of solid substances in different kinds of liquid solvents vary widely. Some substances are essentially insoluble in all known solvents; the materials we classify as macromolecular are typical examples. Most materials are noticeably soluble in one or more solvents. Those substances that we call salts often have very appreciable solubility in water but relatively little solubility in other liquids. Organic compounds, whose molecules contain carbon and hydrogen atoms as their main constituents, are often soluble in organic liquids such as benzene or carbon tetrachloride.

We also often find that the solubility of a given substance in a liquid is sharply dependent on temperature. Most substances are more soluble in a given solvent at high temperatures than at low temperatures, although there are some materials whose solubility is practically temperature-independent and a few others that become less soluble as temperature increases.

By taking advantage of the differences in solubility of different substances, we often find it possible to isolate the components of a mixture in essentially pure form.

In this experiment you will be given a sample containing silicon carbide, potassium nitrate, and copper sulfate. Your goal will be to separate two pure components from the mixture, using water as the solvent. Silicon carbide, SiC, is a black, very hard material; it is the classic abrasive, and completely insoluble in water. Potassium nitrate, KNO_3, and copper sulfate, $CuSO_4 \cdot 5H_2O$, are water-soluble ionic substances, with different solubilities at different temperatures, as indicated in Figure 3.1. The copper sulfate we will use is blue as a crystalline hydrate, as well as in solution. Its solubility increases fairly rapidly with temperature. Potassium nitrate is a white solid, clear and colorless in solution. Its solubility in water increases about twenty-fold between 0°C and 100°C.

Given a mixture containing substantial amounts of both SiC and KNO_3 and a smaller amount of $CuSO_4 \cdot 5H_2O$, we separate out the silicon carbide first. This is done by simply stirring the mixture with water, which dissolves all of the potassium nitrate and copper sulfate in the mixture. The insoluble silicon carbide remains behind and is filtered off.

The solution obtained after filtration contains KNO_3 and $CuSO_4$ in a rather large amount of water. Some of the water is removed by boiling, and then the solution is cooled to 0°C. At that temperature the KNO_3 is not very soluble, and most of it crystallizes from solution. Since $CuSO_4$ is not present in large amount, its solubility is not exceeded and it remains in solution. The solid KNO_3 is separated from the solution by filtration. This procedure, by which a substance can be separated from an impurity, is called fractional crystallization.

The solid potassium nitrate one recovers is contaminated by a small amount of copper sulfate. The purity of the solid can be markedly increased by stirring it with a small amount of water and then filtering off the dissolved $CuSO_4$. The purity can be established by the intensity of the color produced by the copper impurity when treated with ammonia, NH_3.

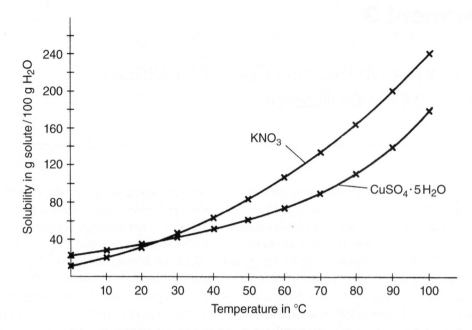

Figure 3.1   The solubility of potassium nitrate (KNO_3) and copper(II) sulfate pentahydrate ($CuSO_4 \cdot 5H_2O$) in water both vary appreciably with temperature. While both solubilities increase with increasing temperature, the solubility of potassium nitrate is lower than that of copper(II) sulfate pentahydrate at low temperatures, but it increases more quickly with temperature, such that in hot water the solubility of KNO_3 is appreciably higher than that of the copper salt.

WEAR YOUR SAFETY GLASSES WHILE PERFORMING THIS EXPERIMENT

Experimental Procedure

Obtain a Büchner funnel, a suction flask, and a sample (about twenty grams) of your unknown solid mixture.

Weigh a 150-mL beaker, a 50-mL beaker, and a piece of filter paper, one item at a time, on a top-loading balance. These, and all other weighings in this experiment, should be made to the nearest tenth of a gram, that is, to ± 0.1 g. (Do not use an analytical balance for any weighings.) Use the masses you obtain as needed in your calculations.

Add your sample to the 150-mL beaker and weigh again. Then add about forty milliliters of deionized water, which will be enough to dissolve the soluble solids. Light your Bunsen burner and adjust the flame so it is blue, quiet, and of moderate size.

A. Separation of SiC

Support the beaker with its contents on a piece of wire gauze on an iron ring. Warm it gently to 50°C (between 40°C and 60°C), while stirring the mixture. When the blue and white solids are all dissolved, pour the contents of the beaker into a Büchner funnel while gentle suction is being applied (see Fig. 3.2 and Appendix IV). Transfer as much of the black solid carbide as you can to the funnel, using your rubber policeman. Transfer the blue filtrate to the (cleaned) 150-mL beaker and add 15 drops of 6 M nitric acid, HNO_3, which will help ensure that the copper sulfate remains in solution in later steps. Reassemble the funnel, apply suction, and wash the SiC on the filter paper with deionized water. Continue the suction for a few minutes to dry the SiC. Turn off the suction, and, using your spatula, lift the filter paper and the SiC crystals from the funnel, and put the paper on the lab bench so that the crystals may dry in the air. When you are finished with the rest of the experiment, weigh the dry SiC on its piece of filter paper.

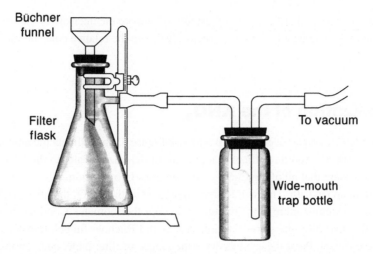

Figure 3.2   To operate the Büchner funnel, put a piece of circular filter paper in the funnel. Thoroughly moisten the filter paper with deionized water from a wash bottle, then immediately turn on the suction. Keep the suction on while filtering the sample.

B. Separation of KNO₃

Heat the blue filtrate in the beaker to the boiling point, and then boil gently until the white crystals of KNO_3 are visible in the liquid. **CAUTION:** **The hot liquid may have a tendency to bump, so do not heat it too strongly.** While the filtrate is being heated, prepare some ice-cold deionized water by putting your wash bottle in an ice-water bath.

When white crystals are clearly apparent in the boiling liquid (the solution may appear cloudy at that point), stop heating and add 12 mL of deionized water to the solution. Stir the mixture with a glass stirring rod to dissolve the solids, including any on the walls; if necessary, warm the solution, but do not boil it.

Cool the solution to room temperature in a water bath, and then to nearly 0°C in an ice bath. White crystals of KNO_3 will come out of solution. Stir the cold slurry of crystals for several minutes. Check the temperature of the slurry with your thermometer. It should be no more than 3°C. Continue stirring until your mixture gets to that temperature or even a bit lower.

Assemble the Büchner funnel. Chill it by adding about one-hundred milliliters of ice-cold deionized water from your wash bottle, and, after about a minute, by drawing the water through with suction. Filter the KNO_3 slurry through the cold Büchner funnel. Your rubber policeman will be helpful when you transfer the last of the crystals. Press the crystals dry with a clean piece of filter paper, and continue to apply suction for another 30 seconds. Turn off the suction. Lift the filter paper and the crystals from the funnel, and put the paper and crystals on the lab bench.

DISPOSAL OF REACTION PRODUCTS.   By this procedure you have separated most of the KNO_3 in your sample from the $CuSO_4$, which is present in the solution in the suction flask. This solution may now be discarded, so dispose of it as directed by your instructor.

Clean and dry your 150-mL beaker and to it add the KNO_3 crystals from the filter paper. Weigh the beaker and its contents.

C. Analysis of the Purity of the KNO₃

The KNO_3 crystals you have prepared contain a small amount of $CuSO_4$ as an impurity. To find the amount of $CuSO_4$ present, weigh out 0.5 g of the crystals into your weighed 50-mL beaker. Dissolve the crystals in 3 mL of deionized water, and then add 3 mL of 6 M ammonia, NH_3. The copper impurity will form a blue solution in the

NH_3. Pour the solution into a small (13×100 mm) test tube. Compare the intensity of the blue color with that in a series of standard solutions prepared by your instructor. Estimate the relative concentration of $CuSO_4 \cdot 5 H_2O$ in your product.

D. Recrystallization of the KNO₃

Given the mass of the KNO_3 you recovered (less 0.5 g), use Figure 3.1 to estimate the amount of water needed to dissolve the solid at 100°C. Add three times that amount of deionized water to the KNO_3. Stir the crystals for a minute or two to ensure that all of the $CuSO_4$ impurity goes into solution.

Cool the mixture in an ice bath. This will recrystallize most of the KNO_3 that dissolved. After stirring for several minutes, check the temperature of the mixture with your thermometer. Continue cooling until the temperature is below 3°C. Then filter the slurry through an ice-cold Büchner funnel, transferring as much of the solid as possible to the funnel. Press the crystals dry with a piece of filter paper and continue to apply suction for a minute or so. Lift the filter paper from the funnel and put it with its batch of crystals on the lab bench.

Transfer the purified crystals to a piece of dry filter paper. Weigh the paper and the crystals.

Determine the amount of $CuSO_4$ impurity in your recrystallized sample as you did with the first batch, and record that value. The recrystallization should have very significantly increased the purity of your KNO_3.

Weigh your dry SiC on its piece of filter paper. Show your samples of SiC and KNO_3 to your instructor for evaluation. Then return the samples.

Name _____ Section _____

Experiment 3

Data and Calculations: Resolution of Matter into Pure Substances,
II. Fractional Crystallization

Mass of 150-mL beaker _____ g   Mass of 50-mL beaker _____ g

Mass of filter paper _____ g

A. Separation of SiC

Mass of sample plus 150-mL beaker   _____ g

Mass of sample   _____ g

Mass of SiC plus filter paper   _____ g

Mass of SiC   _____ g

Percentage of SiC in sample, by mass   _____ %

B. Separation of KNO$_3$

Mass of 150-mL beaker plus KNO$_3$ _____ g

Mass of KNO$_3$ recovered   _____ g

Percentage (by mass) of sample recovered as KNO$_3$ _____ %

(continued on following page)

C. Analysis of Purity of KNO_3

Mass of 50-mL beaker plus KNO_3 _____ g

Mass of KNO_3 used in analysis _____ g

Percentage (by mass) of $CuSO_4 \cdot 5\,H_2O$ present in KNO_3 _____ %

D. Recrystallization of KNO_3

Mass of filter paper plus purified KNO_3 _____ g

Mass of purified KNO_3 obtained _____ g

Percentage (by mass) of sample recovered as pure KNO_3 _____ %

Mass of 50-mL beaker plus purified KNO_3 _____ g

Mass of purified KNO_3 used in analysis _____ g

Percentage (by mass) of $CuSO_4 \cdot 5\,H_2O$ in purified KNO_3 _____ %

Experiment 3

Advance Study Assignment: Resolution of Matter into Pure Substances,
II. Fractional Crystallization

1. Using Figure 3.1, determine

   a.   the number of grams of $CuSO_4 \cdot 5H_2O$ that will dissolve in 100 g of H_2O at 100°C.

   _____ g $CuSO_4 \cdot 5H_2O$

   b.   the number of grams of water required to dissolve 3.9 g of $CuSO_4 \cdot 5H_2O$ at 100°C. [Hint: Your answer to Problem 1(a) gives you the needed conversion factor for (g $CuSO_4 \cdot 5H_2O$) to (g H_2O).]

   _____ g H_2O

   c.   the number of grams of water required to dissolve 37 g KNO_3 at 100°C.

   _____ g H_2O

   d.   the number of grams of water required to completely dissolve a mixture containing 37 g of KNO_3 and 3.9 g of $CuSO_4 \cdot 5H_2O$, assuming that the solubility of one substance is not affected by the presence of another. (This is generally a surprisingly good assumption when the substances have no ions in common!)

   _____ g H_2O

(continued on following page)

2. To the solution in Problem 1(d) at 100°C, an additional 12.0 g of water are added, and the solution is cooled to 0°C.

    a. How much KNO_3 remains in solution?

                                                _____ g KNO_3

    b. How much KNO_3 crystallizes out?

                                                _____ g KNO_3

    c. How much $CuSO_4 \cdot 5\,H_2O$ crystallizes out?

                                                _____ g $CuSO_4 \cdot 5\,H_2O$

    d. What percent of the KNO_3 in the sample is recovered?

                                                _____ %

Experiment 4

Determination of a Chemical Formula

When atoms of one element combine with those of another, the combining ratio is typically an integer or a simple fraction; 1:2, 1:1, 2:1, and 2:3 are ratios one might encounter. The simplest formula of a compound expresses the atom ratio. Some substances with the ratios we list above include $CaCl_2$, KBr, Ag_2O, and Fe_2O_3. When more than two elements are present in a compound, the formula still indicates the atom ratio. Thus the substance with the formula Na_2SO_4 indicates that the sodium, sulfur, and oxygen atoms occur in that compound in the ratio 2:1:4. Many compounds have more complex formulas than those we have noted, but the same principles apply.

To find the formula of a compound we need to find the mass of each of the elements in a weighed sample of that compound. For example, if we resolved a sample of the compound NaOH weighing 40 g into its elements, we would find that we obtained 23 g of sodium, 16 g of oxygen, and 1 g of hydrogen. Since the atomic mass scale tells us that sodium atoms have a relative mass of 23, oxygen atoms have a relative mass of 16, and hydrogen atoms have a relative mass of 1, we would conclude that the sample of NaOH contained equal numbers of Na, O, and H atoms. Since that is the case, the atom ratio Na:O:H is 1:1:1, and so the simplest formula is NaOH. In terms of moles, we can say that that one mole of NaOH, 40 g, contains one mole of Na, 23 g; one mole of O, 16 g; and one mole of H, 1 g; where we define the molar mass to be that mass in grams equal numerically to the sum of the atomic masses in an element or a compound. From this kind of argument we can conclude that the atom ratio in a compound is equal to the mole ratio. We get the mole ratio from chemical analysis, and from that the formula of the compound.

In this experiment we will use these principles to find the formula of the compound with the general formula $Cu_xCl_y \cdot z\,H_2O$, where x, y, and z are integers which, when known, establish the formula of the compound. (In expressing the formula of a compound like this one, where water molecules remain intact within the compound, we retain the formula of H_2O in the formula of the compound.)

The compound we will study, which is called copper chloride hydrate (or hydrated copper chloride), turns out to be ideal for one's first venture into formula determination. It is stable, can be obtained in pure form, has a characteristic blue-green color which changes as the compound is changed chemically, and is relatively easy to decompose into the elements and water. In this experiment you will first drive out the water, which is called the water of hydration, from an accurately weighed sample of the compound. This occurs when you gently heat the sample to a little over 100°C. As the water is driven out, the color of the sample changes from blue-green to a tan-brown color similar to that of dry dirt. The compound formed is anhydrous (not hydrated: that is, free of water) copper chloride. If we subtract its mass from that of the hydrate, we can determine the mass of the water that was driven off, and, using the molar mass of water, find the number of moles of H_2O that were in the sample.

In the next step we need to find either the mass of copper or the mass of chlorine in the anhydrous sample we have prepared. It turns out to be much easier to determine the mass of the copper, and to then calculate the mass of chlorine by difference. We do this by dissolving the anhydrous sample in water, which gives us a green solution containing copper and chloride ions. To that solution we add some aluminum metal wire. Aluminum is what we call an active metal; in contact with a solution containing copper ions, the aluminum metal will react chemically with those ions, converting them to copper metal. The aluminum is said to reduce the copper ions to the metal, and is itself oxidized to aluminum ions in the process. The copper metal appears on the wire as the reaction proceeds, and has its typical red-orange color. When the reaction is complete, we remove the excess Al, separate the copper from the solution, and weigh the dried metal. From its mass we can calculate the number of moles of copper in the sample. We find the mass of chlorine by subtracting the mass of copper from that of the anhydrous copper chloride, and from that value determine the number of moles of chlorine. The mole ratio of $Cu:Cl:H_2O$ gives us the formula of the compound.

Experimental Procedure

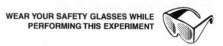

Weigh a clean, dry crucible, without a cover, accurately, on an analytical balance. Place about one gram of the unknown hydrated copper chloride in the crucible. With your spatula, break up any sizeable crystal particles by pressing them against the wall of the crucible. Then weigh the crucible and its contents accurately. Enter your results on the report page.

Place the uncovered crucible on a clay triangle supported by an iron ring. Light your Bunsen burner away from the crucible and adjust the burner so that you have a small flame. Holding the burner in your hand, gently heat the crucible as you move the burner back and forth. Do not overheat the sample. As the sample warms, you will see that the green crystals begin to change to brown around the edges. Continue gentle heating, slowly converting all of the hydrated crystals to the anhydrous brown form. After all of the crystals appear to be brown, continue heating gently, moving the burner back and forth judiciously, for an additional two minutes. Remove the burner, cover the crucible to minimize rehydration, and let it cool for about fifteen minutes. Remove the cover and slowly roll the brown crystals around the crucible. If some green crystals remain, repeat the heating process. Finally, weigh the cool, uncovered crucible and its contents accurately.

Transfer the brown crystals in the crucible to an empty 50-mL beaker. Rinse out the crucible with two 6-mL (taking significant figures into account, that's 6 ± 1 mL) portions of deionized water, and add the rinsings to the beaker. Swirl the beaker gently to dissolve the brown solid. The color will change to green as the copper ions are rehydrated. Measure out about twenty centimeters of 20-gauge aluminum wire (≈ 0.25 g) and form the wire into a loose spiral coil. Put the coil into the solution so that it is completely immersed. Within a few moments you will observe some evolution of H_2, hydrogen gas, and the formation of copper metal on the Al wire. As the copper ions are reduced, the color of the solution will fade. The Al metal wire will be slowly oxidized and enter the solution as aluminum ions. (The hydrogen gas is formed as the aluminum reduces water in the slightly acidic copper solution.)

When the reaction is complete, the solution will be colorless, and most of the copper metal that was produced will be on the Al wire. (This will take roughly thirty minutes, but watch the color of the solution, not the time! Swirling the beaker or stirring its contents periodically will help the reaction reach completion, but be gentle enough that you don't break the wire.) Add 5 drops of 6 M HCl to dissolve any insoluble aluminum salts and clear up the solution. Use your glass stirring rod to remove the copper from the wire as completely as you can. Slide the unreacted aluminum wire up the wall of the beaker with your stirring rod, and, while the wire is hanging from the rod, rinse off any remaining Cu particles with water from your wash bottle. If necessary, complete the removal of the Cu with a drop or two of 6 M HCl added directly to the wire. Put the wire aside; it has done its duty.

In the beaker you now have the metallic copper produced in the reaction, in a solution containing an aluminum salt. In the next step we will use a Büchner funnel to separate the copper from the solution. Weigh accurately a dry piece of filter paper that will fit in the Büchner funnel, and record its mass. Put the paper in the funnel, moisten it with a little water to ensure a good seal, and then apply light suction. Before the filter paper can dry out, decant the solution into the funnel. Wash the copper metal thoroughly with deionized water, breaking up any copper particles with your stirring rod. Transfer the rinsing and residual copper to the filter funnel. Wash any remaining copper into the funnel with water from your wash bottle. *All* of the copper must be transferred to the funnel. Rinse the copper on the paper once again with water. Turn off the suction. Add 10 mL of 95% ethanol to the funnel, and after a minute or so turn on the suction. Draw air through the funnel for about five minutes. With your spatula, lift the edge of the paper, and carefully lift the paper and the copper from the funnel. Dry the paper and copper under a heat lamp for 5 minutes. Allow it to cool to room temperature and then weigh it accurately.

DISPOSAL OF REACTION PRODUCTS. Dispose of the liquid waste and copper produced in this experiment as directed by your instructor.

Experiment 4

Data and Calculations: Determination of a Chemical Formula

Atomic masses: Copper (Cu) _____ g/mol   Chlorine (Cl) _____ g/mol

Hydrogen (H) _____ g/mol   Oxygen (O) _____ g/mol

Mass of crucible _____ g

Mass of crucible and hydrated sample _____ g

Mass of hydrated sample _____ g

Mass of crucible and dehydrated sample _____ g

Mass of dehydrated sample _____ g

Mass of filter paper _____ g

Mass of filter paper and copper _____ g

Mass of copper _____ g

Moles of copper _____ moles

Mass of water evolved _____ g

(continued on following page)

Moles of water   _____ moles

Mass of chlorine in sample (by difference)   _____ g

Moles of chlorine   _____ moles

Mole ratio, chlorine:copper in sample   _____ :1

Mole ratio, water:copper in hydrated sample   _____ :1

Formula of dehydrated sample (round to nearest integer)   _____

Formula of hydrated sample   _____

Experiment 4

Advance Study Assignment: Determination of a Chemical Formula

1. To find the mass of a mole of an element, we look up the atomic mass of the element in a table of atomic masses (see Appendix III or the periodic table). The molar mass of an element is simply the mass in grams of that element that is numerically equal to its atomic mass. For a compound substance, the molar mass is equal to the mass in grams that is numerically equal to the sum of the atomic masses in the formula of the substance. Find the molar mass of

   Cu _____ g/mol   Cl _____ g/mol   H _____ g/mol

   O _____ g/mol H_2O _____ g/mol   CuO _____ g/mol

2. If we can find the ratio of the number of moles of the elements in a compound to one another, we can find the formula of the compound. In a certain compound of copper and oxygen, Cu_xO_y, we find that a sample weighing 0.9573 g contains 0.8503 g Cu.

   a. How many moles of Cu are there in the sample?

   $$\left(\text{Moles of Cu} = \frac{\text{mass of Cu}}{\text{molar mass of Cu}}\right)$$

   _____ moles

   b. How many grams of O are there in the sample? (The mass of the sample equals the mass of Cu plus the mass of O.)

   _____ g

   c. How many moles of O are there in the sample?

   _____ moles

(continued on following page)

d. What is the mole ratio (moles of Cu/moles of O) in the sample?

_____ :1

e. What is the formula of the oxide? (The atom ratio equals the mole ratio and is expressed using the smallest integers possible.)

_____

f. What is the molar mass of the copper oxide?

_____ g/mol

Experiment 5

Identification of a Compound by Mass Relationships

In the previous experiment we showed how we can find the formula of a compound by analysis to determine the elements it contains. When chemical reactions occur, there is a relationship between the masses of the reactants and products that follows directly from the balanced equation for the reaction and the molar masses of the species that are involved. In this experiment we will use this relationship to identify an unknown substance.

Your unknown will be one of the following compounds, all of which are salts:

$$NaHCO_3 \quad Na_2CO_3 \quad KHCO_3 \quad K_2CO_3$$

In the first part of the experiment you will be heating a weighed sample of your compound in a crucible. If your sample is a carbonate, no chemical reaction will occur, but any small amount of adsorbed water will be driven off. If your sample is a hydrogen carbonate, it will decompose by the following reaction, using $NaHCO_3$ as the example:

$$2\,NaHCO_3(s) \rightarrow Na_2CO_3(s) + H_2O(g) + CO_2(g) \tag{1}$$

In this case there will be an appreciable decrease in mass, since some of the products will be driven off as gases. If such a mass decrease occurs, you can be sure that your sample is a hydrogen carbonate.

In the second part of the experiment, you will treat the solid carbonate in the crucible with HCl, hydrochloric acid. There will be considerable effervescence as CO_2 gas is evolved; the reaction that occurs is, using Na_2CO_3 as our example:

$$Na_2CO_3(s) + 2\,H^+(aq) + 2\,Cl^-(aq) \rightarrow 2\,NaCl(s) + H_2O(\ell) + CO_2(g) \tag{2}$$

(Since HCl in solution exists as ions, we write the equation in terms of ions.) You will then heat the crucible strongly to drive off any excess HCl and any water that is present, obtaining pure, dry, solid NaCl as your product.

To identify your unknown, you will need to find the molar masses of the possible reactants and final products. For each of the possible unknowns there will be a different relationship between the mass of the original sample and the mass of the chloride salt that is produced in Reaction 2. If you know your sample is a carbonate, you need only be concerned with the mass relationships in Reaction 2 and should use as the original mass of your unknown the mass of the carbonate after it has been heated. If you have a hydrogen carbonate, the overall reaction your sample undergoes will be the sum of Reactions 1 and 2.

From your experimental data you will be able to calculate the change in mass that occurred when you formed the chloride from the hydrogen carbonate or the carbonate that you started with. That difference divided by the mass of the original salt will be different for each of the possible starting compounds and will not vary with the mass of the sample. Let's call that quantity Q.

$$Q = (\text{mass of chloride} - \text{original mass})/\text{original mass} \tag{3}$$

Your calculation of the theoretical values of Q for each of the possible compounds should allow you to determine the identity of your unknown. Since you already know whether your compound is a carbonate or a hydrogen carbonate, you should then need only work out which of the two possible compounds of that type yours may be.

Experimental Procedure

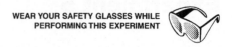

Obtain an unknown sample to use in this experiment.

Clean your crucible and its cover by rinsing them with deionized water and then drying them with a towel. Place the crucible with its cover slightly ajar on a clay triangle. Heat gently with your Bunsen burner flame for a minute or two and then strongly for two more minutes. Allow the crucible and cover to cool to room temperature (this will take about ten minutes; it must cool completely, or your mass measurement will be artificially lowered by convection currents).

Weigh the crucible and cover accurately on an analytical balance. Record the mass on the report page. With a spatula, transfer about half a gram of the unknown to the crucible. Weigh the crucible, cover, and the sample of unknown on the balance. Record the mass.

Put the crucible on the clay triangle, with the cover ajar. Heat the crucible, gently and intermittently, for a few minutes. Gradually increase the flame intensity, to the point where the bottom of the crucible is red hot. Heat for 10 more minutes. Allow the crucible to cool for 10 minutes, and then weigh it, with its cover and contents, on the analytical balance, recording the mass.

At this point the sample in the crucible is a dry carbonate, since the heating process will convert any hydrogen carbonate to carbonate.

Put the crucible on the clay triangle, leaving the cover off. Add about twenty-five drops of 6 M HCl, 1 drop at a time, to the sample. As you add each drop, you will probably observe effervescence as CO_2 is produced. Let the action subside before adding the next drop, to keep the effervescence confined to the lower part of the crucible. We do not want the product to foam up over the edge! When you have added all of the HCl, the effervescence should have ceased, and the solid should be completely dissolved. Heat the crucible gently for a brief period to complete the solution process. If all of the solid is not dissolved, add 6 more drops of 6 M HCl and warm gently.

Place the cover on the crucible in an off-center position, to allow water to escape during the next heating operation. Heat the crucible, gently and intermittently, for about ten minutes, to slowly evaporate the water and excess HCl. If you heat too strongly, spattering will occur and you may lose some sample. When the sample is dry, gradually increase the flame intensity, finally getting the bottom of the crucible to glow red. Then heat at full flame strength for 10 minutes.

Allow the crucible to cool for 10 minutes. Weigh it, with its cover and contents, on the analytical balance, recording the mass.

DISPOSAL OF REACTION PRODUCTS.   Dispose of the sample by dissolving it in water and pouring it down the drain.

Experiment 5

Data and Calculations: Identification of a Compound by Mass Relationships

Atomic masses: Na _____ g/mol   K _____ g/mol   H _____ g/mol

C _____ g/mol   O _____ g/mol   Cl _____ g/mol

Molar masses: $NaHCO_3$ _____ g/mol Na_2CO_3 _____ g/mol   NaCl _____ g/mol

$KHCO_3$ _____ g/mol K_2CO_3 _____ g/mol   KCl _____ g/mol

Mass of crucible and cover plus unknown   _____ g

Mass of crucible and cover   _____ g

Mass of unknown   _____ g

Mass of crucible, cover, and unknown after heating   _____ g

Loss of mass of sample   _____ g

Unknown is a: carbonate   hydrogen carbonate   (Circle your choice)

Mass of crucible, cover, and solid chloride   _____ g

Mass of solid chloride   _____ g

Change in mass when original compound was converted to a chloride   _____ g

Q = change in mass/original mass = _____ = _____

(continued on following page)

Theoretical values of Q, as obtained by Equation 3

Possible sodium compound _____

Possible potassium compound _____

1 mole of compound → _____ moles NaCl

1 mole of compound → _____ moles KCl

_____ g of compound → _____ g NaCl

_____ g of compound → _____ g KCl

Change in mass _____ g $Q =$ _____

Change in mass _____ g $Q =$ _____

Identity of unknown _____   Unknown # _____

Experiment 5

Advance Study Assignment: Identification of a Compound
by Mass Relationships

1. A student attempted to identify an unknown compound by the method used in this experiment. She found that when she heated a sample weighing 0.5032 g, the mass dropped to 0.3176 g. When the product was converted to a chloride, the mass went back up, to 0.3502 g.

    a. Is the sample a carbonate?   Yes / no (Circle one)

        Please provide your reasoning below.

    b. What are the two compounds that might be the unknown?

                                                 _____ or _____

    c. Write the balanced chemical equation for the overall reaction that occurs when each of these two original compounds is converted to a chloride. If the compound is a hydrogen carbonate, use the sum of Reactions 1 and 2. If the sample is a carbonate, use Reaction 2. Write the equation for a sodium salt and then for a potassium salt.

    d. How many moles of the chloride salt would be produced from one mole of the original compound?

                                               _____ moles chloride salt

(continued on following page)

e. How many grams of the chloride salt would be produced from one molar mass of the original compound?

Molar masses: $NaHCO_3$ _____ g/mol Na_2CO_3 _____ g/mol NaCl _____ g/mol

$KHCO_3$ _____ g/mol K_2CO_3 _____ g/mol KCl _____ g/mol

If a sodium salt, _____ g original compound → _____ g chloride

If a potassium salt, _____ g original compound → _____ g chloride

f. What is the theoretical value of Q, as found by Equation 3,

if the student has the Na salt? _____

if the student has the K salt? _____

g. What was the observed value of Q?

_____

h. Which compound did she have as an unknown? _____

Experiment 6

Properties of Hydrates

Most solid chemical compounds will contain some water if they have been exposed to the atmosphere for any length of time. In most cases the water is present in very small amounts and is merely adsorbed on the surface of the crystals. Other solid compounds contain larger amounts of water, chemically bound in the crystal. These compounds are called hydrates and are usually ionic salts. The water that is present in these salts is called water of hydration, and it is usually bound to the cations in the salt.

The water molecules in a hydrate are removed relatively easily. In many cases, simply heating a hydrate to a temperature somewhat above the boiling point of water will drive off the water of hydration. Hydrated barium chloride is typical in this regard; it is converted to anhydrous $BaCl_2$ if heated above 115°C:

$$BaCl_2 \cdot 2H_2O(s) \rightarrow BaCl_2(s) + 2H_2O(g) \text{ at } t \geq 115°C$$

In the dehydration reaction the crystal structure of the solid will change, and the color of the salt may also change. In Experiment 4, when copper chloride hydrate was gently heated, it was converted to the brownish anhydride.

Some hydrates lose water to the atmosphere upon standing in air. This process is called efflorescence. The amount of water lost depends on the amount of water in the air, as measured by its relative humidity, and the temperature. In moist warm air, cobalt(II) chloride ($CoCl_2$) is fully hydrated and exists as $CoCl_2 \cdot 6H_2O$, which is red. In cold, dry air the salt loses most of its water of hydration and is found as anhydrous $CoCl_2$, which is blue. At intermediate humidities and 25°C, we find the stable form is the violet dihydrate, $CoCl_2 \cdot 2H_2O$. In the old days one could obtain inexpensive hygrometers that indicated the humidity by the color of the cobalt chloride they contained.

Some anhydrous ionic compounds will tend to absorb water from the air or other sources so strongly that they can be used to dry liquids or gases. These substances are called desiccants and are said to be hygroscopic. A few ionic compounds can take up so much water from the air that they dissolve in the water they absorb; sodium hydroxide, NaOH, will do this. This process is called deliquescence.

Some compounds evolve water on being heated but are not true hydrates. The water is produced by decomposition of the compound rather than by loss of water of hydration. Organic compounds, particularly carbohydrates, behave this way. Decompositions of this sort are not reversible; adding water to the product will not regenerate the original compound. True hydrates typically undergo reversible dehydration. Adding water to anhydrous $BaCl_2$ will cause formation of $BaCl_2 \cdot 2H_2O$, or if enough water is added you will get a solution containing Ba^{2+} and Cl^- ions. Many ionic hydrates are soluble in water and are usually prepared by crystallization from aqueous solution. If you order barium chloride from a chemical supply house, you will probably get crystals of $BaCl_2 \cdot 2H_2O$, which is a stable, stoichiometrically pure compound. The amount of bound water may depend on the way the hydrate is prepared, but in general the number of moles of water per mole of compound is either an integer or an odd multiple of one-half.

In this experiment you will study some of the properties of hydrates. You will identify the hydrates in a group of compounds, observe the reversibility of the hydration reaction, and test some substances for efflorescence or deliquescence. Finally you will be asked to determine the amount of water lost by a sample of unknown hydrate upon heating. From this amount, given the formula or the molar mass of the anhydrous sample, you will be able to calculate the formula of the hydrate itself.

Experimental Procedure

A. Identification of Hydrates

Place about half a gram of each of the compounds listed below in small, dry test tubes, one compound to a tube. Observe carefully the behavior of each compound when you heat it gently with a burner flame. If droplets of water condense on the cool upper walls of the test tube, this is evidence that the compound may be a hydrate. Note the nature and color of the residue. Let the tube cool and try to dissolve the residue in a few milliliters (mL) of water, warming very gently if necessary. A true hydrate will tend to dissolve in water, producing a solution with a color very similar to that of the original hydrate. If the compound is a carbohydrate, it will give off water on heating and will tend to char. The solution of the residue in water will often be caramel colored.

Nickel chloride	Sucrose
Potassium chloride	Calcium carbonate
Sodium tetraborate (borax)	Barium chloride

B. Reversibility of Hydration

Gently heat a few crystals, ≈ 0.3 g, of cobalt(II) chloride hexahydrate, $CoCl_2 \cdot 6H_2O$, in an evaporating dish until the color change appears to be complete. Dissolve the residue in the evaporating dish in a few milliliters of water from your wash bottle. Heat the resulting solution and boil it to dryness. CAUTION: **Heat only gently once the liquid is gone!** Note any color changes. Put the evaporating dish on the lab bench and let it cool.

C. Deliquescence and Efflorescence

Place a few crystals of each of the compounds listed below on separate watch glasses and put them next to the dish of $CoCl_2$ prepared in Part B. Depending on their composition and the relative humidity (amount of moisture in the air), the samples may gradually lose water of hydration to, or pick up water from, the air. They may also remain unaffected. To establish whether the samples gain or lose mass, weigh each of them on a top-loading balance to ± 0.01 g. Record their masses. Weigh them again after about an hour to detect any change in mass. Observe the samples occasionally during the laboratory period, noting any changes in color, crystal structure, or degree of wetness that may occur.

$Na_2CO_3 \cdot 10H_2O$ (washing soda)	$KAl(SO_4)_2 \cdot 12H_2O$ (alum)
$CaCl_2$	$CuSO_4$

D. Percent by Mass of Water in a Hydrate

Clean a porcelain crucible and its cover with 6 M HNO_3. Any stains that are not removed by this treatment will not interfere with this experiment. Rinse the crucible and cover with deionized water. Put the crucible with its cover slightly ajar on a clay triangle and heat with a burner flame, gently at first and then to redness for about two minutes. Allow the crucible and cover to cool, and then weigh them to ± 0.001 g on an analytical balance. Handle the crucible with clean crucible tongs.

Obtain a sample of unknown hydrate and place about a gram of the sample in the crucible. Weigh the crucible, cover, and sample on the balance. Put the crucible on the clay triangle, with the cover in an off-center position to allow the escape of water vapor. Heat again, gently at first and then strongly, keeping the bottom of the crucible at red heat for about ten minutes. Center the cover on the crucible and let it cool to room temperature. Weigh the cooled crucible along with its cover and contents.

Examine the solid residue. Add water until the crucible is two thirds full and stir. Warm gently if the residue does not dissolve readily. Does the residue appear to be soluble in water?

DISPOSAL OF REACTION PRODUCTS.   Dispose of the residues in this experiment as directed by your instructor.

Name _____ **Section** _____

Experiment 6

Data and Calculations: Properties of Hydrates

A. Identification of Hydrates

	H_2O appears?	Color of residue?	Water soluble?	Hydrate?
Nickel chloride	_____	_____	_____	_____
Potassium chloride	_____	_____	_____	_____
Sodium tetraborate	_____	_____	_____	_____
Sucrose	_____	_____	_____	_____
Calcium carbonate	_____	_____	_____	_____
Barium chloride	_____	_____	_____	_____

B. Reversibility of Hydration

Summarize your observations of $CoCl_2 \cdot 6H_2O$:

Is the dehydration and hydration of $CoCl_2$ reversible?

C. Deliquescence and Efflorescence

	Mass (sample + glass)		Observations	Conclusion(s)
	Initial	Final		
$Na_2CO_3 \cdot 10H_2O$	_____	_____	_____	_____
$KAl(SO_4)_2 \cdot 12H_2O$	_____	_____	_____	_____
$CaCl_2$	_____	_____	_____	_____
$CuSO_4$	_____	_____	_____	_____
$CoCl_2$	_____	_____	_____	_____

(continued on following page)

D. Percent by Mass of Water in a Hydrate

Mass of crucible and cover   _____ g

Mass of crucible, cover, and solid hydrate   _____ g

Mass of crucible, cover, and residue   _____ g

Calculations and Results

Mass of solid hydrate   _____ g

Mass of residue   _____ g

Mass of H_2O lost   _____ g

Percent by mass of H_2O in the unknown hydrate   _____ $\%_{mass}$ H_2O

Formula mass of anhydrous salt (if furnished)   _____

Number of moles of water per mole of unknown hydrate   _____

Unknown #   _____

Experiment 6

Advance Study Assignment: Properties of Hydrates

1. A student is given a sample of a pink manganese(II) sulfate hydrate. She weighs the sample in a dry, covered crucible and obtains a mass of 26.742 g for the crucible, cover, and sample. Earlier she had found that the crucible and cover weighed 23.599 g. She then heats the crucible to drive off the water of hydration, keeping the crucible at red heat for about ten minutes with the cover slightly ajar. She then lets the crucible cool and finds it has a lower mass; the crucible, cover, and contents then weigh 26.406 g. In the process the sample was converted to off-white anhydrous $MnSO_4$.

a. What was the mass of the hydrate sample?

_____ g hydrate

b. What is the mass of the anhydrous $MnSO_4$?

_____ g $MnSO_4$

c. How much water was driven off?

_____ g H_2O

d. What is the percent by mass of water in the hydrate?

$$\%_{mass} \text{ water} = \frac{\text{mass of water in sample}}{\text{mass of hydrate sample}} \times 100\%$$

_____ $\%_{mass} \text{ } H_2O$

(continued on following page)

e. How many grams of water would there be in 100.0 g hydrate? How many moles?

_____ g H_2O; _____ moles H_2O

f. How many grams of $MnSO_4$ are there in 100.0 g hydrate? How many moles? (What percentage of the mass of the hydrate is $MnSO_4$? Convert the mass of $MnSO_4$ to moles. The molar mass of $MnSO_4$ is 151.00 g/mol.)

_____ g $MnSO_4$; _____ moles $MnSO_4$

g. How many moles of water are present per mole of $MnSO_4$?

_____

h. What is the formula of the hydrate?

_____

Experiment 7

Analysis of an Unknown Chloride

One of the important applications of precipitation reactions lies in the area of quantitative analysis. Many substances that can be precipitated from solution are so slightly soluble that the precipitation reaction by which they are formed can be considered to proceed to completion. Silver chloride is an example of such a substance. If a solution containing Ag^+ ion is slowly added to one containing Cl^- ion, the ions will react to form AgCl:

$$Ag^+(aq) + Cl^-(aq) \rightarrow AgCl(s) \tag{1}$$

Silver chloride is so insoluble that essentially all of the Ag^+ added will precipitate as AgCl until all of the Cl^- is used up. When the amount of Ag^+ added to the solution is equal to the amount of Cl^- initially present, the precipitation of Cl^- ion will be, for all practical purposes, complete.*

A convenient method for chloride analysis using AgCl has been devised. A solution of $AgNO_3$ is added to a chloride solution just to the point where the number of moles of Ag^+ added is equal to the number of moles of Cl^- initially present. We analyze for Cl^- by simply measuring how many moles of $AgNO_3$ are required. Surprisingly enough, this measurement is rather easily made by an experimental procedure called a titration.

In the titration, a solution of $AgNO_3$ of known concentration (in moles $AgNO_3$ per liter of solution) is added from a calibrated buret to a solution containing a measured amount of unknown. The titration is stopped when a color change occurs in the solution, indicating that stoichiometrically equivalent amounts of Ag^+ and Cl^- are present. The color change is caused by a chemical reagent, called an indicator, that is added to the solution at the beginning of the titration.

The volume of $AgNO_3$ solution that has been added up to the time of the color change can be measured accurately with the buret, and the number of moles of Ag^+ added can be calculated from the known concentration of the solution.

In the Fajans method for the volumetric analysis of chloride, which we will employ in this experiment, the indicator used is dichlorofluorescein. This compound, in its anionic (negative ion) form, functions as an adsorption indicator. After enough silver nitrate has been added to react with all of the chloride ions, and the concentration of Ag^+ in solution begins to rise, positively charged silver ions are adsorbed onto the surface of the colloidal (very tiny) silver chloride particles. The silver chloride then has a positive charge and attracts (negatively charged) dichlorofluorescein anions onto the surface. When this occurs, the silver chloride takes on a pink color in the presence of the yellowish, fluorescent dichlorofluorescein, indicating that the endpoint of the titration has been reached. Because AgCl gradually darkens in the presence of (ultraviolet) light, your titration solutions will eventually become dark purple in color—more quickly in brighter light. You will not be able to revisit the endpoint color of your titrated solutions for very long after the titration is complete.

In this experiment, you will be obtaining a 100-mL volumetric flask which contains an unknown chloride solution. You will dilute the solution to the 100-mL mark, and after mixing, pipet out 10.00-mL aliquots of the solution into separate flasks for titration with a standardized solution of $AgNO_3$. Using the respective volumes of titrant required to reach the endpoint in each titration, you will obtain three replicate measurements of the concentration of the diluted chloride solution in the volumetric flask. Given the molarity of the $AgNO_3$ titrant,

$$\text{moles } Ag^+ = \text{moles } AgNO_3 = M_{AgNO_3} \times V_{AgNO_3} \tag{2}$$

where the volume of $AgNO_3$ is expressed in liters and the molarity M_{AgNO_3} is in moles per liter of solution. At the endpoint of the titration,

$$\text{moles } Ag^+ \text{ added} = \text{moles } Cl^- \text{ present in unknown} \tag{3}$$

$$\text{moles } Cl^- \text{ present in unknown} = M_{Cl} \times V_{Cl} \tag{4}$$

*See Experiment 26 for a discussion of the principles governing precipitations of this sort.

Experimental Procedure

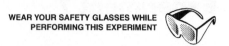

Obtain a buret, a 100-mL volumetric flask containing your concentrated unknown chloride solution, and a 10-mL volumetric pipet. *Being careful not to overshoot,* add deionized water slowly to the volumetric flask up to the mark on the narrow neck, squeezing your wash bottle gently as the bottom of the meniscus becomes just coincident with the ring—this is best seen if the ring on the flask is at the same level as your eyes. Place the stopper in the flask (see Appendix IV), and holding the stopper in place with the thumb, index finger, or palm of the same hand holding the flask, invert the flask 15 times to thoroughly mix the solution. Transfer via pipet (see Appendix IV) 10.00 mL of your diluted chloride solution to a 250-mL Erlenmeyer flask, and add about fifty milliliters of deionized water. Add *1 drop* (not more!) of dichlorofluorescein indicator.

Without diluting it in any way, measure out about one-hundred milliliters of the standardized $AgNO_3$ solution into a clean, *dry* 125-mL Erlenmeyer flask. This will be your total supply for the entire experiment, so do not waste it. Silver nitrate is expensive! Silver solutions will leave long-lasting black stains on skin and clothing, so take care not to spill your titrant. If called for, clean your buret thoroughly with detergent solution, with the aid of a buret brush, and rinse it with deionized water. If your buret starts out wet with anything other than the titrant, pour 2 or 3 successive portions of a few milliliters of the $AgNO_3$ titrant into the buret and tip it back and forth to rinse the inside walls. Allow the $AgNO_3$ to drain out of the buret tip each time. Fill the buret with $AgNO_3$ solution. Open the stopcock momentarily to flush any air bubbles out of the buret tip. Be sure that your stopcock fits snugly and that the buret does not leak. (See Appendix IV for procedures regarding titrations with a buret.)

Read the initial buret level to ± 0.01 mL. You may find it useful when making readings to put a white card, with a thick black stripe or black tape on it, behind the meniscus. If the black line is held just below the level to be read, its reflection in the surface of the meniscus will help you obtain an accurate reading. (Note that if you use such a tool, it is critical that you use it the same way each time you take a measurement: it will change what you read!) Begin to add the $AgNO_3$ titrant solution to the chloride solution while swirling the flask. Tiny colloidal particles of white AgCl will begin to form, giving a slight cloudiness to the yellowish solution. At the beginning of the titration, you can add the titrant (the $AgNO_3$ solution) fairly rapidly, a few milliliters at a time, swirling the flask to mix the solution. You will find that a pinkish color forms in the solution and disappears as you mix by swirling.

Gradually decrease the rate at which you add the $AgNO_3$ solution as the pink color persists longer. At some stage you may find it convenient to set your buret stopcock to deliver titrant slowly, drop by drop while you swirl the flask. The endpoint of the titration is indicated by the pink color no longer going away with swirling. Right at the endpoint you may note that the solution takes on a peach color as the lasting pink and some yellow color are both present. The key to this titration is to focus on the hue (color) of the solution rather than its intensity. If you are careful, you can hit the endpoint within one drop of titrant. When you have reached the endpoint, stop the titration and record the buret level, again to ± 0.01 mL.

Analyze two more 10.00-mL diluted chloride samples, as described above. Note the volume of $AgNO_3$ used in the first titration and refill the buret as necessary, to be certain that you will not go below the 50.00-mL mark at the bottom of the buret.

Calculate the concentration of chloride in the diluted solution within your 100-mL volumetric flask, as independently indicated by each of your three titrations. Also calculate the mean and the standard deviation of these three results. (Consult Appendix IX.)

Take it Further (Optional): Find the mass percent Cl^- or NaCl in a commercial food product, such as a bouillon cube or canned soup. Compare your results with those on the nutrition label.

DISPOSAL OF REACTION PRODUCTS. Silver nitrate is toxic to the environment. Pour all titrated solutions and any $AgNO_3$ remaining in your buret or flask into the waste bottles provided, unless directed otherwise by your instructor.

Name _____ Section _____

Experiment 7

Data and Calculations: Analysis of an Unknown Chloride

Molarity of standard $AgNO_3$ solution _____ M

	I	II	III
Initial buret reading	_____ mL	_____ mL	_____ mL
Final buret reading	_____ mL	_____ mL	_____ mL
Volume of $AgNO_3$ used to titrate sample	_____ mL	_____ mL	_____ mL
Moles of $AgNO_3$ used to titrate sample	_____	_____	_____
Moles of Cl^- present in sample	_____	_____	_____
Volume of diluted chloride solution pipetted	_____ mL	_____ mL	_____ mL
Total moles of Cl^- in 100-mL volumetric flask	_____	_____	_____
Cl^- concentration in diluted solution in volumetric flask	_____ M	_____ M	_____ M
Mean chloride concentration	_____ M		
Standard deviation in chloride concentration	_____ M		
Unknown #	_____		

Name ______________________________ Section ______________________________

Experiment 7

Advance Study Assignment: Analysis of an Unknown Chloride

1. A student performed this experiment and obtained the following concentration values: 0.02813 M, 0.02882 M, and 0.02708 M.

   a. What is the mean concentration?

   ___________ M

   b. What is the standard deviation of these results?

   ___________ M

2. A 10.00-mL diluted chloride sample required 13.89 mL of 0.02104 M $AgNO_3$ to reach the Fajans endpoint.

   a. How many moles of Cl^- ion were present in the sample? (Use Eqs. 2 and 3.)

   ___________ moles Cl^-

   b. What was the concentration of chloride in the diluted solution?

   ___________ M

3. How would the following errors affect the concentration of Cl⁻ obtained in Question 2(b)? Give your reasoning in each case.

   a. The student read the molarity of $AgNO_3$ as 0.02014 M instead of 0.02104 M.

   b. The student was past the endpoint of the titration when he took the final buret reading.

Experiment 8

Verifying the Absolute Zero of Temperature— Determination of the Atmospheric Pressure

Gases differ from liquids and solids in that much of the ordinary physical behavior of gases can be described by simple laws that apply to all gases. Few such relations exist for liquids and solids, so information about their physical properties must be obtained by direct experimentation on the liquid or solid of interest.

Gases are easily compressed relative to liquids and solids. If one doubles the pressure on a gas, the volume decreases by a factor of two. Over rather wide ranges of pressure, the pressure–volume product, PV, remains nearly constant as long as the temperature remains constant. This relationship is called Boyle's Law, after Robert Boyle, who discovered it in 1660. It was one of the first natural laws that scientists recognized.

When a gas is heated at constant volume, the pressure goes up. In 1787, about one-hundred twenty years after Boyle did his work, Jacques Charles found that the pressure increased linearly with Celsius temperature. Boyle was in no position to discover Charles's Law, since in 1660 the idea of temperature was not well developed. In 1848, at the ripe old age of 24, William Thomson (who would later take on the title Lord Kelvin) recognized that one could set up an absolute temperature scale on which the pressure was actually proportional to temperature. This idea was relatively sophisticated and led to the observation that many natural laws, not just the gas laws, are most simply expressed on an absolute temperature scale; the most common of these is now called the Kelvin scale.

The gas we encounter most often is the air around us. Its physical behavior is like that of other gases, so under ordinary conditions it obeys Boyle's and Charles's Laws. The pressure of the air is called the atmospheric pressure. It varies a little from day to day, and at sea level it is equal to the pressure exerted by a column of mercury 76 cm high, or a column of water 10 m high; this is 1×10^5 Pascals, or in common American units, 15 psi (lbf/in^2). Ordinarily we are not aware of this rather substantial pressure because it is nicely balanced by our bodies, but if it decreases suddenly, as it does in a tornado, it will definitely get our attention, since it can, among other things, cause a house to fly apart. The atmospheric pressure can be found by various instruments, but probably the most accurate of these is a mercury barometer, which measures the height to which a column of mercury rises in an evacuated tube immersed in a pool of mercury.

In the first part of this experiment we will try to verify the absolute zero of temperature. We will simply heat a sample of air, keeping its volume constant while we measure the increase in pressure. We assume the validity of Charles's Law, and calculate the temperature at which the pressure would become zero. That is the absolute zero of temperature, and the zero point of the Kelvin temperature scale.

In the rest of the experiment we will measure the atmospheric pressure using a simple apparatus that allows us to find the change in volume when a sample of air is slightly compressed at constant temperature. The changes in volume and pressure are related to the atmospheric pressure in a remarkably simple way.

Experimental Procedure

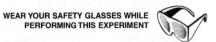
WEAR YOUR SAFETY GLASSES WHILE PERFORMING THIS EXPERIMENT

Obtain a 1000-mL beaker, a 500-mL Erlenmeyer flask with stopper and glass tubing insert, a digital or sensitive mercury thermometer, a glass U-tube manometer, two lengths of rubber tubing, and a disposable Pasteur pipet.

A. Verifying the Absolute Zero of Temperature

In this part of the experiment we will examine how the pressure of a sample of air of fixed volume increases when we raise the temperature. Given that $PV = nRT$ for the air, if we hold the volume and amount of air constant, we can see that the Ideal Gas Law reduces to $P = kT$, where k is a constant equal to nR/V and the pressure is proportional to the absolute (Kelvin) temperature, T. Our goal is to establish the conversion between the Celsius temperature, t, and the Kelvin temperature, T.

Assemble the apparatus shown in Figure 8.1, placing a 1000-mL beaker on a magnetic stirrer if one is available. Put a magnetic stirring bar in the beaker, and clamp a *dry* 500-mL Erlenmeyer flask in the beaker as shown. (If the flask is not dry, add a few milliliters of acetone and shake well. Pour out the acetone and gently blow compressed air into the flask for a minute or two.) Fill the beaker with room temperature water which has been stored overnight in the lab. Start the stirrer. If you don't have a magnetic stirrer available, stirring with a stirring rod should be satisfactory, and indeed you should use one even in addition to the magnetic stirrer. Connect the short piece of rubber tubing to the glass tubing in the stopper, and then moisten the rubber stopper and insert it firmly into the flask. Clamp the U-tube manometer into position. You can rest the bottom of the U-tube on the lab bench surface if that is convenient. Using your Pasteur pipet, add a few milliliters of water to the U-tube manometer, so that the level in the left arm is about four centimeters above the lowest part of the U. Connect the rubber tubing to the left arm of the manometer. Mark the water level in the left arm with a piece of masking tape. Record the levels in the left and right arms to the nearest 0.1 cm. Record the temperature to the nearest 0.1°C. Record the atmospheric pressure in mm Hg.

If the system is tight, and the water in the beaker is at room temperature, with stirring the levels will remain steady for several minutes. If necessary, adjust the masking tape position once the levels hold steady, and then take new readings.

Using the long piece of rubber tubing and a pipet bulb, siphon about one-hundred milliliters of water from the 1000-mL beaker into another container. Using your Pasteur pipet, add a few milliliters of this water to the right arm of the manometer, raising the level in the *left* arm about three centimeters. Add some warm water (at about 40°C) from the tap to a beaker, and very slowly add the warm water to the 1000-mL beaker. The level in the left arm will go down as the air in the flask expands. Add warm water as necessary until the level in the left arm is back at the location you marked with the masking tape. If the level goes too low, add a little room-temperature water to the right arm to bring it to the proper position. Wait a minute or so, with stirring, to see that levels hold steady and the temperature is not changing. Read the two levels and the temperature.

Repeat the steps in the paragraph above at two or three higher temperatures, being sure to wait long enough for the levels and temperature to become steady. At the higher temperatures, the water in the 1000-mL beaker will tend to cool, so you may need to add warm water to make it hold its temperature. Your final level

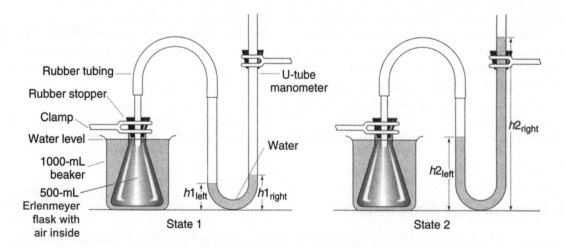

Figure 8.1 The same apparatus, shown above, is used for verifying the absolute zero of temperature and for determining the atmospheric pressure.

in the right arm should be near the top of the manometer. You should have at least four runs for which you have recorded water levels and temperature.

You can process the data you have obtained by making a graph of total pressure, P_{total} versus the Celsius temperature, t. The total pressure of the gas will equal the measured atmospheric pressure plus the difference, Δh, between the level in the right arm of the manometer and the level in the left arm. You have measured pressures in cm of water, whereas our atmospheric pressure is usually in mm Hg. For convenience we will use cm H_2O as our unit, and convert cm Hg to cm H_2O by a conversion factor:

$$P_{atm} \text{ in cm } H_2O = P_{atm} \text{ in } cm \text{ Hg} \times \text{density of Hg}/\text{density of } H_2O$$
$$= 13.57 \times P_{atm} \text{ in } cm \text{ Hg [Note: NOT mm Hg!]} \tag{1}$$

$$\text{In units of cm } H_2O, \quad P_{total} = P_{atm} + \Delta h \tag{2}$$

Using Equations 1 and 2, make a table of temperatures t in °C and the values of P_{total} in cm H_2O. Then using graph paper or a software program like Excel, make a graph of total pressure vs Celsius temperature (see Appendix V). The P vs t graph should be a straight line, with the general equation $y = mx + b$, or, with our variables, $P = mt + b$, where m is the slope of the line and b is a constant. Using your graph, find the equation of the line. Then find the temperature t_0 in °C at which P equals zero. Kelvin realized that if he added minus that temperature to t, defining T on his absolute temperature scale, he would obtain the simple equation $P = mT$ rather than $P = mt + b$. The Advance Study Assignment demonstrates how this is accomplished.

B. Determining the Atmospheric Pressure

In this part of the experiment we will use the same apparatus as in the first part. Disconnect the rubber tubing from the manometer and pour out the water. Take the flask out of the beaker and pour out the water. Clamp the flask in the beaker and firmly insert the stopper. Fill the beaker with room temperature water. Check with your thermometer to see that the water is indeed at room temperature. In this experiment it is imperative that the temperature of the flask remains constant.

Using your Pasteur pipet, add a few milliliters of water to the manometer, so that the levels are about four centimeters above the bottom of the U. Then connect the rubber tubing to the left arm of the manometer. The apparatus at that point should look like the left-hand sketch in Figure 8.1.

You have now isolated the air in the flask in a fixed initial state, which we will call State 1. That state is defined by its volume, pressure, and temperature. The volume, V_1, of the air is essentially that of the flask, V_{flask} (we will ignore the small volume of the air in the tubing). The pressure, P_1, is equal to the atmospheric pressure, P_{atm}, plus the difference between the heights of the liquid levels in the manometer, $h1_{right} - h1_{left}$ (see Appendix IV).

$$P_1 = P_{atm} + h1_{right} - h1_{left} \tag{3}$$

As in the first part of the experiment, we will use cm H_2O as our unit for pressure.

Measure the heights $h1_{right}$ and $h1_{left}$ with a ruler or meter stick and record them to the nearest 0.1 cm. If the temperature is not changing, these levels should hold steady. Wait three minutes and check to see they have not changed. If they do change, add a little warm or cold water to the 1000-mL beaker to bring it to room temperature. Don't proceed until the levels hold steady for at least two minutes.

In the next step of the experiment we compress the air in the flask just a little, by adding water to the right arm of the manometer. Add the water slowly, using your Pasteur pipet. You will see that both levels go up, but the right-hand level goes up faster than the left, since the air in the flask, as its volume is decreased, is going to a higher pressure than before. Continue adding water until the level is near the top of the right arm. Measure the heights of the levels in the right and left arms of the manometer and record them. The levels should not change with time. The air sample is now in State 2 (see Fig. 8.1).

In State 2, the pressure P_2 of the air is equal to P_{atm} plus $h2_{right} - h2_{left}$. The volume V_2 is slightly less than V_1 by an amount equal to the volume of the air in the manometer that was displaced by water as the air was compressed. Let's call that change in volume ΔV, equal to $(h2_{left} - h1_{left})$ times the cross-sectional area, A, of the manometer tube.

$$\Delta V = (h2_{left} - h1_{left}) \times A \tag{4}$$

Summarizing, for States 1 and 2:

$$P_1 = P_{atm} + h1_{right} - h1_{left} \qquad V_1 = V_{flask} \text{ (to be measured)} \tag{5}$$

$$P_2 = P_{atm} + h2_{right} - h2_{left} \qquad V_2 = V_{flask} - \Delta V \tag{6}$$

$$\Delta P = P_2 - P_1$$

Since Boyle's Law is valid for the change in state,

$$P_1 V_1 = P_2 V_2 \text{ (for a gas sample at constant temperature)}$$

or

$$P_1 V_1 = (P_1 + \Delta P)(V_1 - \Delta V) \tag{7}$$

Solving this equation for P_1, we find that

$$P_1 = (V_1 \Delta P - \Delta V \Delta P)/\Delta V \tag{8}$$

We can use Equations 8 and 5 to find P_{atm}, but first we need to calibrate the apparatus by measuring V_{flask} and the cross-sectional area, A, of the manometer.

C. Calibrating the System

The volume V_1 of the apparatus is essentially equal to the volume of the flask, V_{flask}, since it is much larger than the volume of the tubing or the manometer. To determine the volume of the flask, take it out of the beaker and fill it with water up to within one centimeter of the top. Pour the water into a tared beaker and record the mass of the water. We can assume that the density of water is 1.00 g/mL, so the volume of the flask in milliliters is numerically equal to the mass of the water in grams.

The cross-sectional area of the manometer tube is found in a similar way. Fill the manometer with water, using your finger to stopper the tube as necessary. Pour the water into a tared small beaker and record its mass, and hence its volume. Then measure the lengths of the arms of the manometer, and the length of the U as $\pi \times r$, the radius. The length of the manometer is the sum of these three lengths. Use the length and volume to find the cross-sectional area A in cm^2.

Using the data you have obtained and the relevant equations, calculate P_1 and P_{atm}. The units of P_{atm} will be cm H_2O, which you can convert to cm Hg by dividing by 13.57. Report the value of P in both sets of units. Compare the value you obtain with that found with the barometer in the laboratory.

Complete all of your calculations before leaving the lab. When you are finished, turn off the magnetic stirrer and digital thermometer, if you used them, and return all borrowed equipment.

Experiment 8

Data and Calculations: Verifying the Absolute Zero of Temperature— Determination of the Atmospheric Pressure

A. Verifying the Absolute Zero of Temperature

Manometer heights in cm H_2O		Temp, t [°C]	P_{total} [cm H_2O]
$h1_{right}$ _____	$h1_{left}$ _____	t_1 _____	P_1 _____
$h2_{right}$ _____	$h2_{left}$ _____	t_2 _____	P_2 _____
$h3_{right}$ _____	$h3_{left}$ _____	t_3 _____	P_3 _____
$h4_{right}$ _____	$h4_{left}$ _____	t_4 _____	P_4 _____
$h5_{right}$ _____	$h5_{left}$ _____	t_5 _____	P_5 _____

P_{atm} in mm Hg _____ P_{atm} in cm Hg _____ P_{atm} in cm H_2O _____

To convert P_{atm} from cm Hg to cm H_2O, multiply by 13.57.

$$P_{total} = P_{atm} + h_{right} - h_{left} \text{ (for } P \text{ in cm } H_2O)$$

Using graph paper or a spreadsheet, make a graph of P_{total} vs t. (Appendices V, VII, and/or VIII can help you with this.) The graph should be a straight line, with the equation $P = mt + b$, where m is the slope and b is a constant. If you need to find the equation by hand, see Appendix V for the method. If you find the equation using software, print out the graph with the line and the equation shown and include it with this report. In any case, report your result below:

Equation for P_{total} vs t _____

Find the temperature t_0 in °C where P_{total} becomes zero.

$t_0 =$ _____ °C   Let $k = -t_0$,   and $T = t + k$

On the Kelvin scale, $T = 0$ at absolute zero, which is t_0°C on the Celsius scale.

In your equation, substitute $T - k$ for t, and show that, with your value of k, the equation reduces to $P_{total} = mT$.

(continued on following page)

B. Measuring the Atmospheric Pressure

Manometer Heights in cm

$h1_{right}$ _____ $h2_{right}$ _____ t _____ °C

$h1_{left}$ _____ $h2_{left}$ _____

Using Equations 5 and 6, evaluate P_1, P_2, and ΔP.

$P_1 = P_{atm} +$ _____ cm H_2O $P_2 = P_{atm} +$ _____ cm H_2O $\Delta P =$ _____ cm H_2O

C. Calibrating the System

Mass of water in flask _____ g $V_{flask} =$ _____ mL

Evaluate the cross-sectional area, A, of the manometer tube:

Length of left arm _____ cm   Length of right arm _____ cm

Radius of U section, r _____ cm   Length of U section, $\pi \times r$ _____ cm

Total length of manometer = (Length of left arm) + (Length of right arm) + (Length of U Section) = _____ cm

Mass of water in full manometer _____ g = V_{man} in mL = V_{man} in cm^3

Cross-sectional area, $A = V_{man}/$(Total length of manometer) =

_____ cm^2

Change in volume, $\Delta V = (h2_{left} - h1_{left}) \times A$ _____ mL   (Eq. 4)

Using Equation 8, evaluate P_1 and P_{atm}.

$P_1 =$ _____ cm H_2O $P_{atm} =$ _____ cm H_2O

Express P_{atm} in cm Hg and compare your value to that obtained from the laboratory barometer.

$P_{atm} =$ _____ cm Hg $P_{barometer} =$ _____ cm Hg

Name _____ **Section** _____

Experiment 8

For use as needed.

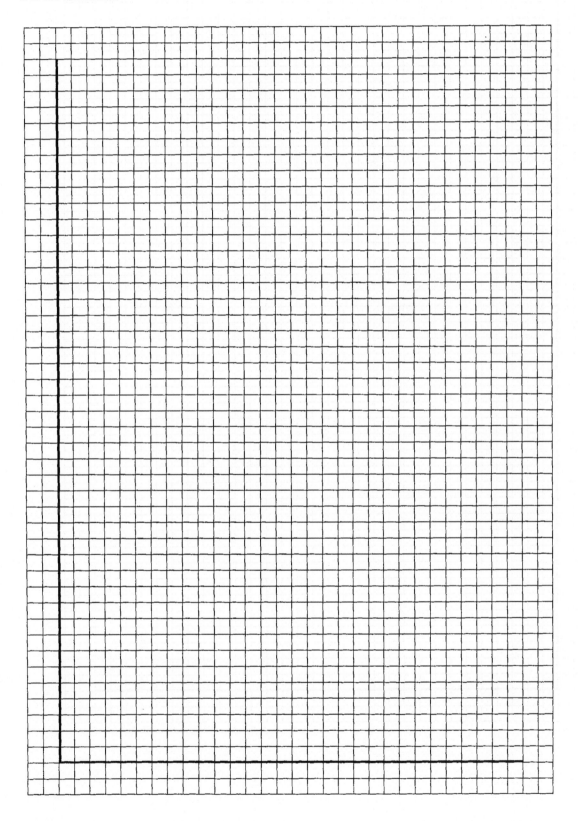

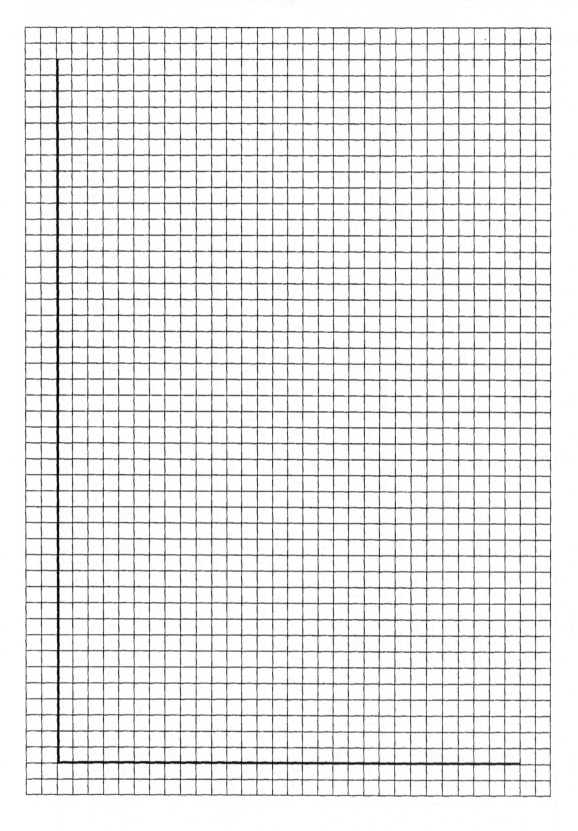

Experiment 8

Advance Study Assignment: Verifying the Absolute Zero of Temperature—
Determination of the Atmospheric Pressure

1. The atmospheric pressure on a spring day in Minnesota was found to be 726 mm Hg.

   a. What would the pressure be in cm H_2O?

   _____ cm H_2O

   b. About how many meters of water would it take to exert that pressure?

   _____ m H_2O

2. In this experiment a student found that when she increased the temperature of a 544-mL sample of air from 22.8°C to 30.9°C, the pressure of the air went from 1021 cm H_2O up to 1049 cm H_2O. Since the air expands linearly with temperature, the equation relating P to t is of the form:

$$P = mt + b \qquad\qquad (9)$$

where m is the slope of the line and b is a constant.

   a. What is the slope of the line? (Find the change in P divided by the change in t.)

   $m =$ _____ cm $H_2O/°C$

   b. Find the value of b. (Substitute known values of P and t into Eq. 9 and solve for b.)

   $b =$ _____ cm H_2O

(continued on following page)

c. Express Equation 9 in terms of the values of m and b.

$$P =$$

d. At what temperature t will P become zero?

$$P = 0 \text{ at } t = \underline{\hspace{4cm}} \text{°C} = t_0 = -k$$

(Please note that you are unlikely to get results anywhere near this good when actually carrying out this experiment, as it involves a very large extrapolation.)

e. The temperature in Part (d) is the absolute zero of temperature. Lord Kelvin suggested that we set up a scale on which that temperature is 0 K. On that scale, $T = t + k$. Show that, on the Kelvin scale, your equation reduces to $P = mT$.

Experiment 9

Molar Mass of a Volatile Liquid

$\mathbf{O}$ne of the important applications of the Ideal Gas Law is found in the experimental determination of the molar masses of gases and vapors. In order to measure the molar mass of a gas or vapor, we need simply determine the mass of a given sample of the gas under known conditions of temperature and pressure under which the gas obeys the Ideal Gas Law,

$$PV = nRT \tag{1}$$

If the pressure P is in atmospheres, the volume V in liters, the temperature T in K, and the amount n in moles, then the gas constant R is equal to 0.0821 L atm/(mole K).

From measured values of P, V, and T for a sample of gas, we can use Equation 1 to find the number of moles of gas in the sample. The molar mass in grams, MM, is equal to the mass m of the gas sample divided by the number of moles, n:

$$n = \frac{PV}{RT} \quad MM = \frac{m}{n} \tag{2}$$

This experiment involves measuring the molar mass of a volatile liquid using Equation 2. A small amount of the liquid is introduced into a weighed flask. The flask is then placed in boiling water, where the liquid vaporizes completely, driving out the air and filling the flask with pure vapor at atmospheric pressure and the temperature of the boiling water. If we cool the flask so that this vapor condenses, we can measure the mass of the vapor and calculate a value for the liquid's molar mass, MM.

Experimental Procedure*

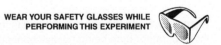

WEAR YOUR SAFETY GLASSES WHILE PERFORMING THIS EXPERIMENT

Obtain a special round-bottomed flask, a tiny stopper, a large stopper with a glass cap in it, and an unknown liquid. Support the flask on an evaporating dish or in a beaker at all times. If you should break or crack the flask, report it to your instructor immediately so that it can be repaired. With the tiny stopper loosely inserted in the neck of the flask, weigh the empty, dry flask on an analytical balance. Use a tared cork ring to support the flask, unless the pan of the balance hangs from a hook above, in which case a copper loop can be used to suspend the flask from the hook supporting the balance pan.

Pour about half of your unknown liquid, about five milliliters, into the flask. **Do not reinsert the tiny stopper!** Add about three-hundred milliliters of water to the 600-mL beaker and assemble the apparatus as shown in Figure 9.1, with the beaker centered on the hotplate and the floating flask kept in place by its neck being under the clamped-in-place cap and large stopper. Add a few boiling chips to the water in the 600-mL beaker and heat the water to the boiling point. Watch the liquid level in your flask: the level should gradually drop as vapor escapes through the cap, even before the water boils. After all the liquid has disappeared, the water starts to boil, and no more vapor comes out of the cap, lower the heat but continue to boil the water gently for 5 to 8 minutes. Measure the temperature of the boiling water. Shut off the hotplate and wait until the water has just stopped boiling and then remove the cap and *immediately* insert the tiny stopper used previously.

Loosen the clamp holding the large stopper in place. Remove the flask from the beaker of water, holding it by the neck, which will be cooler. Immerse the flask in a beaker of cool water to a depth of about five centimeters. After holding the flask in the water for about two minutes to allow it to cool, carefully remove the tiny stopper *for not more than a second or two* to allow air to enter, and then reinsert the stopper. (As the flask

*See W. L. Masterton and T. R. Williams, J. Chem. Educ. *36*, 528 (1959). For an alternate apparatus, see Instructor's Manual.

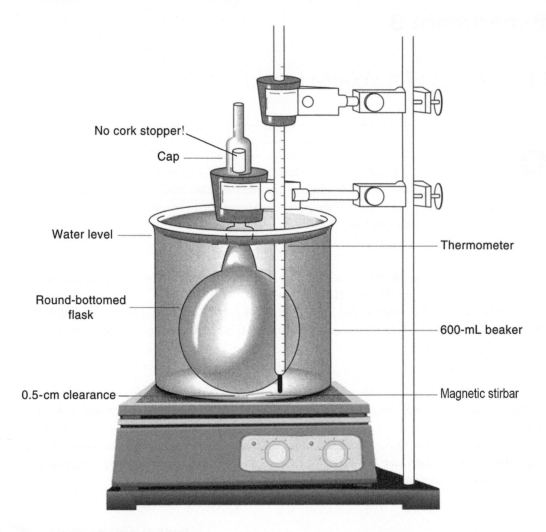

Figure 9.1   In the apparatus used in this experiment, the round-bottom flask floats in place, held down by the cap, rather than being rigidly connected to the large stopper. The thermometer should not touch the sides or bottom of the 600-mL beaker, and the magnetic stirbar must turn freely.

cools, the vapor inside condenses and the pressure drops, which explains why air rushes in when the stopper is removed.)

Dry the flask with a towel to remove the surface water and allow it to come to room temperature. Loosen the tiny stopper again momentarily to equalize any pressure differences, then reweigh the flask. Read the atmospheric pressure from the barometer and record it.

Repeat the procedure using another 5 mL of your liquid sample.

You may be given the volume of the flask by your instructor, or you may be directed to measure its volume by weighing the flask stoppered and full of water on a top-loading balance. *Do not fill the flask with water unless you are specifically told to do so.*

When you have completed the experiment, return the flask: do not attempt to wash or clean it in any way. Also return the stopper and cap, along with any remaining unknown sample, in its original container.

Experiment 9

Data and Calculations: Molar Mass of a Volatile Liquid

	Trial 1	**Trial 2**
Unknown #	_____	
Mass of flask and stopper	_____ g	_____ g
Mass of flask, stopper, and condensed vapor	_____ g	_____ g
Mass of flask, stopper, and water (see directions; only if told to do so!)	_____ g	_____ g
Temperature of boiling-water bath	_____ °C	_____ °C
Atmospheric (ambient) pressure	_____ mm Hg	_____ mm Hg

Calculations and Results

	Trial 1	**Trial 2**
Pressure of vapor, P	_____ atm	_____ atm
Volume of flask (volume of vapor), V	_____ L	_____ L
Temperature of vapor, T	_____ K	_____ K

(continued on following page)

	Trial 1	**Trial 2**
Mass of vapor, m	_____ g	_____ g

	Trial 1	**Trial 2**
Number of moles of vapor, n	_____ mol	_____ mol

	Trial 1	**Trial 2**
Molar mass of unknown, as found by substitution into Equation 2	_____ g/mol	_____ g/mol

Experiment 9

Advance Study Assignment: Molar Mass of a Volatile Liquid

1. A student weighed an empty flask and stopper and found the mass to be 53.256 g. She then added about five milliliters of an unknown liquid and heated the flask (in a boiling-water bath) to 98.8°C. After all the liquid had vaporized, she removed the flask from the bath, stoppered it, and let it cool. Once it had cooled, she momentarily removed the stopper, then replaced it and weighed the flask and condensed vapor, obtaining a mass of 53.870 g. The volume of the flask was known to be 271.1 mL. The absolute atmospheric pressure in the laboratory on that day was 728 mm Hg.

   a. What was the pressure of the vapor in the flask in atm?

   $P =$ _____ atm

   b. What was the temperature of the vapor in K? What was the volume of the flask in liters?

   $T =$ _____ K $V =$ _____ L

   c. What was the mass of condensed vapor that was present in the flask?

   $m =$ _____ grams

   d. How many moles of condensed vapor were present?

   $n =$ _____ moles

   e. What is the mass of one mole of vapor (Eq. 2)?

   $MM =$ _____ g/mole

(continued on following page)

2. How would each of the following procedural errors affect the results to be expected in this experiment? Give your reasoning in each case.

    a. Not all of the liquid was vaporized when the flask was removed from the boiling-water bath.

    b. The flask was not dried before the final weighing with the condensed vapor inside.

    c. The flask had not cooled completely to room temperature in the water bath, but the tiny stopper was not removed again before the final weighing.

    d. The flask was left open to the atmosphere while it was being cooled, and the tiny stopper was inserted just before the final weighing.

    e. The flask was stoppered and removed from the boiling-water bath before the vapor had reached the temperature of the boiling water. All the liquid had vaporized.

Experiment 10

Analysis of an Aluminum-Zinc Alloy*

Some of the more active metals will react readily with solutions of strong acids, producing hydrogen gas and a solution of a salt of the metal. Small amounts of hydrogen are commonly prepared by the action of hydrochloric acid on metallic zinc:

$$Zn(s) + 2\,H^+(aq) \rightarrow H_2(g) + Zn^{2+}(aq) \tag{1}$$

From this equation it is clear that one mole of zinc produces one mole of hydrogen gas in this reaction. If the hydrogen were collected under known conditions, it would be possible to calculate the mass of zinc in a pure sample by measuring the amount of hydrogen it produced on reaction with acid.

Since aluminum reacts spontaneously with strong acids in a manner similar to that shown by zinc,

$$2\,Al(s) + 6\,H^+(aq) \rightarrow 2\,Al^{3+}(aq) + 3\,H_2(g) \tag{2}$$

we could find the amount of aluminum in a pure sample by measuring the amount of hydrogen produced by its reaction with an acid solution. In this case two moles of aluminum would produce three moles of hydrogen.

Since the amount of hydrogen produced by a gram of zinc is not the same as the amount produced by a gram of aluminum,

$$1 \text{ mole Zn} \rightarrow 1 \text{ mole } H_2, \; 65.4 \text{ g Zn} \rightarrow 1 \text{ mole } H_2, \; 1.00 \text{ g Zn} \rightarrow 0.0153 \text{ mole } H_2 \tag{3}$$

$$2 \text{ moles Al} \rightarrow 3 \text{ moles } H_2, \; 54.0 \text{ g Al} \rightarrow 3 \text{ moles } H_2, \; 1.00 \text{ g Al} \rightarrow 0.0556 \text{ mole } H_2 \tag{4}$$

it is possible to react an alloy of zinc and aluminum of known mass with acid, determine the amount of hydrogen gas evolved, and calculate the percentages of zinc and aluminum in the alloy, using Relations 3 and 4. The object of this experiment is to carry out such an analysis.

In this experiment you will react a weighed sample of an aluminum-zinc alloy with an excess of acid and collect the hydrogen gas evolved over water (Fig. 10.1). If you measure the volume, temperature, and total pressure of the gas and use the Ideal Gas Law, taking proper account of the pressure of water vapor in the system, you can calculate the number of moles of hydrogen produced by the sample:

$$P_{H_2}V = n_{H_2}RT, \quad n_{H_2} = \frac{P_{H_2}V}{RT} \tag{5}$$

The volume V and the temperature T of the hydrogen are easily obtained from the data. The pressure exerted by the dry hydrogen P_{H_2} requires more attention. The total pressure P of gas in the flask is, by Dalton's Law, equal to the partial pressure of the hydrogen P_{H_2} plus the partial pressure of the water vapor P_{H_2O}:

$$P = P_{H_2} + P_{H_2O} \tag{6}$$

The gas in the flask is in contact with liquid water, so the gas is saturated with water vapor; the pressure P_{H_2O} under these conditions is equal to the vapor pressure VP_{H_2O} of water at the temperature of the experiment. This value is constant at a given temperature, and is found in Appendix I at the end of this manual. The total gas pressure P in the flask is very nearly equal to the atmospheric pressure P_{atm}.

Substituting these values into (6) and solving for P_{H_2}, we obtain

$$P_{H_2} = P_{atm} - VP_{H_2O} \tag{7}$$

*W. L. Masterton, J. Chem. Educ. *38*, 558 (1961). For a source of alloy samples, see Instructor's Manual.

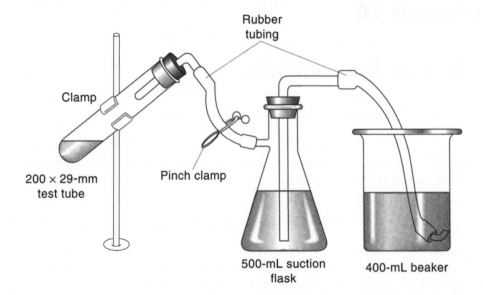

Figure 10.1 This experiment employs the water displacement method to determine the amount of hydrogen gas generated in the chemical reaction.

Using (5), you can now calculate n_{H_2}, the number of moles of hydrogen produced by your weighed sample. You can then calculate the percentages of Al and Zn in the sample by properly applying (3) and (4) to your results. For a sample containing m_{Al} grams Al and m_{Zn} grams Zn, it follows that

$$n_{H_2} = (m_{Al} \times 0.0556) + (m_{Zn} \times 0.0153) \tag{8}$$

For a one-gram sample, m_{Al} and m_{Zn} represent the mass fractions of Al and Zn, that is, % Al / 100 and % Zn / 100. Therefore

$$N_{H_2} = \left(\frac{\% \, Al}{100} \times 0.0556\right) + \left(\frac{\% \, Zn}{100} \times 0.0153\right) \tag{9}$$

where N_{H_2} = number of moles of H_2 produced *per gram* of sample. Since it is also true that

$$\% \, Zn = 100\% - \% \, Al \tag{10}$$

Equation 9 can be written in the form

$$N_{H_2} = \left(\frac{\% \, Al}{100} \times 0.0556\right) + \left(\frac{100 - \% \, Al}{100} \times 0.0153\right) \tag{11}$$

We can solve Equation 11 directly for % Al if we know the number of moles of H_2 evolved per gram of sample. To save time in the laboratory and to avoid arithmetic errors, it is highly desirable to prepare in advance a graph representing N_{H_2} as a function of % Al. Then when N_{H_2} has been determined in the experiment, % Al in the sample can be read directly from the graph. Directions for preparing such a graph are given in Problem 1 in the Advance Study Assignment.

Experimental Procedure

WEAR YOUR SAFETY GLASSES WHILE
PERFORMING THIS EXPERIMENT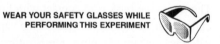

Obtain a suction flask, large (29 × 200 mm) test tube, stopper assemblies, and a sample of Al-Zn alloy. Assemble the apparatus as shown in Figure 10.1.

Take a gelatin capsule from the supply on the lab bench and weigh it on the analytical balance to ± 0.0001 g. Pour your alloy sample out on a piece of paper and add about half of it to the capsule. If necessary, break up the turnings into smaller pieces by simply tearing them. Seal the capsule and weigh it again. The mass of the sample should be between 0.1500 and 0.2500 g. Use care in both weighings, since the sample is small and a small weighing error will produce a large experimental error. Put the remaining alloy back in its container.

Fill the suction flask and beaker about two-thirds full of water. Moisten the stopper on the suction flask and insert it firmly into the flask. Open the pinch clamp and apply suction to the tubing attached to the side arm of the suction flask. Pull water into the flask from the beaker until the water level in the flask is 4 or 5 cm below the side arm. To apply suction, use a suction bulb or a short piece of rubber tubing attached temporarily to the tube that goes through the test tube stopper. Close the pinch clamp to prevent siphoning. The tubing from the beaker to the flask should be full of water, with no air bubbles.

Carefully remove the tubing from the beaker and put the end on the lab bench. As you do this, no water should leak out of the end of the tubing. Pour the water remaining in the beaker into another beaker, letting the 400-mL beaker drain for a second or two. Without drying it, weigh the empty beaker on a top-loading balance to ± 0.1 g. Put the tubing back in this beaker.

Pour 10. mL (that's 10 ± 1 mL) of 6 M HCl, hydrochloric acid, into the large test tube. Gently drop the gelatin capsule into the HCl solution; if it sticks to the tube, poke it down into the acid with your stirring rod. Insert the stopper firmly into the test tube and open the pinch clamp. If a little water goes into the beaker at that point, pour that water out, letting the beaker drain for a second or two.

Within a few minutes the acid will eat through the wall of the capsule and begin to react with the alloy. Removing the test tube from the clamp and agitating it gently can speed things up, **but hold it only above the liquid level,** as the part of the test tube in contact with the acid will become quite hot as the reaction proceeds. **Stop agitating the tube once the reaction picks up speed, or it may get going too fast, and bubble over!** The hydrogen gas that is formed will go into the suction flask and displace water from the flask into the beaker. The volume of water that is displaced will equal the volume of gas that is produced. As the reaction proceeds you will probably observe a dark foam, which contains particles of unreacted alloy. The foam may carry some of the alloy up the tube. Remove the tube from the clamp and tilt and/or rotate it gently to make sure that all of the alloy gets into the acid solution. The reaction should be over within five to ten minutes. At that time the liquid solution will again be clear, the foam will be essentially gone, the capsule will be all dissolved, and there should be no unreacted alloy. When the reaction is over, close the pinch clamp and take the tubing out of the beaker. Weigh the beaker and the displaced water to ± 0.1 g. Measure the temperature of the water and the atmospheric pressure.*

Take it Further (Optional): Using the procedure and apparatus in this experiment, find the percentage by mass of Al in an aluminum soda or beer can.

> **DISPOSAL OF REACTION PRODUCTS.** Pour the used acid solution into the waste container. Reassemble the apparatus and repeat the experiment with the remaining sample of alloy.

---

*The pressure of the gas in the flask will differ slightly from the atmospheric pressure, because the water levels inside and outside the flask are not quite equal. The error arising from this effect is smaller than other experimental errors, so we shall ignore it.

Name _____ **Section** _____

Experiment 10

Data and Calculations: Analysis of an Aluminum-Zinc Alloy

	Trial 1	**Trial 2**
Mass of gelatin capsule	_____ g	_____ g
Mass of alloy sample plus capsule	_____ g	_____ g
Mass of empty beaker	_____ g	_____ g
Mass of beaker plus displaced water	_____ g	_____ g
Atmospheric pressure	_____ mm Hg	
Temperature	_____ °C	
Mass of alloy sample	_____ g	_____ g
Mass of displaced water	_____ g	_____ g
Volume of displaced water (density of water, $\rho = 1.00$ g/mL)	_____ mL	_____ mL
Volume of H_2, V	_____ L	_____ L
Temperature of H_2, T	_____ K	
Vapor pressure of water at T, VP_{H_2O}, from Appendix I	_____ mm Hg	

(continued on following page)

	Trial 1	**Trial 2**
Pressure of dry H_2, P_{H_2}		_____ mm Hg;
		_____ atm
Moles H_2 from sample, n_{H_2}	_____ moles	_____ moles
Moles H_2 per gram of sample, N_{H_2}	_____ moles/g	_____ moles/g
% Al (read from graph)	_____ %	_____ %
Unknown #	_____	

Experiment 10

Advance Study Assignment: Analysis of an Aluminum-Zinc Alloy

1. On the following page, construct a graph of N_{H_2} vs % Al. To do this, refer to Equation 11 and the discussion preceding it. Note that a plot of N_{H_2} vs % Al should be a straight line (why?). To fix the position of a straight line, it is necessary to locate only two points. The most obvious way to do this is to find N_{H_2} when % Al = 0 and when % Al = 100. If you wish you may calculate some intermediate points (for example, N_{H_2} when % Al = 50, or 20, or 70); all these points should lie on the same straight line. To use a spreadsheet, set up Equation 11 for different Al percentages. Graph N_{H_2} vs % Al.

2. A student obtained the following data in this experiment. Fill in the blanks in the data and make the indicated calculations:

Mass of gelatin capsule	0.1134 g	Ambient temperature, t	24°C
Mass of capsule plus alloy sample	0.3218 g	Ambient temperature, T	_____ K
Mass of alloy sample, m	_____ g	Atmospheric pressure	723 mm Hg
Mass of empty beaker	154.3 g	Vapor pressure of H_2O at t (Appendix I)	_____ mm Hg
Mass of beaker plus displaced water	406.1 g		
Mass of displaced water	_____ g	Pressure of dry H_2, P_{H_2} (Eq. 7)	_____ mm Hg
Volume of displaced water (density = 1.00 g/mL)	_____ mL	Pressure of dry H_2	_____ atm

Volume, V, of H_2 = Volume of displaced water = _____ mL = _____ L

Find the number of moles of H_2 evolved, n_{H_2} [Use Equation 5: V in liters, P_{H_2} in atm, T in K, $R = 0.0821$ liter · atm/(mole · K).]

_____ moles H_2

Find N_{H_2}, the number of moles of H_2 per gram of sample (n_{H_2}/m).

_____ moles H_2/g

Find the % Al in the sample from the graph prepared for Problem 1.   _____ % Al

Find the % Al in the sample by using Equation 11.

_____ % Al

Experiment 10

Advance Study Assignment: Analysis of an Aluminum-Zinc Alloy

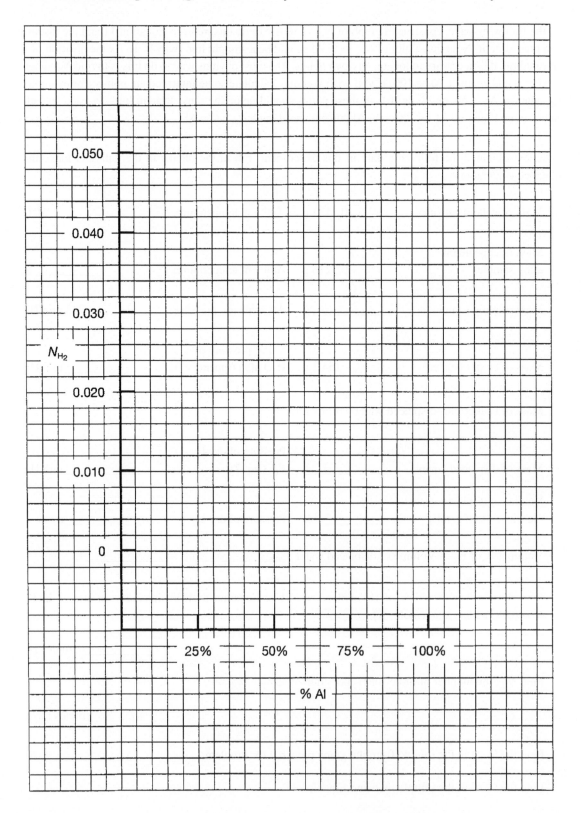

Experiment 11

The Atomic Spectrum of Hydrogen

When atoms are excited, either in an electric discharge or with heat, they tend to give off light. The light is emitted only at certain wavelengths that are characteristic of the atoms in the sample. These wavelengths constitute what is called the atomic spectrum of the excited element and reveal much of the detailed information we have regarding the electronic structure of atoms.

Atomic spectra are interpreted in terms of quantum theory. According to this theory, atoms can exist only in certain states, each of which has an associated, and fixed, amount of energy. When an atom changes its state, it must absorb or emit an amount of energy that is just equal to the difference between the energies of the initial and final states. This energy may be absorbed or emitted in the form of light. The emission spectrum of an atom is obtained when excited atoms fall from higher to lower energy levels. Since there are many such levels, the atomic spectra of most elements are very complex.

Light is absorbed or emitted by atoms in the form of photons, each of which has a specific amount of energy, ϵ. This energy is related to the wavelength of light by the equation

$$\epsilon_{photon} = \frac{hc}{\lambda} \tag{1}$$

where h is Planck's constant, 6.62608×10^{-34} joule seconds, c is the speed of light, 2.997925×10^8 meters per second, and λ is the wavelength, in meters. The energy ϵ_{photon} is in joules and is the energy given off by one atom when it jumps from a higher to a lower energy level. Since energy is conserved (the total amount of energy cannot change, because energy cannot be created or destroyed), the change in energy of the atom, $\Delta\epsilon_{atom}$, must equal the energy of the photon emitted:

$$\Delta\epsilon_{atom} = \epsilon_{photon} \tag{2}$$

where $\Delta\epsilon_{atom}$ is equal to the energy in the upper level minus the energy in the lower one. Combining Equations 1 and 2, we obtain the relation between the change in energy of the atom and the wavelength of light associated with that change:

$$\Delta\epsilon_{atom} = \epsilon_{upper} - \epsilon_{lower} = \epsilon_{photon} = \frac{hc}{\lambda} \tag{3}$$

The amount of energy in a photon given off when an atom makes a transition from one level to another is very small, on the order of 1×10^{-19} joules. This is not surprising since, after all, atoms are very small. To avoid such small numbers, we will work with 1 mole of atoms, much as we do in dealing with energies involved in chemical reactions. To do this we need only to multiply Equation 3 by Avogadro's number, N. Let

$$N\Delta\epsilon = \Delta E = N\epsilon_{upper} - N\epsilon_{lower} = E_{upper} - E_{lower} = \frac{Nhc}{\lambda}$$

Substituting the values for N, h, and c, and expressing the wavelength in nanometers (nm) rather than meters ($1\,\text{meter} = 1 \times 10^9$ nanometers), we obtain an equation relating energy change in kilojoules per mole of atoms to the wavelength of photons associated with such a change:

$$\Delta E = \frac{6.02214 \times 10^{23} \times 6.62608 \times 10^{-34} \text{ J sec} \times 2.997925 \times 10^8 \text{ m/sec}}{\lambda \text{ (in nm)}} \times \frac{1 \times 10^9 \text{ nm}}{1 \text{ m}} \times \frac{1 \text{ kJ}}{1000 \text{ J}}$$

$$\Delta E = E_{upper} - E_{lower} = \frac{1.19627 \times 10^5 \text{ kJ/mole}}{\lambda \text{ (in nm)}} \quad \text{or} \quad \lambda \text{ (in nm)} = \frac{1.19627 \times 10^5}{\Delta E \text{ (in kJ/mole)}} \tag{4}$$

Equation 4 is useful in the interpretation of atomic spectra. Say, for example, we study the atomic spectrum of sodium and find that the wavelength of its strong yellow line is 589.16 nm (see Fig. 11.1). This line is

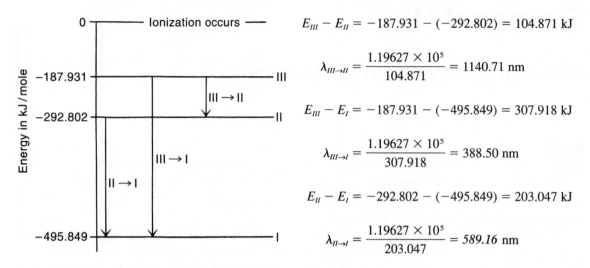

Figure 11.1 Spectral line wavelengths are calculated from the energy levels of the sodium atom.

known to result from a transition between two of the three lowest levels in the atom. The energies of these levels are shown in Figure 11.1. To determine which of the levels give rise to the 589.16-nm line, we note that there are three possible transitions, shown by downward arrows in the figure. We find the wavelengths associated with those transitions by first calculating ΔE ($E_{\text{uppper}} - E_{\text{lower}}$) for each transition. Knowing ΔE we calculate λ by Equation 4. Clearly, the II $\rightarrow$ I transition is the source of the yellow line in the spectrum.

The simplest of all atomic spectra is that of the hydrogen atom. In 1886 Balmer showed that the lines in the spectrum of the hydrogen atom had wavelengths that could be expressed by a rather simple equation. Bohr, in 1913, explained the spectrum on a theoretical basis with his famous model of the hydrogen atom. According to Bohr's theory, the energies allowed to a hydrogen atom are given by the equation

$$\epsilon_n = \frac{-B}{n^2} \tag{5}$$

where B is a constant predicted by the theory and n is an integer, 1, 2, 3, ..., called a quantum number. It has been found that all the lines in the atomic spectrum of hydrogen can be associated with energy levels in the atom which are predicted with great accuracy by Bohr's equation. When we write Equation 5 in terms of a mole of H atoms, and substitute the numerical value for B, we obtain

$$E_n = \frac{-1312.04}{n^2} \text{ kilojoules per mole}, \ n = 1, 2, 3, ... \tag{6}$$

Using Equation 6 you can calculate, very accurately indeed, the energy levels for hydrogen. Transitions between these levels give rise to the wavelengths of light observed in the atomic spectrum of hydrogen. These wavelengths are also known very accurately. Given both the energy levels and the wavelengths, it is possible to determine the actual levels associated with each wavelength. In this experiment your task will be to make determinations of this type for the observed wavelengths in the hydrogen atomic spectrum that are listed in Table 11.1.

Table 11.1

Some Wavelengths in the Spectrum of the Hydrogen Atom, as Measured in a Vacuum					
	Assignment		Assignment		Assignment
Wavelength, λ [nm]	$n_{\text{hi}} \rightarrow n_{\text{lo}}$	Wavelength, λ [nm]	$n_{\text{hi}} \rightarrow n_{\text{lo}}$	Wavelength, λ [nm]	$n_{\text{hi}} \rightarrow n_{\text{lo}}$
97.25	_____	410.29	_____	1005.2	_____
102.57	_____	434.17	_____	1094.1	_____
121.57	_____	486.27	_____	1282.2	_____
389.02	_____	656.47	_____	1875.6	_____
397.12	_____	954.86	_____	4052.3	_____

Experimental Procedure

There are several ways we might analyze an atomic spectrum, given the energy levels of the atom involved. A simple and effective method is to calculate the wavelengths of some of the lines arising from transitions between some of the lower energy levels, and see if they match those that are observed. We shall use this method in our experiment. All the data are good to at least five significant figures, so by using a calculator you should be able to make very accurate determinations. A spreadsheet can be used to good advantage in this experiment. Your instructor may give you some suggestions on how you might proceed.

A. Calculation of the Energy Levels of the Hydrogen Atom

Given the expression for E_n in Equation 6, it is possible to calculate the energy for each of the allowed levels of the H atom starting with $n = 1$. Using your calculator, calculate the energy in kJ/mole of each of the 10 lowest levels of the H atom. Note that the energies are all negative, so that the *lowest* energy will have the *largest* allowed negative value. Enter these values in the table of energy levels, Table 11.2. On the energy level diagram provided just before the Advance Study Assignment (ASA), plot along the y-axis each of the six lowest energies, drawing a horizontal line at the allowed level and writing the value of the energy alongside the line near the y-axis. Write the quantum number associated with the level to the right of the line.

B. Calculation of the Wavelengths of the Lines in the Hydrogen Atomic Spectrum

The lines in the hydrogen spectrum all arise from jumps made by the atom from one energy level to another. The wavelengths in nm of these lines can be calculated by Equation 4, where ΔE is the difference in energy in kJ/mole between any two allowed levels. For example, to find the wavelength of the spectral line associated with a transition from the $n = 2$ level to the $n = 1$ level, calculate the difference, ΔE, between the energies of those two levels. Then substitute ΔE into Equation 4 to obtain this wavelength in nanometers.

Using the procedure we have outlined, calculate the wavelengths in nm of all the lines we have indicated in Table 11.3. That is, calculate the wavelengths of all the lines that can arise from transitions between any two of the six lowest levels of the H atom. Enter these values in Table 11.3.

C. Assignment of Observed Lines in the Hydrogen Atomic Spectrum

Compare the wavelengths you have calculated with those listed in Table 11.1. If you have made your calculations properly, your wavelengths should match, within the error of your calculation, several of those that are observed. On the line opposite each wavelength in Table 11.1, write the quantum numbers of the upper and lower states for each line whose origin you can recognize by comparison of your calculated values with the observed values. On the energy level diagram, draw a vertical arrow pointing down (light is emitted, meaning energy leaves the atom and $\Delta E_{\text{atom}} < 0$) between those pairs of levels that you associate with any of the observed wavelengths. By each arrow write the wavelength of the line originating from that transition.

There are a few wavelengths in Table 11.1 that have not yet been calculated. Enter those wavelengths in Table 11.4. By assignments already made and by an examination of the transitions you have marked on the diagram, deduce the quantum states that are likely to be associated with the as-yet-unassigned lines. This is perhaps most easily done by first calculating the value of the energy change associated with each wavelength, ΔE. Then find two values of E_n whose difference is equal to ΔE. The quantum numbers for the two E_n states whose energy difference is ΔE will be the ones that are to be assigned to the given wavelength. When you have found n_{hi} and n_{lo} for a wavelength, write them in Table 11.1 and Table 11.4; continue until all the lines in the tables have been assigned.

D. The Balmer Series

This is the most famous series in the atomic spectrum of hydrogen. The lines in this series are the only ones in the spectrum that occur in the visible region. Your instructor may have a hydrogen source tube and a spectroscope with which you may be able to observe some of the lines in the Balmer series. There are some questions you should answer relating to this series on the report page.

Experiment 11

Data and Calculations: The Atomic Spectrum of Hydrogen

A. Calculation of the Energy Levels of the Hydrogen Atom

Energies are to be calculated from Equation 6 for the 10 lowest energy states.

Table 11.2

Quantum number, n	Energy, E_n [kJ/mol]	Quantum number, n	Energy, E_n [kJ/mol]
_____	_____	_____	_____
_____	_____	_____	_____
_____	_____	_____	_____
_____	_____	_____	_____
_____	_____	_____	_____

B. Calculation of the Wavelengths of the Lines in the Hydrogen Atomic Spectrum

In the upper half of each box write ΔE, the difference in energy in kJ/mole between $E_{n_{hi}}$ and $E_{n_{lo}}$. In the lower half of the box, write the λ in nm associated with that value of ΔE.

Table 11.3

n_{hi}	6	5	4	3	2	1

n_{lo}

1

2

3

4

5

$$\Delta E = E_{n_{hi}} - E_{n_{lo}}$$

$$\lambda \text{ [nm]} = \frac{1.19627 \times 10^5}{\Delta E}$$

(continued on following page)

C. Assignment of Observed Lines in the Hydrogen Atomic Spectrum

1. As directed in the procedure, assign n_{hi} and n_{lo} for each wavelength in Table 11.1 which corresponds to a wavelength calculated in Table 11.3.

2. List below all wavelengths you cannot yet assign.

Table 11.4

Observed wavelength, λ [nm]	Energy of transition, ΔE [kJ/mol]	Probable transition, $n_{hi} \rightarrow n_{lo}$	Calculated wavelength, λ [nm] (Eq. 4)
_____	_____	_____	_____
_____	_____	_____	_____
_____	_____	_____	_____
_____	_____	_____	_____

D. The Balmer Series

1. When Johann Balmer found his famous series for hydrogen in 1886, he was limited experimentally to wavelengths in the visible and near ultraviolet regions from 250 nm to 700 nm, so all the lines in his series lie in that region. On the basis of the entries in Table 11.3 and the transitions on your energy level diagram, what common characteristic do the lines in the Balmer series have?

What would be the longest possible wavelength for a line in the Balmer series?

$\lambda = $ _____ nm

What would be the shortest possible wavelength that a line in the Balmer series could have? (Hint: What is the largest possible value of ΔE that can be associated with a line in the Balmer series?)

$\lambda = $ _____ nm

Fundamentally, why would any line in the hydrogen spectrum between 250 nm and 700 nm belong to the Balmer series? (Hint: On the energy level diagram note the range of possible values of ΔE for transitions to the $n = 1$ level and to the $n = 3$ level. Could a spectral line involving a transition to the $n = 1$ level have a wavelength in the range indicated?)

The Ionization Energy of the Hydrogen Atom

1. In a normal hydrogen atom at room temperature, the electron is in its lowest energy state, $n = 1$, which is called the ground state of the atom. The maximum electronic energy that a hydrogen atom can have is 0 kJ/mole, at which point the electron would essentially be removed from the atom leaving an H^+ ion. How much energy, in kilojoules per mole, does it take to ionize an H atom?

    _____ kJ/mole

    The ionization energy of hydrogen is often expressed in units other than kJ/mole. What would it be in joules per atom? In electron volts per atom? (1 ev = 1.602×10^{-19} J)

    _____ J/atom; _____ ev/atom

    (The energy level diagram to be completed in Part A is on the following page.)

Experiment 11

Data and Calculations: The Atomic Spectrum of Hydrogen: Energy Level Diagram

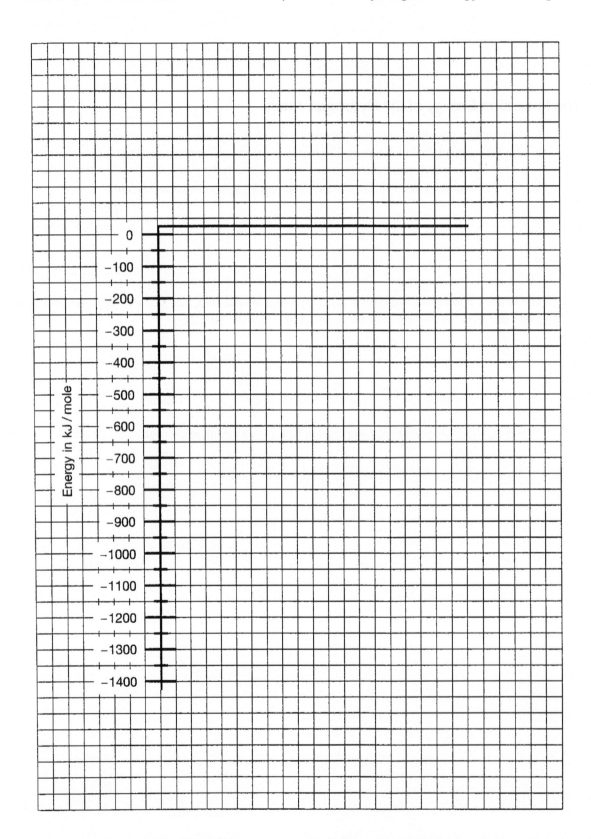

Experiment 11

Advance Study Assignment: The Atomic Spectrum of Hydrogen

1. Found in the gas phase, the boron tetracation (quadruply charged positive, or +4, ion), B^{4+}, has an energy level formula analogous to that of the hydrogen atom, since both species have only one electron. The energy levels of the B^{4+} ion are given by the equation

$$E_n = -\frac{32815}{n^2} \text{ kJ/mole where } n = 1, 2, 3, \ldots$$

   a. Calculate the energies in kJ/mole for the four lowest energy levels of the B^{4+} ion.

   $E_1 = $ _____ kJ/mole

   $E_2 = $ _____ kJ/mole

   $E_3 = $ _____ kJ/mole

   $E_4 = $ _____ kJ/mole

   b. One of the most important transitions for the B^{4+} ion involves a jump from the $n = 2$ to the $n = 1$ level. ΔE for this transition equals $E_2 - E_1$, where these two energies are obtained as in Part (a). Find the value of ΔE in kJ/mole. Find the wavelength in nanometers of the line emitted when this transition occurs; use Equation 4 to make the calculation.

   $\Delta E = $ _____ kJ/mole; $\lambda = $ _____ nm

(continued on following page)

c.   Three of the strongest lines in the B^{4+} ion spectrum are observed at the following wavelengths: (1) 74.99 nm; (2) 26.25 nm; (3) 3.888 nm. Find the quantum numbers of the initial and final states for the transitions that give rise to these three lines. Do this by calculating, using Equation 4, the wavelengths of lines that can originate from transitions involving any two of the four lowest levels. You calculated one such wavelength in Part (b). Make similar calculations with the other possible pairs of levels. When a calculated wavelength matches an observed one, write down n_{high} and n_{low} for that line. Continue until you have assigned all three of the lines. Show your work below.

(1) _____ $\rightarrow$ _____   (2) _____ $\rightarrow$ _____   (3) _____ $\rightarrow$ _____

Experiment 12

The Alkaline Earths and the Halogens — Two Families in the Periodic Table

The periodic table arranges the elements in order of increasing atomic number in horizontal rows of such length that elements with similar properties recur periodically; that is, they fall directly beneath each other in the table. The elements in a given vertical column are referred to as a family or group. The physical and chemical properties of the elements in a given family change gradually as one goes from one element in the column to the next. By observing the trends in properties, the elements can be arranged in the order in which they appear in the periodic table. In this experiment we will study the properties of the elements in two families in the periodic table, the alkaline earths (Group 2) and the halogens (Group 17).

The alkaline earths are all moderately reactive metals and include barium, beryllium, calcium, magnesium, radium, and strontium. (Since beryllium compounds are rarely encountered and are often very poisonous, and radium compounds are highly radioactive, we will not include these two elements in this experiment.) All the alkaline earths exist in their compounds and in solution as M^{2+} cations (Mg^{2+}, Ca^{2+}, etc.). If a solution containing one of these cations is mixed with one containing an anion (CO_3^{2-}, SO_4^{2-}, IO_3^-, etc.), an alkaline earth salt will precipitate if the compound containing those two ions is insoluble.

For example:

$$M^{2+}(aq) + SO_4^{2-}(aq) \rightarrow MSO_4(s) \quad \text{if } MSO_4 \text{ is insoluble} \tag{1a}$$

$$M^{2+}(aq) + 2\ IO_3^-(aq) \rightarrow M(IO_3)_2(s) \quad \text{if } M(IO_3)_2 \text{ is insoluble} \tag{1b}$$

We would expect, and do indeed generally observe, that the salts composed of a given anion and the alkaline earth cations show a smooth trend in solubility, consistent with the order of the cations' corresponding elements in the periodic table. That is, as we go from one end of the alkaline earth family to the other, the solubilities of, say, the sulfate salts either gradually increase or decrease. Similar trends exist for the carbonates, oxalates, and iodates formed by those cations. By determining such trends in this experiment, you will be able to confirm the order of the alkaline earths in the periodic table.

The elemental halogens are also relatively reactive. They include astatine, bromine, chlorine, fluorine, iodine, and the recently discovered element tennessine. We will not study astatine or fluorine in this experiment, since the former is radioactive and the latter is too reactive to be safely employed in labs such as ours. Tennessine is extremely unstable and decays to lighter elements so quickly that nobody has really had a chance to study it! Unlike the alkaline earths, the halogens tend to gain electrons, forming X^- anions (Cl^-, Br^-, etc.). Because of this property, the halogens are oxidizing agents, species that tend to oxidize (remove electrons from) other species. An interesting and simple example of the sort of reaction that may occur arises when a solution containing a halogen (Cl_2, Br_2, I_2) is mixed with a solution containing a (different) halide ion (Cl^-, Br^-, I^-). Taking X_2 to be the halogen, and Y^- to be a halide ion, the following reaction may occur, in which another halogen, Y_2, is formed:

$$X_2(aq) + 2\ Y^-(aq) \rightarrow 2\ X^-(aq) + Y_2(aq) \tag{2}$$

The reaction will occur if X_2 is a better oxidizing agent than Y_2, since then X_2 can produce Y_2 by removing electrons from the Y^- ions. If Y_2 is a better oxidizing agent than X_2, Reaction 2 will not proceed, but will instead be spontaneous in the opposite direction.

In this experiment we will mix solutions of halogens and halide ions to determine the relative oxidizing strengths of the halogens. These strengths show a smooth variation as one goes from one halogen to the next in the periodic table. We will be able to tell if a reaction occurs by the colors we observe. In water, and particularly in some organic solvents, the halogens have characteristic colors. The halide ions are colorless in aqueous solution and insoluble in organic solvents. Bromine (Br_2) in heptane, C_7H_{16} (HEP), is orange, while Cl_2 and I_2 in that solvent have quite different colors.

Say, for example, we mix an aqueous solution of Br_2 with a little heptane, which is lighter than and insoluble in water. The Br_2 is much more soluble in HEP than in water and goes into the HEP layer, giving it an orange color. To that mixture we add a solution containing a halide ion, say Cl^- ion, and mix well. If Br_2 is a better oxidizing agent than Cl_2, it will take electrons from the chloride ions and will be converted to bromide, Br^-, ions; the reaction would be

$$Br_2(aq) + 2\ Cl^-(aq) \rightarrow 2\ Br^-(aq) + Cl_2(aq) \tag{3}$$

If the reaction occurs, the color of the HEP layer will change, since Br_2 will be used up and Cl_2 will form. The color of the HEP layer will go from orange to that of a solution of Cl_2 in HEP. If the reaction does *not* occur, the color of the HEP layer will remain orange. By using this line of reasoning, and by working with the possible mixtures of halogens and halide ions, you should be able to arrange the halogens in order of increasing oxidizing power, which should correspond (directly or inversely) to their order in the periodic table.

One difficulty that you may have in this experiment involves terminology rather than actual chemistry. You must learn to distinguish the halogen *elements* from the halide *ions*, since the two kinds of species are not at all the same, even though their names are similar:

Elemental halogens	**Halide ions**
Bromine, Br_2	Bromide ion, Br^-
Chlorine, Cl_2	Chloride ion, Cl^-
Iodine, I_2	Iodide ion, I^-

The *halogens* are molecular substances and oxidizing agents, and all have odors. They are only slightly soluble in water and are much more soluble in HEP, where they have distinct colors. The *halide ions* exist in solution only in water, have no color or odor, and are *not* oxidizing agents. They do not dissolve in HEP.

Given the solubility properties of the alkaline earth cations, and the oxidizing power of the halogens, it is possible to develop a systematic procedure for determining the presence of any Group 2 cation and any Group 17 anion in a solution. In the last part of this experiment you will be asked to set up such a procedure and use it to establish the identity of an unknown solution containing a single alkaline earth halide.

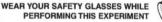
WEAR YOUR SAFETY GLASSES WHILE PERFORMING THIS EXPERIMENT

Experimental Procedure

A. Determining Relative Solubilities of Some Salts of the Alkaline Earths

To each of four small (13×100 mm) test tubes add 1.0 mL (that's 1.0 ± 0.1 mL: 12 drops should reliably fall in that range) of 1.0 M Na_2SO_4. Then add 1.0 mL (again, 12 drops) of 0.10 M solutions of the nitrate salts of barium, calcium, magnesium, and strontium to those tubes, one solution to a tube. Stir each mixture with your glass stirring rod, rinsing the rod in a beaker of deionized water between stirs. Record your results on the solubilities of the sulfates of the alkaline earths in the table on the report page, noting whether a precipitate forms, and any characteristics (such as color, amount, size of particles, and settling tendencies) that might distinguish it.

Rinse out the test tubes, and to each add 1.0 mL of 1.0 M Na_2CO_3. Then add 1.0 mL of the solutions of the alkaline earth salts, one solution to a tube, as before. Record your observations on the solubility properties of the carbonates of the alkaline earth cations. Rinse out the tubes, and test for the solubilities of the oxalates of these cations, using 0.25 M $(NH_4)_2C_2O_4$ as the precipitating reagent. Finally, determine the relative solubilities of the iodates of the alkaline earths, using 1.0 mL of 0.10 M KIO_3 as the test reagent.

B. Relative Oxidizing Powers of the Halogens

Place a few milliliters of bromine-saturated water in a small test tube, and add 1.0 mL of heptane (HEP). Mix the two layers well (stopper the tube and shake it, or withdraw the lower layer into a pipet and squirt it back into the tube through the upper layer), until the bromine color is mostly in the HEP layer. **CAUTION:** **Avoid breathing** the halogen vapors. **Don't use your finger to stopper the tube, or otherwise touch the solutions, since halogens can give you bad chemical burns.** Repeat the experiment using chlorine water and iodine water with separate samples of HEP, noting any color changes as the bromine, chlorine, and iodine are extracted from the water layer into the HEP layer.

To each of three small test tubes add 1.0 mL of bromine water and 1.0 mL of HEP. Then add 1.0 mL of 0.10 M NaCl to the first test tube, 1.0 mL of 0.10 M NaBr to the second, and 1.0 mL of 0.10 M NaI to the third. Mix the two layers in each tube well. Note the color of the HEP phase above each solution. If the color is not that of Br_2 in HEP, a reaction indeed occurred, and Br_2 oxidized that anion, producing the halogen. In such a case, Br_2 is a stronger oxidizing agent than the halogen that was produced.

Rinse out the tubes, and this time add 1.0 mL of chlorine water and 1.0 mL of HEP to each tube. Then add 1.0 mL of the 0.10 M solutions of the sodium halide salts, one solution to a tube, as before. Mix the two layers well and then note the color of the HEP layer. Depending on whether the color is that of Cl_2 in HEP or not, decide whether Cl_2 is a better oxidizing agent than Br_2 or I_2. Again, rinse out the tubes, and add 1.0 mL of iodine water and 1.0 mL of HEP to each. Test each tube with 1.0 mL of a sodium halide salt solution, and determine whether I_2 is able to oxidize Cl^- or Br^- ions. Record all your observations in the table on the report page.

C. Identification of an Alkaline Earth Halide

Your observations on the solubility properties of the alkaline earth cations should allow you to develop a method for determining which of those cations is present in a solution containing one Group 2 cation and no other cations. The method will involve testing samples of the solution with one or more of the reagents you used in Part I. Indicate on the report page how you would proceed.

In a similar way you can determine which halide ion is present in a solution containing only one such anion and no others. There you will need to test a solution of an oxidizing halogen with your unknown to see how the halide ion is affected. From the behavior of the halogen-halide ion mixtures you studied in Part II, you should be able to easily identify the particular halide that is present. Describe your method on the report page. Obtain an unknown solution of an alkaline earth halide, and then use your procedures to determine the cation and anion that it contains.

Optional D. Microscale Procedure for Determining Relative Solubilities of Some Salts of the Alkaline Earths

Your instructor may have you carry out Part I of this experiment using a microscale approach. This method requires much smaller amounts of the reagents. Plastic well plates are employed as containers, and reagents are measured out with small Beral pipets.

Using Beral pipets, add 4 drops of 0.10 M $Ba(NO_3)_2$, barium nitrate, to wells A1–A4, 4 drops to each well. Similarly, add 4 drops of 0.10 M $Ca(NO_3)_2$, calcium nitrate, to wells B1–B4; 4 drops of 0.10 M $Mg(NO_3)_2$, magnesium nitrate, to wells C1–C4, and 4 drops of 0.10 M $Sr(NO_3)_2$, strontium nitrate, to wells D1–D4.

Then, with another Beral pipet, add 4 drops of 1.0 M Na_2SO_4, sodium sulfate, to wells A1–D1. In the table on the report page, record your results on the solubilities of the sulfates of the alkaline earths. Note whether a precipitate formed, and any characteristics, such as amount, size of particles, and cloudiness, which might distinguish it.

With a different Beral pipet, add 4 drops of 1.0 M Na_2CO_3, sodium carbonate, to wells A2–D2. Record your observations on the solubilities of the carbonates of the alkaline earths. Then carry out the same sort of tests with 0.25 M $(NH_4)_2C_2O_4$, ammonium oxalate, in wells A3–D3, and finally with 0.10 M KIO_3, potassium iodate, in wells A4–D4. Record all of your observations in the table on the report page. ■

Take it Further (Optional):   Dissolve a sample of limestone in acid, and determine whether it contains Mg^{2+} as well as Ca^{2+} ions. If both are present, estimate the relative concentrations of the two ions.

DISPOSAL OF REACTION PRODUCTS: Dispose of the reaction products from this experiment as directed by your instructor.

Experiment 12

Observations and Analysis: The Alkaline Earths and the Halogens

A or D. Determining Relative Solubilities of Some Salts of the Alkaline Earths

	1 M Na_2SO_4	1 M Na_2CO_3	0.25 M $(NH_4)_2C_2O_4$	0.1 M KIO_3
$Ba(NO_3)_2$				
$Ca(NO_3)_2$				
$Mg(NO_3)_2$				
$Sr(NO_3)_2$				

Key: P = precipitate forms; S = no precipitate forms.

Note any distinguishing characteristics of a precipitate, such as amount and degree of cloudiness.

Consider the relative solubilities of the Group 2 cations in the various precipitating reagents. On the basis of the trends you observed, list the four alkaline earths in the order in which they should appear in the periodic table. *Start with the one which forms the most soluble oxalate.*

most soluble _____ _____ _____ _____ least soluble

Why did you arrange the elements as you did? Is the order consistent with the properties of the cations in all of the participating reagents?

B. Relative Oxidizing Powers of the Halogens

1. Color of the halogens in solution:

	Br_2	Cl_2	I_2
HEP	_____	_____	_____
Water	_____	_____	_____

(continued on following page)

2. Reactions between halogens and halides:

	Br^-	Cl^-	I^-
Br_2			
Cl_2			
I_2			

State your observations with each mixture, noting the initial and final colors of the HEP layer, and which halogen ends up in the HEP layer. Key: R = reaction occurs; NR = no reaction occurs.

Rank the halogens in order of their increasing oxidizing power.

weakest _____ _____ _____ strongest

Is this their order in the periodic table?

C. Identification of an Alkaline Earth Halide

Procedure for identifying the Group 2 cation:

Procedure for identifying the Group 17 anion:

Observations on unknown alkaline earth halide solution:

Cation present _____

Anion present _____

Unknown # _____

Experiment 12

Advance Study Assignment: The Alkaline Earths and the Halogens

1. As pure elements, all of the halogens are diatomic molecular species. Their melting points are: F_2, 54 K; Cl_2, 172 K; Br_2, 266 K; and I_2, 387 K. Using the periodic table, predict as best you can the molecular formula of elemental astatine, At, the only radioactive element in this family. Also predict whether it will be a solid, liquid, or gas at room temperature.

   Elemental formula: _____   Phase at room temperature: _____

2. Substances A_2, B_2, and C_2 can all act as oxidizing agents. In solution, A_2 is green, B_2 is yellow, and C_2 is red. In the reactions in which they participate, they are reduced to A^-, B^-, and C^- ions, all of which are colorless. When a solution of C_2 is mixed with one containing A^- ions, the color changes from red to green.

   Which species is oxidized? _____

   Which is reduced? _____

   When a solution of C_2 is mixed with one containing B^- ions, the color remains red.

   Is C_2 a better oxidizing agent than A_2? _____

   Is C_2 a better oxidizing agent than B_2? _____

   Arrange A_2, B_2, and C_2 in order of increasing strength as an oxidizing agent.

   _____ < _____ < _____
   Weakest oxidizing agent   Strongest oxidizing agent

(continued on following page)

3. You are given a colorless unknown solution that contains one of the following salts: NaA, NaB, or NaC. In solution, each salt dissociates completely into the Na^+ ion and the anion A^-, B^-, or C^-, whose properties are given in Problem 2. The Na^+ ion is effectively inert. Given the availability of solutions of A_2, B_2, and C_2, develop a simple procedure for determining which salt is present in your unknown.

Experiment 13

The Geometric Structure of Molecules— An Experiment Using Molecular Models

Many years ago it was observed that in many of its compounds the carbon atom formed four chemical linkages to other atoms. As early as 1870, structural formulas of carbon compounds were drawn as shown:

$$
\begin{array}{cc}
\text{H} & \text{H}\quad\text{H} \\
| & |\quad\ | \\
\text{H}-\text{C}-\text{H} & \text{C}=\text{C} \\
| & |\quad\ | \\
\text{H} & \text{H}\quad\text{H} \\
\text{methane} & \text{ethylene}
\end{array}
$$

Although such drawings as these imply that the atom-atom linkages, indicated by valence strokes, lie in a plane, chemical evidence, particularly the existence of only one substance with the chemical formula CH_2Cl_2, requires that the linkages be directed toward the corners of a tetrahedron, at the center of which is the carbon atom. Were the molecules truly flat, one would expect two distinct (geometric) isomers to exist, with the following structural formulae:

$$
\begin{array}{cc}
\text{Cl} & \text{H} \\
| & | \\
\text{H}-\text{C}-\text{Cl} & \text{Cl}-\text{C}-\text{Cl} \\
| & | \\
\text{H} & \text{H}
\end{array}
$$

The physical significance of the chemical linkages between atoms, expressed by the lines or valence strokes in molecular structure diagrams, became evident soon after the discovery of the electron. In a classic paper, G. N. Lewis suggested, on the basis of chemical evidence, that the single bonds in structural formulas involve two electrons and that an atom tends to hold eight electrons in its outermost or valence shell.

Lewis's proposal that atoms generally have eight electrons in their outer shells proved to be extremely useful and has come to be known as the octet rule. It can be applied to many atoms, but is particularly important in the treatment of covalent compounds of atoms in the second row of the periodic table. For atoms such as carbon, oxygen, nitrogen, and fluorine, the eight valence electrons occur in pairs that occupy tetrahedral positions around the central atom core. Some of the electron pairs do not participate directly in chemical bonding and are called unshared or nonbonding pairs; however, the structures of compounds containing such unshared pairs reflect the tetrahedral arrangement of the four pairs of valence shell electrons. In the H_2O molecule, which obeys the octet rule, the four pairs of electrons around the central oxygen atom occupy essentially tetrahedral positions; there are two unshared nonbonding pairs and two bonding pairs that are shared by the O atom and the two H atoms. The H—O—H bond angle is nearly but not exactly tetrahedral since the properties of shared and unshared pairs of electrons are not exactly alike.

$$
\begin{array}{c}
\overset{..}{\underset{.\ .}{\text{O}}} \\
\diagup\quad\diagdown \\
\text{H}\qquad\text{H}
\end{array}
$$

Most molecules obey the octet rule. Essentially, all organic molecules obey the rule, and so do most inorganic molecules and ions. For species that obey the octet rule it is possible to predict electron-dot, or Lewis, structures. The previous drawing of the H_2O molecule is an example of a Lewis structure. Here are several others:

$$
\begin{array}{c}
:\ddot{Cl}: \\
| \\
H-\underset{|}{\overset{|}{C}}-H \\
:\ddot{Cl}:
\end{array}
\qquad
\begin{array}{c}
H-\ddot{N}-H \\
| \\
H
\end{array}
\qquad
:\ddot{O}-H^{-}
\qquad
\begin{array}{c}
H \quad H \\
| \quad | \\
H-\underset{|}{\overset{|}{C}}-\underset{|}{\overset{|}{C}}-H \\
H \quad H
\end{array}
\qquad
\begin{array}{c}
H-C=C-H \\
| \quad | \\
H \quad H
\end{array}
$$

In each of the previous structures there are eight electrons around each atom (except for H atoms, which always have two electrons). There are two electrons in each bond. When counting electrons in these structures, consider *both* of the shared electrons in a bond between two atoms as belonging to (being shared by) *each* of the atoms participating in the bond. In the CH_2Cl_2 molecule just shown, for example, each Cl atom has eight electrons, including the two in the single bond to the C atom. The C atom also has eight electrons, two from each of the four bonds to that atom. The bonding and nonbonding electrons in Lewis structures are all from the *outermost* shells of the atoms involved, and are the so-called valence electrons of those atoms. For the main group elements, the number of valence electrons in an atom is equal to the last digit in the group number of the element in the periodic table. Carbon, in Group 14, has four valence electrons in its atoms; hydrogen, in Group 1, has one; chlorine, in Group 17, has seven valence electrons. In an octet rule structure the valence electrons from all the atoms are arranged in such a way that each atom, except hydrogen, has eight electrons.

Often it is easy to construct an octet rule structure for a molecule. Given that an oxygen atom has six valence electrons (Group 16) and a hydrogen atom has one, it is clear that one O and two H atoms have a total of eight valence electrons; the octet rule structure for H_2O, which we discussed earlier, follows by inspection. Structures like that of H_2O, involving only single bonds and nonbonding electron pairs, are common. Sometimes, however, there is a "shortage" of electrons; that is, it is not possible to construct an octet rule structure in which all the electron pairs are either in single bonds or are nonbonding. C_2H_4 is a typical example of such a species. In such cases, octet rule structures can often be made in which two atoms are bonded by two pairs, rather than one pair, of electrons. The two pairs of electrons form a double bond. In the C_2H_4 molecule, shown above, the C atoms each get four of their electrons from the double bond. The assumption that electrons behave this way is supported by the fact that the $C=C$ double bond is both shorter and stronger than the $C-C$ single bond in the C_2H_6 molecule (see previous example). Double bonds, and triple bonds, occur in many molecules, usually between C, O, N, and/or S atoms.

Lewis structures can be used to predict molecular and ionic geometries. All that is needed is to assume that the four pairs of electrons around each atom are arranged tetrahedrally, or at least nearly so. We have seen how that assumption leads to the correct geometry for H_2O. Applying the same principle to the species whose Lewis structures we listed earlier, we would predict, correctly, that the CH_2Cl_2 molecule would be tetrahedral (roughly, anyway), that NH_3 would be pyramidal (with the nonbonding electron pair sticking up from the pyramid made from the atoms), that the bond angles in C_2H_6 are all tetrahedral, and that the C_2H_4 molecule is planar (the two bonding pairs in the double bond being in a sort of banana bonding arrangement above and below the plane of the molecule). In describing molecular geometry we indicate the positions of the atomic nuclei, not the electrons. The NH_3 molecule is pyramidal, not tetrahedral.

It is also possible to predict polarity from Lewis structures. Polar molecules have their center of positive charge at a different point than their center of negative charge. This separation of charges produces a dipole moment in the molecule. Covalent bonds between different kinds of atoms are polar; all heteronuclear diatomic molecules are polar. In some molecules, the polarity due to one bond may be canceled out by that due to others. Carbon dioxide, CO_2, which is linear, is a nonpolar molecule. Methane, CH_4, which is tetrahedral, is also nonpolar. Among the species whose Lewis structures we have listed, we find that H_2O, CH_2Cl_2, NH_3, and OH^- are polar. C_2H_6 and C_2H_4 are nonpolar.

For some molecules with a given molecular formula, it is possible to satisfy the octet rule with different atomic arrangements. A simple example would be

$$
\begin{array}{c}
H \quad H \\
| \quad | \\
H-\underset{|}{\overset{|}{C}}-\underset{|}{\overset{|}{C}}-\ddot{O}-H \\
H \quad H
\end{array}
\qquad \text{and} \qquad
\begin{array}{c}
H \qquad\quad H \\
| \qquad\quad | \\
H-\underset{|}{\overset{|}{C}}-\ddot{O}-\underset{|}{\overset{|}{C}}-H \\
H \qquad\quad H
\end{array}
$$

These two molecules are called isomers of each other, and the phenomenon is called isomerism. Although the molecular formulas of both substances are the same, C_2H_6O, their properties differ markedly because of their different atomic arrangements.

Isomerism is very common, particularly in organic chemistry, and when double bonds are present, isomerism can occur in very small molecules:

The first two isomers result from the fact that there is no rotation around a double bond, although such rotation can occur around single bonds. The third isomeric structure cannot be converted to either of the first two without breaking bonds.

With certain molecules, given a fixed atomic geometry, it is possible to satisfy the octet rule with more than one bonding arrangement. The classic example is benzene, whose molecular formula is C_6H_6:

These two individual bonding depictions are called resonance structures, and molecules such as benzene, which have two or more resonance structures, are said to exhibit resonance. The actual bonding in such molecules is thought to be an average of the bonding present in the resonance structures. The double-headed arrow between resonance structures is NOT meant to indicate that the molecule spends some of its time in each structure but rather that the one actual structure of the molecule is the *average* of the resonance structures. Thus, the bond between carbon atoms in benzene doesn't ever have the characteristics of a single bond, nor those of a double bond, but instead lies somewhere in-between, slightly stronger than one might expect for a "one-and-a-half" bond. The stability of molecules exhibiting resonance is found to be higher than that anticipated for any single resonance structure.

Although the conclusions we have drawn regarding molecular geometry and polarity can be obtained from Lewis structures, it is much easier to draw such conclusions from models of molecules and ions. The rules we have cited for octet rule structures transfer readily to models. In many ways the models are easier to construct than are the drawings of Lewis structures on paper. In addition, the models are three dimensional and hence much more representative of the actual species. Using the models, it is relatively easy to see both geometry and polarity, as well as to deduce Lewis structures. In this experiment you will assemble models for a sizeable number of common chemical species and interpret them in the ways we have discussed.

Experimental Procedure

IN THIS EXPERIMENT YOUR INSTRUCTOR MAY ALLOW YOU TO WORK WITHOUT SAFETY GLASSES

In this experiment you may work in pairs during the first portion of the laboratory period.

The models you will use consist of drilled wooden balls, short sticks, and springs, or their equivalent. The balls represent atomic nuclei surrounded by the inner electron shells. The sticks and springs represent valence electron pairs, and fit in the holes in the wooden balls. The model (molecule or ion) consists of wooden balls (atoms) connected by sticks or springs (chemical bonds). Some sticks may be connected to only one atom (nonbonding pairs).

In this experiment we will deal with atoms that obey the octet rule; such atoms have four electron pairs around the central core and will be represented by balls with four tetrahedral holes in which there are four sticks or springs. The only exception will be hydrogen atoms, which share two electrons in covalent compounds, and which will be represented by balls with a single hole in which there is a single stick.

In assembling a molecular model of the kind we are considering, it is possible, indeed desirable, to proceed in a systematic manner. We will illustrate the recommended procedure by developing a model for a molecule with the formula CH_2O:

1. Determine the total number of valence electrons in the species. This is easily done once you realize that the number of valence electrons on an atom of a main group element is equal to the last digit of number of the group to which the atom belongs in the periodic table. For CH_2O,

C   Group 14   H   Group 1   O   Group 16

Therefore each carbon atom in a molecule or ion contributes four electrons, each hydrogen atom one electron, and each oxygen atom six electrons. The total number of valence electrons equals the sum of the valence electrons on all of the atoms in the species being studied. For CH_2O this total would be $4 + (2 \times 1) + 6$, or 12 valence electrons. If we are working with an ion, we add one electron for each negative charge or subtract one for each positive charge on the ion.

2. Select wooden balls and sticks to represent the atoms and electron pairs in the molecule. You should use four-holed balls for the carbon atom and the oxygen atom, and one-holed balls to represent the hydrogen atoms. Since there are 12 valence electrons in the molecule and electrons occur in pairs, you will need six sticks to represent the six electron pairs. The sticks will serve both as bonds between atoms and as nonbonding electron pairs.

3. Connect the balls with some of the sticks. (Assemble a skeleton structure for the molecule, joining atoms by single bonds.) In some cases this can only be done in one way. Usually, however, there are various possibilities, some of which are more reasonable than others. In CH_2O the model can be assembled by connecting the two H atom balls to the C atom ball with two of the available sticks, and then using a third stick to connect the C atom and O atom balls. It is generally the case that atoms with valence electron counts closest to four have the most other atoms connected to them, which is why we choose to connect the H atoms to the C atom rather than the O atom.

4. The next step is to use the sticks that are left over in such a way as to fill all the remaining holes in the balls. (Distribute the electron pairs so as to give each atom eight electrons and so satisfy the octet rule.) In the model we have assembled, there is one unfilled hole in the C atom ball, three unfilled holes in the O atom ball, and three available sticks. An obvious way to meet the required condition is to use two sticks to fill two of the holes in the O atom ball, and then use two springs instead of two sticks to connect the C atom and O atom balls. The completed model is shown in Figure 13.1.

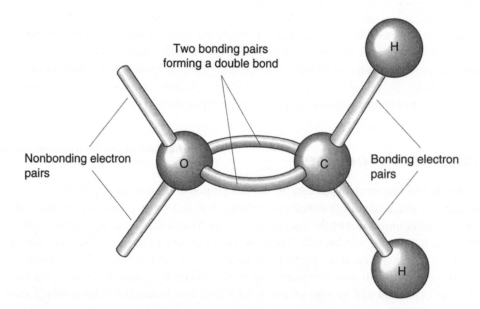

Figure 13.1 Formaldehyde, CH_2O, looks like this when constructed using a molecular model kit of the type typically used in this experiment.

5. Interpret the model in terms of the atoms and bonds represented. The sticks and spatial arrangement of the balls will closely correspond to the electronic and atomic arrangement in the molecule. Given our model, we would describe the CH_2O molecule as being planar with single bonds between carbon and hydrogen atoms and a double bond between the C and O atoms. The H—C—H angle is approximately tetrahedral. There are two nonbonding electron pairs on the O atom. Since all bonds are polar and the molecular symmetry does not cancel the polarity in CH_2O, the molecule is polar. The Lewis structure of the molecule is given below:

$$\ddot{\underset{\displaystyle ..}{O}}\!\!=\!\!C\!\!\underset{\displaystyle \diagdown H}{\overset{\displaystyle \diagup H}{}}$$

(The compound having molecules with the formula CH_2O is well known and is called formaldehyde. The bonding and structure in formaldehyde are as given by the model.)

6. Investigate the possibility of the existence of isomers or resonance structures. It turns out that in the case of CH_2O one can easily construct an isomeric form that obeys the octet rule, in which the central atom is oxygen rather than carbon. It is found that this isomeric form of CH_2O does not exist in nature. As a general rule, carbon atoms almost always form a total of four bonds; put another way, nonbonding electron pairs on carbon atoms are very rare. Another useful rule of a similar nature is that if a species contains several atoms of one kind and one of another, the atoms of the same kind will assume equivalent positions in the species. In SO_4^{2-}, for example, the four O atoms are all equivalent, and are bonded to the S atom and not to one another. Resonance structures are reasonably common. For resonance to occur, however, the atomic arrangement must remain fixed for two or more possible electronic structures. (The sticks and springs can move, but the balls cannot.) For CH_2O there are no resonance structures.

You are now ready to try this on your own!

A. Using the procedure we have outlined, construct and report on models of the molecules and ions listed here and/or other species assigned by your instructor. Draw the complete Lewis structure for each molecule, showing nonbonding as well as bonding electrons. Given the structure you come up with, describe the geometry of the molecule or ion, and state whether the species is polar. Finally, draw the Lewis structures of any likely isomers or resonance forms.

CH_4	H_3O^+	N_2	C_2H_2	SCN^-
CH_2Cl_2	HF	P_4	SO_2	NO_3^-
CH_4O	NH_3	C_2H_4	SO_4^{2-}	HNO_3
H_2O	H_2O_2	$C_2H_2Br_2$	CO_2	$C_2H_4Cl_2$

B. Assuming that stability requires that each atom obey the octet rule, predict the stability of the following species (can you build a structure that obeys the octet rule?):

$$PCl_3 \qquad H_3O \qquad CH_2 \qquad CO$$

C. When you have completed parts A and B, see your laboratory instructor, who will check your results and assign you a set of unknown species. Working now by yourself, assemble models for each species as in the previous section, and report on the geometry and bonding in each of the unknown species on the basis of the model you construct. Also consider and report the polarity and the Lewis structures of any isomers and resonance forms for each species.

Name _____ Section _____

Experiment 13

Observations and Analysis: The Geometric Structure of Molecules

A. Species	Lewis structure	Molecular geometry	Polar?	Isomers or resonance structures
CH_4				
CH_2Cl_2				
CH_4O				
H_2O				
H_3O^+				
HF				
NH_3				
H_2O_2				
N_2				
P_4				
C_2H_4				

(continued on following page)

A. Species	Lewis structure	Molecular geometry	Polar?	Isomers or resonance structures
$C_2H_2Br_2$				
C_2H_2				
SO_2				
SO_4^{2-}				
CO_2				
SCN^-				
NO_3^-				
HNO_3				
$C_2H_4Cl_2$				

B. Stability predicted for PCl_3 _____ H_3O _____ CH_2 _____ CO _____

C. Unknowns

_____

_____

_____

_____

Experiment 13

Advance Study Assignment: The Geometric Structure of Molecules

You are asked by your instructor to construct a model of the $NHCl_2$ molecule. Being of a conservative nature, you proceed as directed in the section on Experimental Procedure.

1. First you need to find the number of valence electrons in $NHCl_2$. For counting purposes with Lewis structures, the number of valence electrons in an atom of a main group element is equal to the last digit in the group number of that element in the periodic table.

    N is in Group _____ H is in Group _____ Cl is in Group _____

    In $NHCl_2$ there is a total of _____ valence electrons.

2. The model consists of balls and sticks.

    a. How many holes should be in the ball you select for the N atom? _____

    b. How many holes should be in the ball you select for the H atom? _____

    c. How many holes should be in the balls you select for the Cl atoms? _____

    The electrons in the molecule are paired, and each stick represents a valence electron pair.

    d. How many sticks do you need? _____

3. Assemble a skeleton structure for the molecule, connecting the balls and sticks to make one unit. Use the rule that N atoms form three bonds, whereas Cl and H atoms usually form only one. Draw a sketch of the skeleton below:

4. a. How many sticks did you need to make the skeleton structure? _____

    b. How many sticks are left over? _____

(continued on following page)

If your model is to obey the octet rule, each ball must have four sticks in it (except for hydrogen atom balls, which need and can only have one). (Each atom in an octet rule species is surrounded by four pairs of electrons.)

c. How many holes remain to be filled? _____

Fill them with the remaining sticks, which represent nonbonding electron pairs. Draw the complete Lewis structure for $NHCl_2$ using lines for bonds and pairs of dots for nonbonding electrons.

5. Describe the geometry of the model, which is that of $NHCl_2$. _____ Is the $NHCl_2$ molecule polar? _____ Why?

Would you expect $NHCl_2$ to have any isomeric forms? _____ Explain your reasoning.

6. Would $NHCl_2$ have any resonance structures? _____ If so, draw them below.

Experiment 14

Heat Effects and Calorimetry

Heat is a transfer of random kinetic energy, sometimes called thermal energy: the transfer occurs spontaneously from an object at a high temperature to an object at a lower temperature. If two objects are in contact, they will, given sufficient time, both reach the same temperature.

Heat is ordinarily measured in a device called an adiabatic calorimeter. An adiabatic calorimeter is simply a container with insulating walls, made so that essentially no thermal energy is exchanged between its contents and the surroundings. Within the calorimeter chemical reactions may occur or thermal energy may transfer from one part of the contents to another, but (ideally) no thermal energy flows into or out of the calorimeter from or to the surroundings.

Specific Heat Capacity

When thermal energy flows into a substance, the temperature of that substance increases. The quantity of heat (thermal energy transfer), q, required to cause a temperature change Δt of any substance is proportional to the mass m of the substance and the temperature change, as shown in Equation 1. The proportionality constant is called the specific heat capacity, abbreviated c, of that substance.

$$q = (\text{specific heat capacity}) \times m \times \Delta t = c \times m \times \Delta t \tag{1}$$

The specific heat capacity can be considered to be the amount of thermal energy required to raise the temperature of one gram of the substance by 1°C (if you make m and Δt in Equation 1 both equal to 1, then q will equal c). Quantities of thermal energy transfer are measured in either joules or calories. To raise the temperature of 1 g of water by 1°C, 4.18 joules of thermal energy must be transferred into the water. The specific heat capacity of water is therefore 4.18 joules/g°C. Since 4.18 joules equals 1 calorie, we can also say that the specific heat capacity of water is 1 calorie/g°C. Ordinarily heat into or out of a substance is determined by the effect that it has on a known amount of water. Because water plays such an important role in these measurements, the calorie, which was the unit of heat most commonly used until recently, was actually defined based on the specific heat capacity of water.

The specific heat capacity of a metal can be readily determined with an adiabatic calorimeter: a device that thermally insulates its contents from the outside world. A weighed amount of metal is heated to some known temperature and is then quickly transferred into a calorimeter containing a measured amount of water at a known temperature. Thermal energy is transferred from the metal to the water, and the two equilibrate at some temperature between the initial temperatures of the metal and the water.

Assuming that no thermal energy is lost from the calorimeter to the surroundings, and that a negligible amount of thermal energy is absorbed by the calorimeter walls, the amount of thermal energy that flows from the metal as it cools is equal to the amount absorbed by the water.

In thermodynamic terms, the heating of the metal is equal in magnitude but opposite in direction, and hence in sign, to that of the water. For the heat q,

$$q_{H_2O} = -q_{metal} \tag{2}$$

If we now express heat in terms of Equation 1 for both the water and the metal M, we get

$$q_{H_2O} = c_{H_2O}\, m_{H_2O}\, \Delta t_{H_2O} = -c_M\, m_M\, \Delta t_M = -q_M \tag{3}$$

In this experiment we measure the masses of water and metal and their initial and final temperatures. (Note that $\Delta t_M < 0$ and $\Delta t_{H_2O} > 0$, since $\Delta t = t_{final} - t_{initial}$.) Given the specific heat capacity of water, we can find the positive specific heat capacity of the metal by Equation 3. We will use this procedure to obtain the specific heat capacity of an unknown metal.

The specific heat capacity of a metal is generally, if somewhat roughly, correlated in a simple way to its molar mass. Dulong and Petit discovered many years ago that about twenty-five joules were required to raise the temperature of one mole of many metals by 1°C. This relation, shown in Equation 4, is known as the Law of Dulong and Petit:

$$MM[g/mol] \approx \frac{25}{c\,[J/g°C]} \qquad (4)$$

where MM is the molar mass of the metal. Once the specific heat capacity of a metal is known, its approximate molar mass can be calculated by Equation 4. The Law of Dulong and Petit was one of the few rules available to early chemists in their studies of molar masses.

Heat of Reaction

When a chemical reaction occurs in aqueous solution, the situation is similar to that which is present when a hot metal sample is put into water. With such a reaction there is an exchange of thermal energy between the reaction mixture and the solvent, water. As in the specific heat capacity experiment, the heating of the reaction mixture is equal in magnitude but opposite in sign to that of the water. The heat, q, associated with the reaction mixture is also equal to the enthalpy change, ΔH, for the reaction, so we obtain the equation

$$q_{reaction} = \Delta H_{reaction} = -q_{H_2O} \qquad (5)$$

By measuring the mass of the water used as solvent, and by observing the temperature change that the water undergoes, we can find q_{H_2O} by Equation 1 and ΔH by Equation 5. If the temperature of the water goes up, energy has left the reaction mixture as thermal energy, so the reaction is *exo*thermic; q_{H_2O} is *positive* and ΔH is *negative*. If the temperature of the water goes down, the reaction mixture has gained thermal energy from the water and the reaction is *endo*thermic. In this case q_{H_2O} is *negative* and ΔH is *positive*. Both exothermic and endothermic reactions are observed.

One of the simplest reactions that can be studied in solution occurs when a solid is dissolved in water. As an example of such a reaction, consider the dissolution of solid sodium hydroxide, NaOH(s), in water:

$$NaOH(s) \rightarrow Na^+(aq) + OH^-(aq); \quad \Delta H = \Delta H_{solution} \qquad (6)$$

When this reaction occurs, the temperature of the solution becomes much higher than that of the NaOH and water that were used. If we dissolve a known amount of NaOH in a measured amount of water in a calorimeter, and measure the temperature change that occurs, we can use Equation 1 to find q_{H_2O} for the reaction and use Equation 5 to obtain ΔH. Noting that ΔH is directly proportional to the amount of NaOH used, we can easily calculate $\Delta H_{solution}$ for either a gram or a mole of NaOH. In the second part of this experiment you will measure $\Delta H_{solution}$ for an unknown ionic solid.

Chemical reactions often occur when solutions are mixed. For example, a precipitate may form, in a reaction opposite in direction to that in Equation 6. A very common reaction is that of neutralization, which occurs when an acidic solution is mixed with one that is basic. In the last part of this experiment you will measure the heat effect when a solution of HCl, hydrochloric acid, is mixed with one containing NaOH, sodium hydroxide, which is basic. The heat effect is quite large, and is the result of the reaction between H^+ ions in the HCl solution with OH^- ions in the NaOH solution:

$$H^+(aq) + OH^-(aq) \rightarrow H_2O(\ell); \quad \Delta H = \Delta H_{neutralization} \qquad (7)$$

Experimental Procedure

A. Specific Heat Capacity

Obtain a calorimeter, a sensitive thermometer, a sample of metal in a large stoppered test tube, and a sample of unknown solid. (The thermometer is expensive, so be careful when handling it.)

The calorimeter consists of two nested expanded polystyrene coffee cups fitted with a Styrofoam cover. There are two holes in the cover: one for a thermometer and the other for a glass stirring rod with a perpendicular loop on one end. Assemble the experimental setup as shown in Figure 14.1.

Fill a 400-mL beaker two-thirds full of water and begin heating it to the boiling point. While the water is heating, weigh your sample of unknown metal in the large stoppered test tube to the nearest 0.1 g on a top-loading or triple-beam balance. Pour the metal into a dry container and weigh the empty test tube and stopper. Replace the metal in the test tube and put the *loosely* stoppered tube into the hot water in the beaker. The water level in the beaker should be high enough that the top of the metal is below the water surface. Continue heating the metal in the water for at least 10 minutes after the water begins to boil, to ensure that the metal attains the temperature of the boiling water. Add small amounts of water as necessary to maintain the water level.

While the water is boiling, weigh the calorimeter to ± 0.1 g. Place about forty milliliters of water in the calorimeter and weigh again. Insert the stirrer and thermometer into the cover and put it on the calorimeter. If a glass thermometer is being used, the thermometer bulb should be completely immersed.

Measure the temperature of the water in the calorimeter to $\pm 0.1°C$. Take the test tube out of the beaker of boiling water, remove the stopper, and pour the metal into the water in the calorimeter. Be careful that no water adhering to the outside of the test tube runs into the calorimeter when you are pouring the metal. Replace the calorimeter cover and agitate the water as best you can with the glass stirrer. Record to $\pm 0.1°C$ the maximum temperature reached by the water. Repeat the experiment, using about fifty milliliters of water in the calorimeter. Be sure to dry your metal before reusing it; this can be done by heating the metal briefly in the test tube in boiling water and then pouring the metal onto a paper towel to drain. You can dry the hot test tube with a little compressed air.

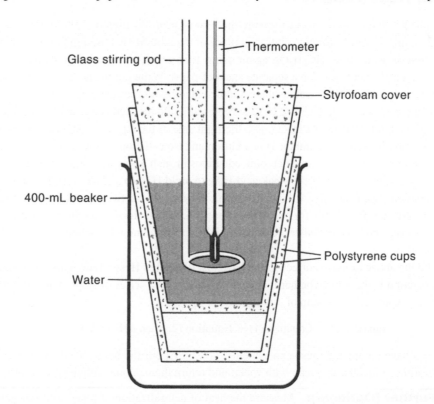

Figure 14.1 The calorimeter employed in this experiment is assumed to be adiabatic (it does not allow thermal energy exchange between its contents and the surroundings).

B. Heat of Solution

Place about fifty milliliters of deionized water in the calorimeter and weigh as in the previous procedure. Measure the temperature of the water to $\pm 0.1°C$. The temperature should be within a degree or two of room temperature. In a small beaker, weigh out about five grams of the solid compound assigned to you. Weigh the beaker and then the beaker plus solid, each to ± 0.1 g. Add the compound to the calorimeter. Stirring continuously and occasionally swirling the calorimeter, determine to $\pm 0.1°C$ the maximum or minimum temperature reached as the solid dissolves. Check to make sure that *all* the solid is dissolved. A temperature change of at least 5 degrees should be obtained in this experiment. If necessary, repeat the experiment, increasing the amount of solid used.

DISPOSAL OF REACTION PRODUCTS. Dispose of the solution in Part B as directed by your instructor.

C. Heat of Neutralization

Rinse out your calorimeter with deionized water, pouring the rinse into the sink. In a graduated cylinder, measure out 25.0 mL of 2.00 M HCl; pour that solution into the calorimeter. Rinse out the cylinder with deionized water, and measure out 25.0 mL of 2.00 M NaOH; pour that solution into a dry 50-mL beaker. Measure the temperature of the acid and of the base to $\pm 0.1°C$, making sure to rinse and dry your thermometer before immersing it in the solutions. Put the thermometer back in the calorimeter cover. Pour the NaOH solution into the HCl solution and put on the cover of the calorimeter. Stir the reaction mixture, and record the maximum temperature that is reached by the neutralized solution.

Calculate the heat of neutralization. It is necessary to slightly modify Equation 1 when calculating q_{H_2O}. Use a value of 1.02 g/mL as an average density for all solutions in this experiment; thus the mass of the solution is obtained by multiplying 50.0 mL by 1.02 g/mL. We will continue to use 4.18 J/g°C as the specific heat capacity.

Optional ## D. Hess's Law

Using the same procedure as in Part C, measure the heat of neutralization of 2.00 M acetic acid (HAc). You should use 25.0 mL of the acetic acid solution in place of the 2.00 M HCl used in Part C. Calculate the molar heat of solution of acetic acid, $HC_2H_3O_2$, again using 1.02 g/mL as the average density for all solutions. Record your data and calculations on a separate sheet of paper. Write out the net ionic equation for the strong acid–strong base neutralization as well as the net ionic equation for the acetic acid–strong base neutralization reaction. (Acetic acid is a weak acid; use HAc to represent undissociated acetic acid.) In each case, show the value of the measured molar heat of reaction on the right side of the equation. Remember that the heat for an exothermic reaction is negative. Combine (via addition and cancellation of any species that appears on opposite sides of the final equation) the two net ionic equations in such a way as to end up with an equation representing the dissociation reaction of the weak acid HAc to yield $H^+(aq)$ and $Ac^-(aq)$. If you needed to reverse one of the equations, the heat value will change signs. As you add the two equations you should add the two heat values, to give the final value for the desired reaction. Just as the dissociation of water to form $H^+(aq)$ and $OH^-(aq)$ is expected to be endothermic, so should the corresponding dissociation reaction for a weak acid, such as acetic acid.

If you did the above exercise correctly, you have just illustrated Hess's Law. This law states that the value of ΔH for a reaction is the same whether it occurs directly, or in a series of steps. Thus, if a thermochemical equation can be expressed as the sum of two equations,

$$\text{Equation (3)} = \text{Equation (1)} + \text{Equation (2), then } \Delta H_3 = \Delta H_1 + \Delta H_2 \ \blacksquare$$

When you have completed these experiments, you may pour the neutralized solutions down the sink. Rinse the calorimeter and thermometer with water, and return them, along with the metal sample.

Take it Further (Optional): Measure the heat of neutralization of household vinegar in its reaction with NaOH solution. Assuming that household vinegar is $5.0\%_{mass}$ acetic acid (or, better, using the actual concentration indicated on the vinegar, if so labeled), find the molar heat of neutralization of acetic acid.

Name _____   Section _____

Experiment 14

Data and Calculations: Heat Effects and Calorimetry

A. Specific Heat Capacity

	Trial 1		Trial 2
Mass of stoppered test tube plus metal	_____ g	$\rightarrow$	_____ g
Mass of test tube and stopper	_____ g	$\rightarrow$	_____ g
Mass of calorimeter	_____ g	$\rightarrow$	_____ g
Mass of calorimeter and water	_____ g		_____ g
Mass of water	_____ g		_____ g
Mass of metal	_____ g	$\rightarrow$	_____ g
Initial temperature of water in calorimeter	_____ °C		_____ °C
Initial temperature of metal (assume 100°C unless directed to do otherwise)	_____ °C	$\rightarrow$	_____ °C
Equilibrium temperature of metal and water in calorimeter	_____ °C		_____ °C
Δt_{water} $(t_{final} - t_{initial})$	_____ °C		_____ °C
Δt_{metal}	_____ °C		_____ °C
q_{H_2O}	_____ J		_____ J
Specific heat capacity of the metal (Eq. 3)	_____ J/g°C		_____ J/g°C
Approximate molar mass of the metal	_____		_____
Unknown metal #			_____

B. Heat of Solution

Mass of calorimeter plus water	_____ g
Mass of beaker	_____ g

(continued on following page)

Mass of beaker plus solid _____ g

Mass of water, m_{H_2O} _____ g

Mass of solid, m_s _____ g

Initial temperature _____ °C

Final temperature _____ °C

q_{H_2O} for the reaction (Eq. 1) ($c = 4.18$ J/g°C) _____ joules

ΔH for the reaction (Eq. 5) _____ joules

The quantity you have just calculated is approximately* equal to the heat of solution of your sample. Calculate the heat of solution per gram of solid sample.

$$\Delta H_{solution} = \underline{\hspace{3cm}} \text{ joules/g}$$

The solution reaction is endothermic/exothermic. (Circle the correct answer.) Give your reasoning.

Solid unknown # _____

Optional   Formula of compound used (if furnished) _____ Molar mass _____ g/mol

Heat of solution per mole of compound _____ kJ/mol ■

C. Heat of Neutralization

Initial temperature of HCl solution _____ °C

Initial temperature of NaOH solution _____ °C

Final temperature of neutralized mixture _____ °C

Change in temperature, Δt [use the average (arithmetic mean) of the initial temperatures of HCl and NaOH] _____ °C

q_{H_2O} (assume 50.0 mL of solution and use average density of 1.02 g/mL) _____ J

Total ΔH for the neutralization reaction _____ J

ΔH per mole of H$^+$ and OH$^-$ ions reacting _____ kJ/mol

*The value of ΔH will be approximate for several reasons. One is that we do not include the amount of heat absorbed by the solute. This effect is smaller than the likely experimental error, and thus we will ignore it.

Experiment 14

Advance Study Assignment: Heat Effects and Calorimetry

1. A metal sample weighing 147.90 g and at a temperature of 99.5°C was placed in 49.73 g of water in an adiabatic calorimeter at 23.0°C. At equilibrium, the temperature of the water and metal was 51.8°C.

   a. What was Δt for the water? ($\Delta t = t_{final} - t_{initial}$)

   _____ °C

   b. What was Δt for the metal?

   _____ °C

   c. How much thermal energy flowed into the water? (Take the specific heat capacity of the water to be 4.18 J/g°C.)

   _____ joules

   d. Calculate the specific heat capacity of the metal, using Equation 3.

   _____ joules/g°C

   e. What is the approximate molar mass of the metal? (Use Eq. 4.)

   _____ g/mol

2. When 4.89 g of NaOH was dissolved in 47.92 g of water in an adiabatic calorimeter at 23.7°C, the temperature of the solution went up to 50.1°C.

   a. Is this dissolution reaction exothermic? _____ Why?

   b. Calculate q_{H_2O}, using Equation 1.

   _____ joules

   c. Find ΔH for the reaction as it occurred in the calorimeter (Eq. 5).

   $\Delta H =$ _____ joules

(continued on following page)

d.   Find ΔH for the dissolution of 1.00 g NaOH in water.

$$\Delta H = \underline{\hspace{2cm}} \text{ kJ/g}$$

e.   Find ΔH for the dissolution of 1 mole of NaOH in water.

$$\Delta H = \underline{\hspace{2cm}} \text{ kJ/mol}$$

f.   Given that NaOH exists as Na^+ and OH^- ions in solution, write the equation for the reaction that occurs when NaOH is dissolved in water.

g.   Given the following heats of formation, ΔH_f, in kJ per mole, as obtained from a table of ΔH_f data, calculate ΔH for the reaction in Part (f). Compare your answer with the result you obtained in Part (e). NaOH(s), -425.6; Na^+(aq), -240.1; OH^-(aq), -230.0

$$\Delta H = \underline{\hspace{2cm}} \text{ kJ/mol}$$

Experiment 15

The Vapor Pressure and Heat of Vaporization of a Liquid*

If we pour a liquid into an open container, which we then close, we find that some of the liquid will evaporate into the air in the container. After a short time, the vapor will establish a partial pressure, which remains constant so long as the temperature does not change. That pressure is called the vapor pressure of the liquid, and at a given temperature it remains the same, regardless of whatever other gases may be present in the container and whatever the total pressure in the container may be.

In the air we breathe, there is usually some water vapor, but with a partial pressure well below the vapor pressure of water (unless it is foggy or raining). The relative humidity is defined as the partial pressure of water vapor in a given air sample, divided by the vapor pressure of water at that temperature. On a hot, sticky day the relative humidity is high, perhaps 80 or 90%, and the air is nearly saturated with water vapor; most people (but not all!) feel uncomfortable under such conditions.

If we raise the temperature of a liquid, its vapor pressure increases. At the "normal boiling point" of the liquid, the vapor pressure is equal to 1 atm. That would be the pressure at 100°C in a closed container in which there is just water and its vapor, with no air.

At high temperatures, the vapor pressure can become very large, reaching about 5 atm for water at 150°C and 15 atm at 200°C. A liquid that boils at a low temperature, say -50°C, will have a very large vapor pressure at room temperature, and confining it at 25°C in a closed container may well cause the container to explode. Failure to recognize this fact has caused several scientists of the authors' acquaintance to have serious accidents.

There is an equation relating the vapor pressure of a liquid to the Kelvin temperature and the enthalpy of vaporization of the liquid, $-\Delta H_{vap}$. It is called the Clausius-Clapeyron Equation and has the form:

$$\ln VP = \frac{-\Delta H_{vaporization}}{RT} + \text{a constant} \tag{1}$$

If we plot ln VP vs the reciprocal of the Kelvin temperature, we should obtain a straight line, the slope of which should equal $-\Delta H_{vap}/R$. In the equation, ln VP is the natural logarithm of the vapor pressure, and the units of ΔH_{vap} and R must match. VP may be in any pressure unit, such as atm, mm Hg, or Pascals.

There are many methods for measuring vapor pressure, most of which are not intuitively obvious. In this experiment we will use one of the simplest, and yet most accurate, methods. We will inject a measured sample of air, free of the liquid under study, into a graduated pipet filled with the liquid under study. We measure the volume of the air-vapor bubble that forms, which will be larger than that of the injected air, since some of the liquid vaporizes. If we raise the temperature, the volume of the bubble will increase: in part because the air in the bubble expands as the temperature rises, but also because more and more liquid vaporizes as the temperature increases. The total pressure in the bubble remains equal to the ambient pressure, so if we can calculate the partial pressure of air in the bubble, we can find the vapor pressure of the liquid at that temperature by simple subtraction.

WEAR YOUR SAFETY GLASSES WHILE PERFORMING THIS EXPERIMENT

Experimental Procedure

Obtain a "device" (a half-sealed 0.5-mL graduated pipet segment: see Fig. 15.1), a Pasteur pipet, a digital thermometer, and a sample of a liquid unknown, whose density will be given to you.

*The general method used in this experiment is one described by DeMuro, Margarian, Mkhikian, No, and Peterson, *J Chem Ed* 76:1113–1116, 1999.

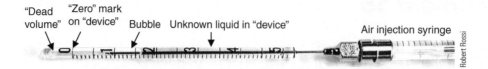

Figure 15.1 The "device" used in this experiment will have an air bubble of known volume injected into it by your instructor.

A. Finding the Vapor Pressure of the Liquid at Room Temperature

Accurately determine the mass of the empty, dry device on an analytical balance and record the result on the report page. Then place the device, open end up, in a small (25- or 50-mL) Erlenmeyer flask (so you don't warm it by holding it with your hand while it is being filled). Using your Pasteur pipet, carefully fill the device completely with your unknown liquid. Dry off the outside of the device and accurately determine its mass again. Be careful around the top of the device—dry only up to the top edge, and do not draw liquid out of the device by drying the top face. You want the liquid level in the device roughly flush with the rim, but the outside of the device must be dry. The surface tension of the liquid will keep it from escaping the device, even when it is set on its side, so weigh *only* the device, setting it directly on the balance—don't weigh your Erlenmeyer flask. The accuracy of your results in this experiment hinges strongly on the quality of this and your next mass measurement—so make both very carefully.

Now, request your instructor's help. Using a syringe, your instructor will inject 0.200 mL of ambient air at the closed end of the device (holding it horizontally, or even with the sealed end tilted slightly up), forming an air bubble. (Surface tension will keep the liquid in the device, even if it is inverted.) After the syringe is removed, the device will no longer be completely full at the open end, so gently return the device to the Erlenmeyer flask, open end up, and add more unknown with your Pasteur pipet, until the liquid level is back up at the rim of the device. Do not worry, the bubble should not escape so long as the device does not experience any sudden motion. Do not heat up the bubble by holding the part of the device containing it with your hand. You need to keep the bubble at room temperature in order to get good results!

Carefully dry and reweigh your device on the same analytical balance you used before. Be sure that the liquid level in the device is again even with the rim, as close as possible to the way it was when you first weighed the device (without the bubble in it). This is your second crucial mass measurement. Record it, and return the device to the Erlenmeyer flask.

Read and record the ambient air temperature and the ambient, atmospheric (laboratory) pressure.

Now fill a medium (18 × 150 mm) test tube to within about 3 cm of the top with your unknown liquid, and also add a bit more to the top of the device, so that the liquid is at or above its rim. (It is no longer important that the outside of the device be dry.) Tilt the test tube to a 45° angle and gently drop in your device, *open end down and sealed end up,* so that it slides down the wall of the liquid-filled test tube, to the bottom. Swirl your test tube and the device inside until the bubble moves all the way to the very top of the device. Read and record the reading on the device as accurately as possible. *(This is a crucial step, so take your time, and read to the nearest 0.002 mL. Note that the numbers on the pipet are in 0.1-mL increments, so each gradation indicates 0.01 mL.)*

Next, insert the temperature probe of the digital thermometer into the unknown liquid, holding it in position with a loose-fitting stopper such that **the tube is not completely sealed,** and record the unknown liquid's temperature. Do not hold the liquid-filled part of the test tube with your hand while taking this measurement, or you will warm it up.

From the data you have obtained so far, you can calculate the vapor pressure of your unknown at the ambient temperature in the lab. Using the mass measurements, you can find the mass of the liquid in the full pipet and the mass of liquid driven out by the bubble. The volume of the bubble is equal to the mass of liquid driven out divided by the liquid's density. This volume is larger than that of the air injected, because some of the unknown liquid vaporizes into the bubble. You may safely assume the bubble always remains at atmospheric pressure: the pressure on it due to the mass of the liquid is negligible. The partial pressure of the air in the bubble is equal, by Boyle's Law, to the atmospheric pressure times the initial volume of the air, divided by the volume of the bubble. The vapor pressure of the liquid is obtained by subtracting the partial pressure of the air from the atmospheric pressure. You can (and should) perform these calculations while waiting for water bath temperatures throughout the rest of the experiment. *If the initial volume of the bubble at room temperature calculates to less than 0.210 mL, consult with your instructor. This value is critical, so do not leave lab (or even return your device) before having calculated it.*

B. Finding the Vapor Pressure of the Liquid at Other Temperatures

We will use a water bath to study how the vapor pressure changes as a function of temperature. Set up the bath as shown in Figure 15.2. Use a 1000-mL beaker that you can heat on a hot plate. Fill the beaker to near the top with warm water from the tap (between 40° and 45°C). Clamp the test tube with the device and digital thermometer in it as shown in Figure 15.2, so that the water level in the bath is as high as the liquid level in the test tube. Place your lab thermometer in the water in the beaker.

The temperature of the water bath should be at 40 ± 3°C for the first measurement. Heat or cool the bath if necessary, with stirring, to get to within a few degrees of that value. It should stabilize within a few minutes. When the temperature on the digital thermometer holds steady for at least 30 seconds, compare it against that of the water bath; the two temperatures should be close to one another when steady. Read the meniscus level in the graduated pipet as carefully as you can, and record it. Also record the temperature of the unknown liquid, as indicated by the digital thermometer.

Now heat the water in the large beaker, with stirring. When the temperature of the bath is roughly 5°C above the first value, stop heating. The temperature will continue to rise for a few minutes as you stir, but will finally level off. When the temperatures on the digital thermometer and the bath thermometer become about equal, and the temperature on the digital thermometer has remained constant for at least 30 seconds, read the level of the meniscus as accurately as you can, and record its value and the temperature of the unknown liquid as indicated by the digital thermometer.

Repeat the measurements you have just made at two higher temperatures, each about five degrees higher than the previous temperature actually measured by the digital thermometer. Each time stop heating at that point, let the temperature level off, and when the bath and device temperatures are similar and both the device temperature and bubble volume are steady for at least 30 seconds, record them. Eventually, you will heat the sample to the point that the bubble volume exceeds the volume of the device, and it escapes. But if this appears likely to happen prematurely, because you have mistakenly overheated the bath, *immediately* remove the test tube from the bath, then add some cold water to the bath and stir. Put the test tube back in position, and then proceed to make a measurement.

The data you have now obtained will allow you to calculate the vapor pressure of the liquid at each temperature you used. It is important to realize that there is a unique "dead volume" above the zero mark in each device (see Fig. 15.1) that the volume reading on the device does not take into account. (Making a device that measured the bubble volume without this complication would require custom glassware, and would be very expensive.) You calculated this "dead volume" in Part A, and on the report page you need to add this value to the volume measurements made in Part B to obtain the actual volume of the bubble at each temperature. You must also calculate the volume that the

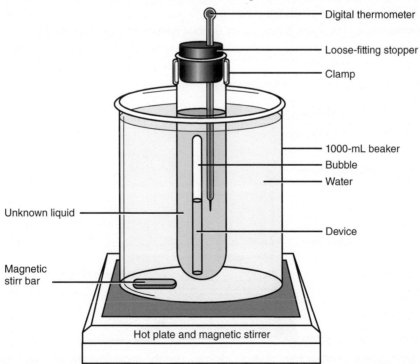

Figure 15.2 After taking a reading at room temperature, the device is heated to progressively higher temperatures in the apparatus above, and readings are taken, until the bubble escapes the device.

air sample alone would occupy at each temperature and atmospheric pressure (use Charles's Law, or the Ideal Gas Law. The moles of *air* present in the bubble remains constant, so the air volume effectively scales with the ratio of the Kelvin temperatures.). The partial pressure of air in the bubble will equal the atmospheric pressure multiplied by the ratio of the air volume to the total bubble volume at each temperature. Calculate the partial pressure of air and the liquid vapor pressure at each temperature as you wait for boiling during the next part of this experiment.

C. Direct Measurement of the Boiling Point

In the last part of this experiment you will measure the boiling point of your liquid at atmospheric pressure. Pour the liquid sample from your test tube into a larger test tube. Determine the boiling point of the liquid using the apparatus shown in Figure 15.3. The thermometer bulb should be just *above* the liquid surface. Heat the water in the bath until the liquid in the tube boils gently, with the vapor condensing *at least 5 cm below* the top of the test tube, but well above the thermometer bulb. Boiling chips may help in keeping the liquid boiling smoothly. As the boiling proceeds, liquid will condense onto and droplets will begin falling from the thermometer bulb. Shortly after that the temperature should become reasonably steady. Record the temperature. The liquid you are using may be flammable and toxic, so you should not inhale the vapor unnecessarily. *Do not* heat the water bath so strongly that condensation of vapors occurs only at the top of the test tube.

Calculating the Heat of Vaporization of the Liquid

When you have completed the calculations described in the previous sections, you should have a table of the vapor pressure of the liquid at each of the temperatures you used. To find the molar heat of vaporization of the liquid, we use Equation 1. First make a table listing ln *VP* (the natural logarithm of the vapor pressure), the temperature T in Kelvin, and $1/T$, as found from your data. Make a graph of ln *VP* vs $1/T$, using the graph paper provided or spreadsheet software such as Excel or Google Sheets (see Appendices VII and VIII). The slope of the line, which should be straight, is equal to $-\Delta H/R$, the molar heat of vaporization divided by the gas constant, R. If you want to express ΔH in joules/mole, R will equal 8.314 J/mole K.

Compare the value of the boiling point as measured directly with that which you can calculate from your graph. The value of $1/T$ at the boiling point will occur where ln *VP* equals ln $P_{atmospheric}$.

Optional   Finally, make a graph of the vapor pressure as a function of the temperature in °C. Use a spreadsheet or another sheet of graph paper. Connect the points with a smooth curve. Comment on how the vapor pressure (*VP*) varies with the Celsius temperature (*t*). ■

Include your graph(s) with your report. When you are finished with the experiment, pour the liquid from the test tube back into its container and return it along with the digital thermometer, the device, and the Pasteur pipet.

Figure 15.3 To determine the boiling point of your unknown, use the apparatus shown above. Note that the stopper **does not seal the test tube**, and that the bulb of the thermometer is *above* the liquid itself. The temperature of the liquid's vapor condensing onto the thermometer bulb provides a better boiling point value than does the temperature of the boiling liquid itself.

Name _____   Section _____

Experiment 15

Data and Calculations: The Vapor Pressure and Heat of Vaporization of a Liquid

A. Finding the Vapor Pressure of the Liquid at Room Temperature

Mass of empty, dry device   _____ g

Mass of device, completely filled with unknown liquid   _____ g

Mass of device filled with liquid and bubble   _____ g

Volume reading on device when at room temperature   _____ mL

Atmospheric (ambient) pressure (P_{atm})   _____ mm Hg

Ambient temperature (t)   _____ °C

Mass of liquid driven out by addition of the bubble   _____ g

Volume of liquid driven out by addition of the bubble (use the density of the liquid)
(Note: This is the volume of the air bubble after it has been saturated with the liquid's vapor.)

_____ mL

Partial pressure of air in the bubble ($P_{air} = P_{atm} \times 0.200$ mL / volume of bubble)

_____ mm Hg

Vapor pressure of the liquid at room temperature ($VP = P_{atm} - P_{air}$)

_____ mm Hg

Difference between volume reading on device and volume of the bubble (based on readings at room temperature—it will be unique to your device, but the same at all temperatures: this is the "dead volume")

_____ mL

(continued on following page)

B. Finding the Vapor Pressure of the Liquid at Other Temperatures

Device temperature (t) [°C]	Volume reading on device [mL]
_____	_____
_____	_____
_____	_____

Calculations:

Absolute temperature (T) [K]	Actual volume of the bubble [mL]	Volume just the air in the bubble would occupy at this temperature [mL]	Partial pressure of air in the bubble [mm Hg]	Partial pressure of vapor in the bubble (VP) [mm Hg]
_____	_____	_____	_____	_____
_____	_____	_____	_____	_____
_____	_____	_____	_____	_____

Absolute temperature (T) [K]	Partial pressure of vapor in the bubble (VP) [mm Hg]	Reciprocal of absolute temperature ($1/T$) [K^{-1}]	Natural logarithm of vapor pressure (ln VP)
_____	_____	_____	_____
_____	_____	_____	_____
_____	_____	_____	_____
_____	_____	_____	_____

Calculated heat of vaporization of unknown liquid (attach your plot and show your work!)

_____ kJ/mol

C. Direct Measurement of the Boiling Point

Observed boiling point of unknown liquid   _____ °C

Unknown #   _____

Experiment 15

Data and Calculations: Vapor Pressure and Heat of Vaporization of Liquids

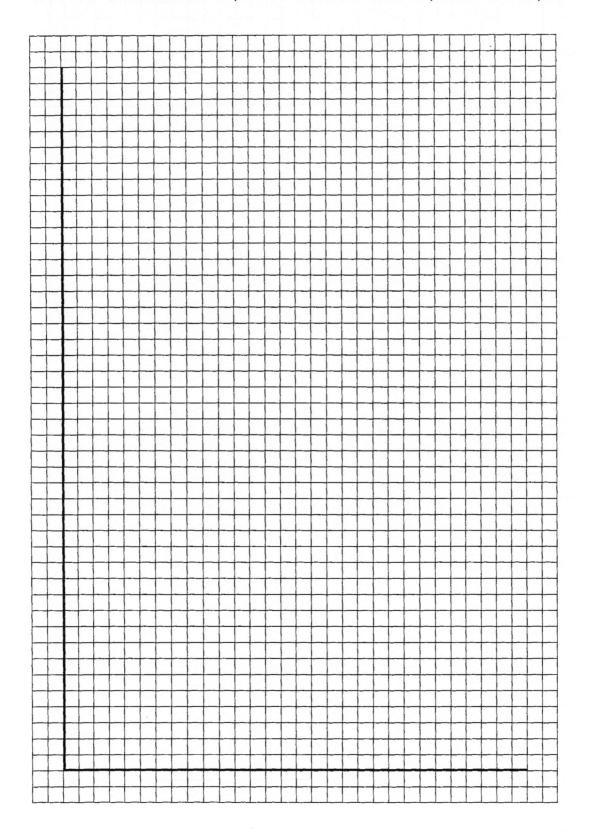

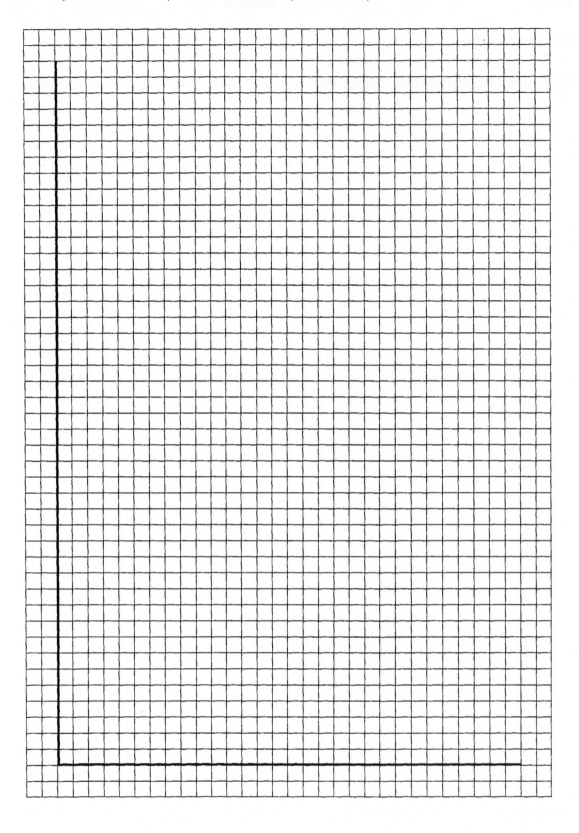

Experiment 15

Advance Study Assignment: Vapor Pressure and Heat of Vaporization
of Liquids

1. In an experiment to measure the vapor pressure of cyclohexane, the following data were obtained:

Mass of empty device	2.4033 g	P_{atm} = 731 mm Hg
Mass full of cyclohexane	2.7870 g	t = 23.7°C
Mass after injecting 0.200 mL of air	2.6093 g	density of cyclohexane = 0.778 g/mL
Bubble meniscus reading	0.182 mL	

a. How many grams of cyclohexane were in the full device?

_____ g

b. How many grams were driven out by the air?

_____ g

c. What is the volume of the cyclohexane driven out, and thus the
volume of the bubble (calculate this using the density)?

_____ mL

d. What is the "dead volume" for this device (see Fig. 15.1)? This is
the difference between the bubble meniscus reading and the actual
volume of the bubble, which you just calculated.

_____ mL

e. Find the partial pressure of air in the bubble.
(Note that the mass and moles of air in the bubble remain the same, but the bubble's volume grows larger
than 0.200 mL since some vapor enters the bubble. The partial pressure of air in the bubble is equal to
the fraction of the bubble's volume that the air is responsible for {0.200 mL divided by the volume you
just calculated} multiplied by the total pressure of the bubble, which is the atmospheric pressure.)

_____ mm Hg

f. What is the partial pressure of vapor in the bubble? The vapor pressure of the cyclohexane and the
partial pressure of the air must add up to the total pressure of the bubble, which is the atmospheric
pressure. (This is the vapor pressure of cyclohexane at 23.7°C!)

_____ mm Hg

(continued on following page)

2. When the device is heated to 46.4°C, the meniscus reading increases to 0.270 mL.

    a. What is the new volume of the bubble? Note that the *difference* in meniscus readings gives you the *difference* in the actual volumes of the two bubbles. Adding the "dead volume" to meniscus readings gives actual bubble volumes.

                              _____ mL

    b. Why does the volume of the bubble increase? (Two reasons)

    c. What volume would the 0.200 mL of air in the bubble expand to at 46.4°C? The pressure and the moles of air remain the same, but you must use the ratio of the two *absolute* temperatures.

                              _____ mL

    d. What is the partial pressure of air in the bubble at 46.4°C? The partial pressure of air in the bubble is again equal to the fraction of the bubble's volume that the air is responsible for, multiplied by the total pressure of the bubble, which is still the atmospheric pressure.

                              _____ mm Hg

    e. What is the vapor pressure of cyclohexane at 46.4°C?

                              _____ mm Hg

Experiment 16

The Structure of Crystals—
An Experiment Using Models

If we examine the crystals of an ordinary substance like table salt, using a magnifying glass or a microscope, we find many cubic particles, among others, in which planes at right angles are present. This is the situation with many common solids. The regularity we see implies a deeper regularity in the arrangement of atoms or ions in the solid. Indeed, when we study crystals by X-ray diffraction, we find that the atomic nuclei are present in remarkably symmetrical arrays, which continue in three dimensions for thousands or millions of units. Substances having a regular arrangement of atom-size particles in the solid are called crystalline, and the solid material consists of crystals. This experiment deals with some of the simpler arrays in which atoms or ions occur in crystals, and what these arrays can tell us about such properties as atomic sizes, densities of solids, and the efficiency of packing of particles.

Many crystals are complex, almost beyond belief. We will limit ourselves to the simplest crystals, those which have cubic structures. They are by far the easiest to understand, yet they exhibit many of the interesting properties of more complicated structures. We will further limit our discussion to substances containing only one or two kinds of atoms, so we will be working with the crystals of some elements and some binary compounds.

The atoms in crystals occur in planes. Sometimes a crystal will cleave readily along such planes. This is the reason the cubic structure of salt is so apparent. Let us begin our study of crystals with a simple example, a two-dimensional crystal in which all of the atoms lie in a square array in a plane, as shown in the following manner:

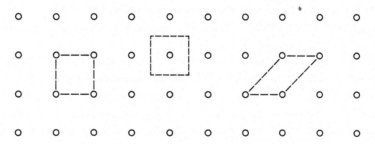

The first thing that you need to realize is that the array goes on essentially forever, in both directions. In this array O is always the same kind of atom, so all the atoms are equivalent. The distance, d_o, between atoms in the vertical direction is the same as in the horizontal, so if you knew this distance you could generate the array completely. In crystal studies we select a small section of the array, called the unit cell, that represents the array in the sense that by moving the unit cell repeatedly in either the x- or y-direction, a distance d_o at a time, you could generate the entire array. In the sketch we have used dotted lines to indicate several possible unit cells. All the cells have the same area, and by moving them up or down or across we could locate all the sites in the array. Ordinarily we select the unit cell on the left, because it includes four of the atoms and has its edges along the natural axes, x and y, for the array, but that is not necessary. In fact, the middle cell has the advantage that it clearly tells us that the number of O atoms in the cell is equal to 1. The number of atoms in whatever cell we choose must also equal 1, but that is not so apparent in the cell on the left. However, once you realize that only one-fourth of each atom on the corners of the cell actually belongs to that cell, because it is shared by three other cells, the number of atoms in the whole cell becomes equal to $1/4 \times 4$, or 1, which is what we got by drawing the cell in a different position.

When we extend the array to three dimensions, the same ideas regarding unit cells apply. The unit cell is the smallest portion of the array that could be used to generate the array. With cubic cells the unit cell is usually chosen to have edges parallel to the x-, y-, and z-axes we could put on the array.

Experimental Procedure

In this experiment we will mix the discussion with the experimental procedure, since one supports and illustrates the other. We will start with the simplest possible cubic crystal, deal with its properties, and then go on to the next more complex example, and the next, and so on. Each kind of crystal will be related in some ways to the earlier ones but will have its own properties as well. So get out your model set and let's begin.

Work in pairs, and complete each section before going on to the next, unless directed otherwise by your instructor.

A. Simple Cubic (SC) Crystals

You can probably guess the form of the array in the crystal structure we call simple cubic (SC). It is shown on the report page for this experiment.

The unit cell is a cube with an edge length equal to the distance from the center of one atom to the center of the next. The length of the cell edge is usually given the symbol d_o. The volume of the unit cell is d_o^3, which is very small, since d_o is on the order of half a nanometer. Using X-rays we can readily measure d_o to four significant figures. The number of atoms in the unit cell in a simple cubic crystal is equal to 1. Can you see why? Only one-eighth of each corner atom actually belongs to the cell, since it is shared equally by eight cells.

Using your model set, assemble three attached unit cells having the SC structure shown in the sketch. Use the short bonds (#6) and the gray atoms. Each bond goes into a square face on the atom. An actual crystal with this structure would have many such cells, in three dimensions.

If you extended the model you have made further, you would find that each atom would be connected to six others. We say that the coordination number of the atoms in this structure is 6. Only the closest atoms are considered to be bonded to each other. In the other models we will be making, *only* those atoms that are *bonded*, by covalent or ionic bonds, will be connected to one another, so the number of bonds to an atom in a large model will be equal to the coordination number.

Although the model has an open structure to help us see relationships better, in an actual crystal we consider that the atoms that are closest are touching. It is on this assumption that we determine atomic radii. In this SC crystal, if we know d_o we can find the atomic radius r of the atoms, since, if the atoms are touching, d_o must equal $2r$. Another property we can calculate knowing d_o is the density, given the nature of the atoms in the crystal. From d_o we can easily find the volume V of the unit cell. Since there is one atom per cell, a mole will contain Avogadro's number of cells. Given the molar mass of the element, we can find the mass m of a cell. The density is simply m/V. With more complex crystals we can make these same calculations, but must take account of the fact that the number of atoms or ions per unit cell may not be equal to one.

Essentially no elements crystallize in an SC structure. The reason is that SC packing is inefficient, in that the atoms are farther apart than they need be. There is, as you can see, a big hole in the middle of each unit cell that is begging for an atom to go there, and atoms indeed do. Before we go into that, let's calculate the fraction of the volume of the unit cell that is actually occupied by atoms. Calculate that fraction on the report page, which steps you through how to determine it.

That fraction is indeed pretty small; only 52% of the cell volume is occupied by atoms. Most of the empty space is in that hole. You can calculate by simple geometry that an atom having a radius equal to 73% of that of the atom on a corner would fit into the hole. Or, putting it another way, an atom bigger than that would have to push the corner atoms back to fit in. If the particles on the corners were anions, and the atom in the hole were a cation, the cation in the hole would push the anions apart, decreasing the repulsion between them and maximizing the attraction between anions and cations. We will have more to say about this when we deal with binary salts, but for now you need to note that if $r_+/r_- > 0.732$, a cation will not fit in the cubic hole formed by anions on the corners of an SC cell.

B. Body-Centered Cubic (BCC) Crystals

In a body-centered cubic (BCC) crystal, the unit cell still contains the corner atoms present in the SC structure, but in the center of the cell there is another atom of the same kind. The unit cell is shown on the report page.

Using your model set, assemble a BCC crystal. Use the blue balls. Put the short bonds in the eight holes in the triangular faces on one blue ball. Attach a blue ball to each bond, again using the holes in a triangular face. Those eight blue balls define the unit cell in this structure. Add as many atoms to the structure as are available, so that you can see how the atoms are arranged. In a BCC crystal each atom is bonded to eight others, so the coordination number is 8. Verify this with your model. (There are no bonds along the edges of the unit cell, since the corner atoms are not as close to one another as they are to the central atom.)

The BCC lattice is much more stable than the SC one, in part at least because of the higher coordination number. Many metals crystallize in a BCC lattice—including sodium, chromium, tungsten, and iron, when these metals are at room temperature.

There are several properties of the BCC structure that you should note. The number of atoms per unit cell is two, one from the corner atoms and one from the atom in the middle of the cell, where it is unshared. As with SC cells, there is a relationship between the unit cell size and the atom radius. Given that in sodium metal the cell edge is 0.429 nm, calculate the radius of a sodium atom on the report page. When you are finished, calculate the density of sodium metal.

The fraction of the volume that is occupied by atoms in a BCC crystal is quite a bit larger than with SC crystals. Using the same kind of procedure that we did with the SC structure, we can show that in BCC crystals 68% of the cell volume is occupied.

C. Close-Packed Structures

Although many elements form BCC crystals, still more prefer structures in which the atoms are close-packed. In such structures there are layers of atoms in which each atom is in contact with six others, as in the following sketch:

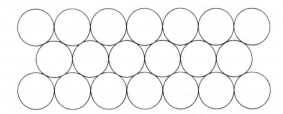

This is the way that billiard balls lie in a rack, or the honeycomb cells are arranged in a bees' nest. It is the most efficient way to pack spheres, with 74% of the volume in a close-packed structure filled with atoms. It turns out that there is more than one close-packed crystal structure. The layers all have the same structure, but they can be stacked on one another in two different ways. This is certainly not apparent at first sight. The first close-packed crystal we will examine is cubic, amazingly enough, considering all those triangles in the layers of atoms.

D. Face-Centered Cubic (FCC) Crystals

In the face-centered cubic (FCC) unit cell there are atoms on the corners and an atom at the center of each face. There is no atom in the center of the cell (see the sketch on the report page, where we show the bonding on only the three exposed faces).

Constructing an FCC unit cell, using bonds between closest atoms, is not as simple as you might think, so let's do it the easy way. This time select the large gray balls, the ones with the most holes. Make the bottom face of the unit cell by attaching four gray balls to a center ball, using short bonds, and holes that are in rectangular faces. You should get the structure shown on the left in the following figure:

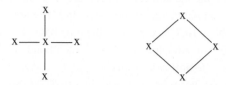

To make the middle layer of atoms, connect four of the balls in a square, again using short bonds and the holes in rectangular faces, as in the sketch on the right, above. The top layer is made the same way as the bottom one. Then connect the layers to one another, again using holes in the rectangular faces. You should then have an FCC cubic unit cell. Because the atoms in the middle of adjacent faces are as close to one another as they are to the corner atoms, there should be bonds between them. There are quite a few bonds in the final cell, 32 in all. Put them all in. As with the BCC structure there are no bonds along the edges of the cell.

The FCC, or cubic close-packed, structure is a common one. Among the metals with this type of crystal structure at room temperature are copper, silver, nickel, and calcium.

In close-packed structures the coordination number has the largest possible value. It can be found by looking at a face-centered atom, say on the top face of the cell. That atom has eight bonds to it and would also have bonds to atoms in the cell immediately above, four of them. So its coordination number, like that of every atom in the cell, is 12.

Having seen how to deal with other structures, you should now be able to find the number of atoms in the FCC unit cell and the radius of an atom as related to d_o. From these relationships and the measured density, you can calculate Avogadro's number. On the report page, find these quantities, using copper metal, which has a unit cell edge d_o equal to 0.361 nm and a density of 8.92 g/cm^3.

In the center of the unit cell there is a hole. It is smaller than a cubic hole but of significant size nonetheless. It is called an octahedral hole, because the six atoms around it define an octahedron. In an ionic crystal, the anions often occupy the sites of the atoms in your cell, and there is a cation in the center of the octahedral hole. On the report page, find the maximum radius, r_+, of the cation that would just fit in an octahedral hole surrounded by anions of radius r_-. You should find that the r_+/r_- ratio turns out to be 0.414.

There is another kind of hole in the FCC lattice that is important. See if you can find it; there are eight of them in the unit cell. If you look at an atom on the corner of the cell, you can see that it lies in a tetrahedron. In the center of the tetrahedron there is a tetrahedral hole, which is small compared to an octahedral or cubic hole. However, in some crystals small cations are found in some or all of the tetrahedral holes, again in a close-packed anion lattice. The cation-anion radius ratio at which cations would just fit into a tetrahedral hole is not so easy to find; r_+/r_- turns out to be 0.225.

The close-packed layers of atoms in the FCC structure are not parallel to the unit cell faces, but rather are perpendicular to the cell diagonal. If you look down the cell diagonal, you see six atoms in a close-packed triangle in the layer immediately behind the corner atom, and another layer of close-packed atoms below that, followed by another corner atom. The layers are indeed close-packed, and, as you go down the diagonal of this and succeeding cells, the layers repeat their positions in the order ABCABC…, meaning that atoms in every fourth layer lie below one another.

Clearly there is another way we could stack the layers. The first and second will always be in the same relative positions, but the third layer could be below the first one if it were shifted properly. So we can have a close-packed structure in which the order of the layers is ABABAB…. The crystal obtained from this arrangement of layers is not cubic, but hexagonal. It too is a common structure for metals. Cadmium, zinc, and manganese adopt this crystal structure at room temperature. As you might expect, the stability of this structure is very similar to that of FCC crystals. We find that simply changing the temperature often converts a metal from one form to another. Calcium, for example, is FCC at room temperature, but if heated to 450°C it converts to close-packed hexagonal.

E. Crystal Structures of Some Common Binary Compounds

We have now dealt with all of the possible cubic crystal structures for metals. It turns out that the structures of binary ionic compounds are often related to these metal structures in a very simple way. In many ionic crystals the anions, which are large compared with cations, are essentially in contact with each other, in either an SC or FCC structure. The cations go into the cubic, or octahedral, or tetrahedral holes, depending on the cation-anion radius ratios that we calculated. The idea is that the cation will tend to go into a hole in which it will not quite fit. This increases the unit cell size from the value it would have if the anions were touching, which reduces the repulsion energy due to anion-anion interaction and increases to the maximum the cation-anion attraction energy, producing the most stable possible crystal structure. According to the so-called radius-ratio rule, large cations go into cubic holes, smaller ones into octahedral holes, and the smallest ones into tetrahedral holes.

The deciding factor for which hole is favored is given by the radius ratio:

If $r_+/r_- > 0.732$	cations go into cubic holes
If $0.732 > r_+/r_- > 0.414$	cations go into octahedral holes
If $0.414 > r_+/r_- > 0.225$	cations go into tetrahedral holes

1. The Sodium Chloride (NaCl) Crystal

To apply the radius-ratio rule to NaCl, we simply need to find r_+/r_-, using the data in Table 16.1, found on the first report page. Since $r_{Na^+} = 0.095$ nm, and $r_{Cl^-} = 0.181$ nm, the radius ratio is $0.095/0.181$, or 0.525. That value is less than 0.732 and greater than 0.414, so the sodium ions should go into octahedral holes. We saw the octahedral hole in the center of the FCC unit cell, so Na^+ ions go there, and the Cl^- ions have the FCC structure. Actually, there are 12 other octahedral holes associated with the cell, one on each edge, which would be apparent if we had been able to make more cells. An Na^+ ion goes into each of these holes, giving the classic NaCl structure shown on the report page. (Again, only the ions and bonds on the exposed faces are shown.)

Make a model of the unit cell for NaCl, using the large gray balls for the Cl^- ions and the blue ones for Na^+ ions. Clearly the Na^+ ion at the center of the cell is in an octahedral hole, but so are all of the other sodium ions, because if you extend the lattice, every Na^+ ion will be surrounded by six Cl^- ions. The coordination number of each Na^+ ion in NaCl is six. The coordination number of the Cl^- ions in NaCl is also six. The NaCl crystal is FCC in Cl^- and also in Na^+, because you could put Na^+ ions on the corners of the unit cell and maintain the same structure. The unit cell extends from the center of one Cl^- ion to the center of the next Cl^- along the cell edge; or, from the center of one Na^+ ion to the center of the next Na^+ ion.

On the report page, find the number of Na^+ and of Cl^- ions in the unit cell.

2. The Cesium Chloride (CsCl) Crystal

Cesium chloride has the same type of formula as NaCl, 1:1. The Cs^+ ion, however, is larger than Na^+, and has a radius equal to 0.169 nm. This makes r_+/r_- equal to 0.933. Because this value is greater than 0.732, we would expect that in the CsCl crystal the Cs^+ ions will fill cubic holes, and this is what is observed. The structure of CsCl will look like that of the BCC unit cell you made earlier, except that the ion in the center will be Cs^+ and those on the corners Cl^-. If you put gray balls in the center of each BCC unit cell you made from the blue balls, you would have the CsCl structure. This structure is *not* BCC, because the corner and center atoms are not the same. Rather it consists of two interpenetrating simple cubic lattices, one made from Cl^- ions and the other from Cs^+ ions.

3. The Zinc Sulfide (ZnS) Crystal

Zinc sulfide is another 1:1 compound, but its crystal structure is not that of NaCl or of CsCl. In ZnS, the Zn^{2+} ions have a radius of 0.074 nm and the S^{2-} ions a radius of 0.184 nm, making r_+/r_- equal to 0.402. By the radius-ratio rule, ZnS should have close-packed S^{2-} ions, with the Zn^{2+} ions in tetrahedral holes. In the ZnS unit cell, we find that this is indeed the case; the S^{2-} ions are FCC, and alternate tetrahedral holes in the unit cell are occupied by Zn^{2+} ions, which themselves form a tetrahedron.

To make a model for ZnS, first assemble an SC unit cell, using blue balls (b) and the long bonds. Use gray balls (g) for the zinc ions, and assemble the unit shown in the following figure, using the short bonds and the holes in triangular faces:

Attach this unit to two corner atoms on the diagonal of the bottom face. Make another unit and attach it to the two corner atoms in the upper face that lie on the face diagonal that is not parallel to the bottom face diagonal. The gray balls should then form a tetrahedron. Then attach the four other FCC blue balls to the gray ones. In this structure the coordination number is four (the long bonds do not count, they just keep the unit cell from falling apart). If the gray balls in the unit cell are replaced with blue ones, so that all atoms are of the same element, we obtain the diamond crystal structure.

These are the three common cubic structures of 1:1 compounds. The radius-ratio rule allows us to predict which structure a given compound will have. It does not always work, but it is correct most of the time. On the report page, use the rule to predict the cubic structures that crystals of the following substances will have: KI, CuBr, and TlBr.

4. The Calcium Fluoride (CaF) Crystal

Calcium fluoride is a 1:2 compound, so it cannot have the structure of any crystal we have discussed so far. The radii of Ca^{2+} and F^- are 0.099 and 0.136 nm, respectively, so r_+/r_- is 0.727. This places CaF_2 on the boundary between compounds with cations in cubic holes and those with cations in octahedral holes. It turns out that in CaF_2, the F^- ions have a simple cubic structure, with half of the cubic holes filled by Ca^{2+} ions. This produces a crystal in which the Ca^{2+} ions lie in an FCC lattice, with 8 F^- ions in a cube inside the unit cell.

Use your model set to make a unit cell for CaF_2. Use gray balls (g) for the Ca^{2+} ions and red ones (r) for F^-. The cations and anions are linked by short bonds that go into holes in nonadjacent triangular faces. The Ca^{2+} ions on each face diagonal are linked to F^- ions as shown in the following diagram:

To complete the bottom face, attach two red and two gray balls to the initial line, so that the four red balls form a square and the gray ones an FCC face. The top of the unit cell is made the same way. Attach the two assemblies through four gray balls, which lie at the centers of the other faces. When you are done, the red balls should form a cube inside an FCC unit cell made from gray balls. Extend the model into an adjacent unit cell to show that every other cube of F^- ions has a Ca^{2+} ion at its center.

When you have completed this part of the experiment, your instructor may assign you a model with which to work. Report your results as directed. Then disassemble the models you made and pack the components in the box.

Experiment 16

Data and Calculations: The Structure of Crystals

Table 16.1

Atom	Molar mass, g/mole	Atomic radius, nm	Ionic radius, nm
Na	22.99	calculated below	0.095
Cu	63.55	calculated below	0.096
Cl	35.45	0.099	0.181
Cs	132.9	0.262	0.169
Zn	65.38	0.133	0.074
S	32.06	0.104	0.184
K	39.10	0.231	0.133
I	126.9	0.133	0.216
Br	79.90	0.114	0.195
Tl	204.4	0.171	0.147

Some Cubic Unit Cells

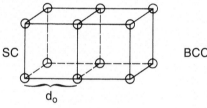

SC

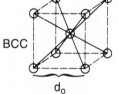

BCC

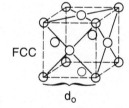

FCC

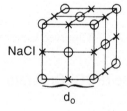

NaCl

A. Simple Cubic (SC) Crystals

Fraction of volume of unit cell occupied by atoms

Volume of unit cell in terms of d_o _____

Number of atoms per unit cell _____

Radius r of atom in terms of d_o _____

Volume of atom in terms of $d_o \left(V = \dfrac{4\pi}{3} r^3 \right)$ _____

(Volume of atom)/(volume of cell) _____ = _____

(continued on following page)

B. Body-Centered Cubic (BCC) Crystals

a. Radius of a sodium atom

In BCC, atoms touch along the cube diagonal
(see your model and the Advance Study Assignment {ASA})

Length of cube diagonal if $d_o = 0.429$ nm   _____ nm

$4 \times r_{Na}$ = length of cube diagonal r_{Na} = _____ nm

b. Density of sodium metal

Length of unit cell, d_o, in cm (1 cm = 10^7 nm)   _____ cm

Volume of unit cell, V _____ cm³

Number of atoms per unit cell   _____

Number of unit cells per mole Na   _____

Mass of a unit cell, m _____ g

Density of sodium metal, m/V (Observed value = 0.97 g/cm³)   _____ g/cm³

D. Face-Centered Cubic (FCC) Crystals

a. Number of atoms per unit cell

Number of atoms on corners _____, shared by _____ cell(s) → _____ × _____ = _____

Number of atoms on faces _____, shared by _____ cell(s) → _____ × _____ = _____

Total atoms per unit cell _____

b. Radius of a Cu atom

In FCC, atoms touch along face diagonal (see your model and the Advance Study Assignment {ASA})

Length of face diagonal in Cu, where $d_o = 0.361$ nm   _____ nm

Number of Cu atom radii on face diagonal _____ r_{Cu} = _____ nm

(continued on following page)

c.   Avogadro's number, N

Length of cell edge, d_o, in cm, in Cu metal _____ cm

Volume of a unit cell in Cu metal _____ cm^3

Volume of a mole of Cu metal (V = mass/density) _____ cm^3

Number of unit cells per mole Cu _____

Number of atoms per unit cell _____

Number of atoms per mole, $N_{calculated}$ _____

d.   Size of an octahedral hole

Following is a sketch of the front face of a FCC cell with the NaCl structure. The ions are drawn so that they are just touching their nearest neighbors. The cations, shaded in gray, are in octahedral holes.

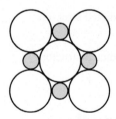

Show the length of d_o on the sketch. (See your model.)

What is the relationship between r_- and d_o?

$r_- = $ _____ $\times d_o$

What is the equation relating r_+, r_-, and d_o?

$d_o = $ _____

What is the relationship between r_+ and d_o?

$r_+ = $ _____ $\times d_o$

What is the value of the radius ratio r_+/r_-? Express the ratio to three significant figures.

$r_+/r_- = $ _____

(continued on following page)

E. Crystal Structures of Some Common Binary Compounds

1. The Sodium Chloride (NaCl) Crystal

Number of Cl^- ions in the unit cell (FCC) _____

Number of Na^+ ions on edges of cell _____, shared by _____ cell(s)

Number of Na^+ ions in center of cell _____, shared by _____ cell(s)

Total number of Na^+ ions in unit cell _____

3. The Zinc Sulfide (ZnS) Crystal

Application of the radius-ratio rule (necessary data is in Table 16.1, at the start of the report pages):

KI $\quad$ r_+/r_- _____ Predicted structure _____ Observed structure = NaCl

CuBr $\quad$ r_+/r_- _____ Predicted structure _____ Observed structure = ZnS

TlBr $\quad$ r_+/r_- _____ Predicted structure _____ Observed structure = CsCl

Report on unknown (if assigned)

Experiment 16

Advance Study Assignment: The Structure of Crystals

1. Many substances crystallize in a cubic structure. The unit cell for such crystals is a cube having an edge with a length equal to d_o.

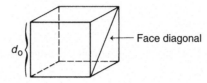

a. What is the length, in terms of d_o, of the face diagonal, which runs diagonally across one face of the cube? (Hint: Use the Pythagorean Theorem.)

_____

b. What is the length, again in terms of d_o, of the cube diagonal, which runs from one corner, through the center of the cube, to the other corner? (Hint: Make a right triangle having a face diagonal and an edge of the cube as its sides, with the hypotenuse equal to the cube diagonal. You will again employ the Pythagorean Theorem.)

_____

2. In an FCC structure, the centers of the atoms are found on the corners of the cubic unit cell and at the center of each face. The unit cell has an edge whose length is the distance from the center of one corner atom to the center of another corner atom on the same edge. The atoms on the diagonal of any face are touching. One of the faces of the unit cell is shown in the following:

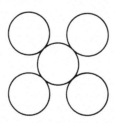

a. Show the distance d_o on the sketch. Draw the boundaries of the unit cell.

b. What is the relationship between the length of the face diagonal and the radius of the atoms in the cell?

Face diagonal = _____

(continued on following page)

c.   How is the radius of the atoms related to d_o?

$r =$ _____

d.   Silver metal crystals have an FCC structure. The unit cell edge in silver is 0.4086 nm long. What is the radius of a silver atom?

_____nm

Experiment 17

Classification of Chemical Substances

Depending on the kind of bonding present in a chemical substance, the substance may be called ionic, molecular, or metallic.

In a solid ionic compound there are ions; the large electrostatic forces between the positively and negatively charged ions are responsible for the bonding which holds these particles together.

In a molecular substance the bonding is caused by the sharing of electrons by atoms. When the stable aggregates resulting from covalent bonding contain relatively small numbers of atoms, they are called molecules. If the aggregates are very large and include essentially all the atoms in a macroscopic particle, the substance is called macromolecular.

Metals are characterized by a kind of bonding in which the electrons are much freer to move than in other kinds of substances. The metallic bond is stable but is probably less localized than other bonds.

The terms *ionic, molecular, macromolecular,* and *metallic* are somewhat arbitrary, and some substances have properties that would place them in a borderline category, somewhere between one group and another. It is useful, however, to consider some of the general characteristics of typical ionic, molecular, macromolecular, and metallic substances, since many common substances can be readily assigned to one of these categories or another.

Ionic Substances

Ionic substances are all solids at room temperature. They are typically crystalline but may exist as fine powders as well as clearly defined crystals. While many ionic substances are stable up to their melting points, some decompose on heating. It is common for an ionic crystal to release loosely bound water of hydration at temperatures below 200°C. Anhydrous (dehydrated) ionic compounds have high melting points, usually above 300°C but below 1000°C. They are not readily volatilized and boil only at very high temperatures (see Table 17.1).

When molten, ionic compounds will conduct an alternating electric current. In the solid state they do not conduct electricity. The conductivity in the molten liquid is attributed to the freedom of motion of the ions, which arises when the crystal lattice is no longer present.

Ionic substances are frequently but not always appreciably soluble in water. The solutions produced conduct alternating electric current rather well. The conductivity of a solution of a slightly soluble ionic substance is often several times that of the solvent water. Ionic substances are usually not nearly so soluble in other liquids as they are in water. For a liquid to be a good solvent for ionic compounds it must be highly polar, containing molecules with well-defined positive and negative regions with which the ions can interact.

Molecular Substances

All gases and essentially all liquids at room temperature are molecular in nature. If the molar mass of a molecular substance is greater than about one hundred grams per mole, it may be a solid at room temperature. The melting points of molecular substances are usually below 300°C; these substances are relatively volatile, but a good many will decompose before they boil. Most molecular substances do not conduct electric current either when solid or when molten.

Organic compounds, which contain primarily carbon and hydrogen, often in combination with other nonmetals, are essentially molecular in nature. Since there are a great many organic substances, it is true that most substances are molecular. If an organic compound decomposes on heating, the residue is frequently a

Table 17.1

			Physical Properties of Some Representative Chemical Substances			
	Melting point, M.P. [°C]	Boiling point, B.P. [°C]	Solubility in		Electrical conductance	Classification
Substance			Water	Hexane		
NaCl	801	1413	Sol	Insol	High in melt and in soln	Ionic
MgO	2800	—	Sl sol	Insol	Low in sat'd soln	Ionic
CoCl$_2$	Sublimes	1049	Sol	Insol	High in soln	Ionic
CoCl$_2 \cdot 6 H_2O$	86	Dec	Sol	Insol	High in soln	Ionic hydrate, loses H$_2$O at 110°C
C$_{10}$H$_8$	70	255	Insol	Sol	Zero in melt	Molecular
C$_6$H$_5$COOH	122	249	Sl sol	Sl sol	Low in sat'd soln	Molecular-ionic
FeCl$_3$	282	315	Sol	Sl sol	High in soln and in melt	Molecular-ionic
SnI$_4$	144	341	Dec	Sol	~Zero in melt	Molecular
SiO$_2$	1600	2590	Insol	Insol	Zero in solid	Macromolecular
Fe	1535	3000	Insol	Insol	High in solid	Metallic

Key: Sol = soluble, having a solubility of at least 0.1 mole/L; Sl sol = slightly soluble, having significant solubility but less than 0.1 mole/L; Insol = (essentially) insoluble; Dec = decomposes (changes chemical form); sat'd = saturated; soln = solution

black carbonaceous (carbon-containing) material. Reasonably large numbers of inorganic substances are also molecular; those that are solids at room temperature include some of the binary compounds of elements in Groups 14, 15, 16, and 17.

Molecular substances are usually soluble in at least a few organic solvents, with the solubility being enhanced if the substance and the solvent are similar in molecular structure.

Some molecular compounds are markedly polar, which tends to increase their solubility in water and other polar solvents. Such substances may ionize appreciably in water, or even in the melt, so that they become conductors of electricity. Usually the conductivity is considerably lower than that of an ionic material. Most polar molecular compounds in this category are organic, but a few, including some transition metal compounds, are inorganic.

Macromolecular Substances

Macromolecular substances are all solids at room temperature. They have very high melting points, usually above 1000°C, and low volatility. They are typically highly resistant to thermal decomposition. They do not conduct electric current and are often good insulators. They are not soluble in water or any organic solvents. They are frequently chemically inert, and inorganic ones may be used as abrasives or refractories (materials that resist extremely high temperatures).

Metallic Substances

The properties of metals appear to derive mainly from the freedom of movement of their bonding electrons. Metals are good electrical conductors in both the solid form and the melt, and have characteristic luster and solid malleability. Most metals are solid at room temperature, but their melting points range widely, from below 0°C to over 2000°C. They are not soluble in water or organic solvents. Some metals are prepared as gray or black powders, which may not appear to be electrical conductors. However, if you measure the conductance while the powder is under pressure, its metallic character is revealed.

Experimental Procedure

In this experiment you will investigate the properties of several substances with the purpose of determining whether they are molecular, ionic, macromolecular, or metallic. In some cases, the classification will be straightforward. In others, the assignment to a class will not be so easy and you may find that the substance exhibits characteristics associated with more than one class.

As the earlier discussion indicates, there are several properties we can use to find out to which class a substance belongs. In this experiment we will use the melting point; the solubility in water and organic solvents; and the electrical conductivity of the aqueous solution, the solid, and the melt in making the classification.

The substances to be studied in the first part of the experiment are on the laboratory tables along with two organic solvents, one polar and one nonpolar. You need only carry out enough tests on each substance to establish the class to which it belongs, so you will not need to perform every test on every substance. You may, however, carry out any extra tests you wish, if only to satisfy your curiosity. Follow the directions for each test as given below, recording your findings in Table 17.2 on the report page.

A. Melting Point

Approximate melting points of substances can be determined by heating a small amount of the substance (a sample the size of a pea) in a test tube. Substances with low melting points, less than 100°C, will melt readily when warmed gently over a Bunsen burner flame in a small (13×100 mm) test tube, or when the tube is immersed in boiling water. If the sample melts between 100° and 300°C, it will take more than gentle warming over a Bunsen burner flame, but it will melt before the test tube imparts a yellow-orange color to the flame. Above 300°C, the flame will take on an increasingly yellow-orange color; up to 500°C you can still use a test tube and a strong burner flame, but at 550°C the pyrex tube will begin to soften. In this experiment we will not attempt to measure any melting points above 500°C. Stop heating as soon as the test tube itself begins to glow (you will need to take it out of the flame periodically to check; even the slightest glow from the glass indicates you are over 500°C).

While heating a sample, keep the tube **loosely** stoppered with a cork. **Do not breathe** any vapors that are given off, and **do not continue to heat** a sample after it has melted. As you heat the sample, watch for evidence of decomposition, sublimation, or evolution of water.

B. Solubility and Conductance of Solutions

In testing for solubility, again use a sample about the size of a pea, this time in a medium (18 mm $\times$ 150 mm) test tube. Use about two milliliters of solvent, enough to fill the tube to a depth of about one centimeter. Stir well, using a clean stirring rod. Some samples will dissolve completely almost immediately; some are only slightly soluble and may produce a cloudy suspension; others are completely insoluble. Make solubility tests with deionized water and the two organic solvents provided, and record your results. Use fresh deionized water in your wash bottle for the aqueous solubility tests.

Conductance measurements need only be made on aqueous solutions. We will use a portable ohmmeter for this purpose. Your instructor may take the measurements for you, or may have you take them. An ohmmeter measures the electrical resistance of a sample in ohms, Ω. A solution with a high resistance has a low electrical conductance, and vice versa. Some of your solutions will have a low resistance, on the order of 1000 Ω or less; these are good conductors. Deionized water has a high resistance; with your meter it will probably have a resistance of 50,000 Ω or greater. Small amounts of contaminants can lower the resistance of a solution markedly, particularly if the main solute shows high resistance.

Measure the resistance of any of the aqueous solutions containing soluble or slightly soluble substances. Between tests, rinse the electrodes in a beaker filled with deionized water. For our purposes, a solution with a resistance less than 2000 Ω is a good conductor, denoted "G." Between 2000 and 20,000 Ω it is a weak conductor, denoted "W." Above 20,000 Ω we will consider it to be essentially nonconducting, and denote it with an "N." Record the resistances you observe in ohms. Then note, with a G, W, or N, whether the solution is a good, weak, or poor conductor.

C. Electrical Conductance of Solids and Melts

Some substances conduct electricity in the solid state. If the sample contains large crystals, the conductance test is easy. Select a crystal, put it on the lab bench, and touch it with the two wires on the ohmmeter probe. Metals have a very low resistance. In powder form most substances, metals included, appear to have essentially infinite resistance. However, under pressure, metal powders, unlike those of other substances, show good conductance. To test a powder for conductance, put a penny on the lab bench. On it place a small rubber washer from the box on the bench. Fill the hole in the washer with the powder, and put another penny on top of the washer. Put the whole sandwich between the jaws of a pair of insulated pliers. Touch the electrodes from the ohmmeter probe to the pennies, one electrode to each penny, and squeeze the pliers. If the powder is a metal, the resistance will gradually fall from infinity to a small value. Make sure any drop in resistance is not caused by the pennies touching each other. Record your results.

To check the conductance of a melt, put a pea-size sample in a dry, medium test tube and melt it. Heat the electrodes on the probe for a few seconds in the Bunsen flame and touch them to the melt. Heat gently to ensure that no solid is crystallized on the electrodes. Many melts are good conductors. After testing a melt, clean the electrodes by washing them with water or an organic solvent, or, if necessary, scraping them off with a spatula.

Having made the tests we have described, you should be able to assign each substance to its class, or, possibly, to one or both of two classes. Make this classification for each substance, and give your reasons for doing so.

When you have classified each substance, report to your laboratory instructor, who will assign you two unknowns for characterization.

DISPOSAL OF REACTION PRODUCTS. All residues from your tests should be discarded in the waste container unless directed otherwise by your instructor.

Name _____ Section _____

Experiment 17

Observations and Analysis: Classification of Chemical Substances

Table 17.2

| Substance # | Approximate melting point [°C] (< 100, 100–300, 300–500, > 500) | Solubility | | | Electrical resistance, Ω | | | Classification and reason |
		H₂O	Nonpolar organic	Polar organic	Solution in H₂O	Solid	Melt	
I								
II								
III								
IV								
V								
VI								
Unknown # ___								
___								

Key: Sol = soluble; Sl sol = slightly soluble; Insol = insoluble; G = good electrical conductor, $R < 2000\ \Omega$; W = weak electrical conductor, $2000\ \Omega < R < 20{,}000\ \Omega$; N = nonconductor, $R > 20{,}000\ \Omega$.

Experiment 17

Advance Study Assignment: Classification of Chemical Substances

1. List the properties of a substance that would definitely establish that it is an ionic material.

2. If we classify substances as ionic, molecular, macromolecular, or metallic, in which (if any) categories are *all* the members

   a. soluble in water?

   b. electrical conductors in the melt?

   c. solids at room temperature?

   d. insoluble in all common solvents?

3. A given substance is a white, granular solid at 25°C that does not conduct electricity. It melts at 250°C, then decomposes to a black goo at 312°C. As a liquid it does not conduct electricity, not even an alternating current. It does not dissolve in water, but dissolves readily in hexane to produce a solution that does not conduct electricity. What would the proper classification of the substance be, based on this information?

4. A dull gray-white solid melts at 1650°C. It is not electrically conductive as either a solid or a liquid, and is not soluble in either water or any organic solvent. Classify the substance as best you can from these properties.

Experiment 18

Some Nonmetals and Their Compounds— Preparations and Properties

Some of the most commonly encountered chemical substances are nonmetallic elements or their simple compounds. O_2 and N_2 in the air, CO_2 produced by combustion of oil, coal, or wood, and H_2O in rivers, lakes, and air are typical of such substances. Substances containing nonmetallic atoms, whether elementary or compound, are all molecular, reflecting the covalent bonding that holds their atoms together. They are often gases at room temperature, due to weak intermolecular forces. However, with high molecular masses, hydrogen bonding, or macromolecular structures they can be liquids, like H_2O and Br_2, and solids, like I_2 and graphite.

Several of the common nonmetallic elements and some of their gaseous compounds can be prepared by simple reactions. In this experiment you will prepare some typical examples of such substances and examine a few of their characteristic properties.

WEAR YOUR SAFETY GLASSES WHILE PERFORMING THIS EXPERIMENT

Experimental Procedure

In several of the experiments you perform, you will prepare gases. In general we will not describe in detail what should be observed, so perform each preparation carefully and report what you actually observe, not what you think we expect you to observe!

There are several tests we will make on the gases you prepare, and the way each of these should be carried out is summarized below.

Test for Odor
To determine the odor of a gas, first pass your hand across the end of the tube, bringing the gas toward your nose. If you don't detect an odor, sniff near the end of the tube, first at some distance and then gradually closer. Do not just put the tube at the end of your nose immediately and take a deep whiff! Some of the gases you will make have no odor, and some will have very potent ones. Some of the gases are quite toxic and, even though we will be making only small amounts, caution in testing for odor is important.

CAUTION: **The following gases prepared in this experiment are toxic: Br_2, SO_2, NO_2, NH_3, and H_2S. Do not inhale these gases unnecessarily when testing their odors. One small sniff will be sufficient and will not be harmful. It is good for you to know these odors, in case you encounter them in the future.**

Test for Support of Combustion
A few gases will support combustion, but most will not. To make the test, ignite a wood splint with a Bunsen burner flame, blow out the flame, and put the glowing, but not burning, splint into the gas in the test tube. If the gas supports combustion, the splint will glow more brightly, or may make a small popping noise. If the gas does not support combustion, the splint will go out almost instantly. You can use the same splint for all the combustion tests.

Test for Acid-Base Properties
Many gases are acids. This means that if the gas is dissolved in water it will produce some H^+ ions. A few gases are bases; in aqueous solution such gases produce OH^- ions. Other gases do not react with water and are neutral. It is easy to establish the acidic or basic nature of a gas by using a chemical indicator. One of the most common acid-base indicators is litmus, which is red in acidic solution and blue in basic solution. To test whether a gas is an acid, moisten a piece of blue litmus paper with water from your wash bottle, and put the

paper down in the test tube in which the gas is present. If the gas is an acid, the paper will turn red. Similarly, to test if the gas is a base, moisten a piece of red litmus paper, and hold it down in the tube. A color change to blue will occur if the gas forms a basic solution. Since you may have used an acid or a base in making the gas, do not touch the walls of the test tube with the paper. The color change will occur fairly quickly and smoothly over the surface of the paper if the gas is acidic or basic. It is not necessary to use a new piece of litmus for each test. Start with a piece of blue and a piece of red litmus. If you need to regenerate the blue paper, hold it over an open bottle of 6 M NH_3, whose vapor is basic. If you need to make red litmus, hold the moist paper over the acidic vapor of an open bottle of 6 M acetic acid.

A. Preparation and Properties of Nonmetallic Elements: O_2, N_2, Br_2, I_2

1. Oxygen, O_2

Oxygen can be easily prepared in the laboratory by the decomposition of H_2O_2, hydrogen peroxide, in aqueous solution. Hydrogen peroxide is not very stable and will break down to water and oxygen gas on addition of a suitable catalyst, such as the Fe^{3+} ion:

$$2\ H_2O_2(aq) \xrightarrow{\ Fe^{3+}\ } O_2(g) + 2\ H_2O(\ell) \tag{1}$$

Add 5 mL of $3\%_{mass}$ aqueous H_2O_2 to a small (13×100 mm) test tube. (This will be a depth of about five centimeters in such a tube.) Add 5 drops of concentrated $Fe(NO_3)_3$ solution to the test tube and swirl. You may wish to hold your finger over the end of the tube to help confine the O_2 that is generated. Test the evolved gas for odor. Test the gas for any acid-base properties, using moist blue and red litmus paper. Test the gas for support of combustion. Record your observations.

2. Nitrogen, N_2

Sodium sulfamate in aqueous solution will react with nitrite ion to produce nitrogen gas:

$$NO_2^-(aq) + NH_2SO_3^-(aq) \rightarrow N_2(g) + SO_4^{2-}(aq) + H_2O(\ell) \tag{2}$$

Add about one milliliter of 1.0 M KNO_2, potassium nitrite, to a small test tube. Then add ten to twelve drops of 0.5 M $NaNH_2SO_3$, sodium sulfamate, and place the test tube in a hot-water bath made from a 250-mL beaker half full of water. Bubbles of nitrogen should form within a few moments. Confine the gas for a few seconds with a stopper. Cautiously test the evolved gas for odor. Carry out the tests for acid-base properties and for support of combustion. If you need to generate more N_2 to complete the tests, add sodium sulfamate solution as necessary, five to ten drops at a time. Report your observations.

3. Iodine, I_2

The halogen elements are most readily prepared from their sodium or potassium salts. The reaction involves an oxidizing agent, which can remove electrons from the halide ions, freeing the halogen. The reaction that occurs when a solution of potassium iodide, KI, is treated with 6 M HCl and a little MnO_2 is

$$2\ I^-(aq) + 4\ H^+(aq) + MnO_2(s) \rightarrow I_2(aq) + Mn^{2+}(aq) + 2\ H_2O(\ell) \tag{3}$$

At 25°C, I_2 is a solid, with relatively low solubility in water, and appreciable volatility. I_2 can be extracted from aqueous solution into organic solvents, particularly heptane, C_7H_{16} (HEP). The solid, vapor, and solutions in water and HEP all have characteristic colors.

Put 2 drops of 1.0 M KI into a *medium* (18×150 mm) test tube. Add 6 drops 6 M HCl and a tiny amount of manganese dioxide, MnO_2. Swirl the mixture and note any changes that occur. Put the test tube into a hot-water bath. After a minute or two, a noticeable amount of I_2 vapor should be visible above the liquid. Remove the test tube from the water bath and add 10. mL of deionized water. Stopper the tube and shake. Note the color of I_2 in the solution. Cautiously sniff the vapor above the solution. Decant the liquid into another medium test tube and add 3 mL of heptane, C_7H_{16}. Stopper and shake the tube. Observe the color of the HEP layer and the relative solubility of I_2 in water and HEP. Record your observations.

4. Bromine, Br$_2$

Bromine can be made by the same reaction used to make iodine, substituting bromide ion for iodide. Bromide ion is less easily oxidized than iodide ion. Bromine at 25°C is a liquid. Br$_2$ can be extracted from aqueous solution into heptane. Its liquid, vapor, and solutions in water and HEP are colored.

To two or three drops of 1.0 M NaBr in a medium test tube, add 6 drops of 6 M HCl and a small amount of MnO$_2$, about the size of a small pea. Swirl to mix the reagents, and observe any changes. Heat the tube in a hot-water bath for a minute or two. Try to detect Br$_2$ vapor above the liquid by observing its color. Add 10. mL of water to the tube, stopper, and shake. Note the color of Br$_2$ in the solution. Sniff the vapor, cautiously. Decant the liquid into another medium test tube. Add 3 mL of HEP, stopper, and shake. Note the color of Br$_2$ in HEP. Record your observations.

B. Preparation and Properties of Some Nonmetallic Oxides: CO$_2$, SO$_2$, NO, and NO$_2$

1. Carbon Dioxide, CO$_2$

Carbon dioxide, like several nonmetallic oxides, is easily made by treating an oxyanion with an acid. With carbon dioxide the oxyanion is CO$_3{}^{2-}$, carbonate ion, which is present in solutions of carbonate salts, such as Na$_2$CO$_3$. The reaction is

$$CO_3{}^{2-}(aq) + 2\,H^+(aq) \rightarrow H_2CO_3(aq) \rightarrow CO_2(g) + H_2O(\ell) \tag{4}$$

Carbon dioxide is not very soluble in water, and on acidification, carbonate solutions will tend to effervesce as CO$_2$ is liberated. Nonmetallic oxides in solution are often acidic but never basic.

To 1.0 mL of 1.0 M Na$_2$CO$_3$ in a small test tube add 6 drops of 3 M H$_2$SO$_4$. Test the gas for odor, acidic properties, and ability to support combustion.

2. Sulfur Dioxide, SO$_2$

Sulfur dioxide is readily prepared by acidification of a solution of sodium sulfite, containing sulfite ion, SO$_3{}^{2-}$. As you can see, the reaction that occurs is very similar to that with carbonate ion.

$$SO_3{}^{2-}(aq) + 2\,H^+(aq) \rightarrow H_2SO_3(aq) \rightarrow SO_2(g) + H_2O(\ell) \tag{5}$$

Sulfur dioxide is considerably more soluble in water than is carbon dioxide. Some effervescence may be observed on acidification of concentrated sulfite solutions; effervescence will increase if the solution is heated.

To 1.0 mL of 1.0 M Na$_2$SO$_3$ in a small test tube add 6 drops of 3 M H$_2$SO$_4$. Cautiously test the evolved gas for odor. Test its acidic properties and its ability to support combustion. Put the test tube into a hot-water bath for a few seconds to see if effervescence occurs when the solution is hot.

3. Nitrogen Dioxide, NO$_2$, and Nitric Oxide, NO

If a solution containing nitrite ion, NO$_2{}^-$, is treated with acid, two oxides are produced: NO$_2$ and NO. In solution these gases are combined in the form of N$_2$O$_3$, which is colored. When these gases come out of solution, the mixture contains NO and NO$_2$; the latter is colored. NO is colorless and reacts readily with oxygen in the air to form NO$_2$. The preparation reaction is

$$2\,NO_2{}^-(aq) + 2\,H^+(aq) \rightarrow N_2O_3(aq) + H_2O(\ell) \rightarrow NO(g) + NO_2(g) + H_2O(\ell) \tag{6}$$

To 1.0 mL of 1.0 M KNO$_2$ in a small test tube, add 6 drops of 3 M H$_2$SO$_4$. Swirl the mixture for a few seconds and note the color of the solution. Warm the tube in a hot-water bath for a few seconds to increase the rate of gas evolution. Note the color of the gas that is given off. Cautiously test the odor of the gas. Test its acidic properties and its ability to support combustion. Record your observations.

C. Preparation and Properties of Some Nonmetallic Hydrides: NH_3, H_2S

1. Ammonia, NH_3

A solution of 6 M NH_3 in water is a common laboratory reagent. You may have been using it in this experiment to make your litmus paper turn blue. Ammonia gas can be made by simply heating 6 M $NH_3(aq)$. It can also be prepared by addition of a strongly basic solution to a solution of an ammonium salt, such as NH_4Cl. The latter solution contains NH_4^+ ion. On treatment with OH^- ion, as in a solution of NaOH, the following reaction occurs:

$$NH_4^+(aq) + OH^-(aq) \rightarrow NH_3(aq) + H_2O \rightarrow NH_3(g) + H_2O(\ell) \tag{7}$$

The odor of NH_3 is characteristic. NH_3 is very soluble in water, so effervescence is not observed, even on heating concentrated solutions.

Add 1.0 mL of 1.0 M NH_4Cl to a small test tube. Add 1.0 mL of 6 M NaOH. Swirl the mixture and **cautiously** test the odor of the evolved gas; **too large a whiff of ammonia can knock anyone unconscious!** Put the test tube into a hot-water bath for a few moments to increase the amount of NH_3 in the gas phase. Test the gas with moistened blue and red litmus paper. Test the gas for support of combustion.

2. Hydrogen Sulfide, H_2S

Hydrogen sulfide can be made by treating some solid sulfides, particularly FeS, with an acid such as HCl or H_2SO_4. With FeS the reaction is

$$FeS(s) + 2 H^+(aq) \rightarrow H_2S(g) + Fe^{2+}(aq) \tag{8}$$

This reaction was used for many years to make H_2S in the laboratory in courses in qualitative analysis. In this experiment we will employ the method currently used in such courses for H_2S generation. This involves the decomposition of thioacetamide, CH_3CSNH_2, which occurs in solution on treatment with acid and heat. The reaction is

$$CH_3CSNH_2(aq) + 2 H_2O(\ell) \rightarrow H_2S(g) + CH_3COO^-(aq) + NH_4^+(aq) \tag{9}$$

The odor of H_2S is notorious, and potent. H_2S is moderately soluble in water and, since it is produced reasonably slowly in Reaction 9, there will be little if any effervescence.

To 1.0 mL of 1.0 M thioacetamide in a small test tube add 6 drops of 3 M H_2SO_4. Put the test tube in a boiling-water bath for about one minute. There may be some cloudiness due to formation of free sulfur. **Carefully** smell the gas in the tube; H_2S is toxic! Test the gas for acidic and basic properties and for the ability to support combustion.

Optional D. Identification of an Unknown Solution

In this experiment you prepared nine different species containing nonmetallic elements. In each case the source of the species was in solution. In this part of the experiment we will give you an unknown solution that can be used to make one of the nine species. It will be a solution used in preparing one of the species, but it will not be an acid or a base. Identify by suitable tests the species that can be made from your unknown and the substance that is present in your unknown solution. ■

DISPOSAL OF REACTION PRODUCTS. Most of the chemicals used in this experiment can be discarded down the sink drain. Pour the solutions from Sections 3 and 4 of Part A, containing I_2 and Br_2 in heptane, into a waste container, unless directed otherwise by your instructor.

Name _____   Section _____

Experiment 18

Observations and Analysis: Some Nonmetals and Their Compounds

A. Preparation and Properties of Nonmetallic Elements: O_2, N_2, Br_2, I_2

Element prepared	Degree of effervescence	Odor	Acid-base tests	Support of combustion test
1. O_2	_____	_____	_____	_____
2. N_2	_____	_____	_____	_____

	Odor	Of vapor	Color	In HEP
			In H_2O	
3. I_2	_____	_____	_____	_____
4. Br_2	_____	_____	_____	_____

B. Preparation and Properties of Some Nonmetallic Oxides: CO_2, SO_2, NO, and NO_2

	Oxide prepared	Degree of effervescence	Odor	Acid test	Color	Support of combustion test
1.	CO_2	_____	_____	_____	_____	_____
2.	SO_2	_____	_____	_____	_____	_____
3.	$NO_2 + NO$	_____	_____	_____	soln: _____	_____
					gas: _____	_____

C. Preparation and Properties of Some Nonmetallic Hydrides: NH_3, H_2S

	Hydride prepared	Degree of effervescence	Odor	Acid-base tests	Support of combustion test
1.	NH_3	_____	_____	_____	_____
2.	H_2S	_____	_____	_____	_____

Optional **D. Properties of an Unknown Solution**

Nonmetal species that can be made from unknown _____

Identity of unknown solution _____

Unknown # _____

Experiment 18

Advance Study Assignment: Some Nonmetals and Their Compounds

In this experiment nine species are prepared and studied. For each species, list the reagents used in its preparation and the reaction that occurs.

	Reagents used	Reaction
1. O_2		
2. N_2		
3. I_2		
4. Br_2		
5. CO_2		
6. SO_2		
7. $NO_2 + NO$		
8. NH_3		
9. H_2S		

Experiment 19

Molar Mass Determination by Depression of the Freezing Point

The most common liquid we encounter in our daily lives is water. In this experiment we will study the equilibria that can exist between pure water and its aqueous solutions, and ice, the solid form of water. (Water is one of a very few substances for which we have a separate name for the solid.)

If we take some ice cubes from the refrigerator and put them into a glass of water from the tap, we find that the water temperature falls and some ice melts. This occurs because heat will always tend to flow from a higher to a lower temperature. Heat from the water flows into the ice. It takes heat, called the heat of fusion, to melt ice. If there is enough ice present, the water temperature will ultimately fall to the freezing point of pure water, T_f°, which is 0°C, and stay there. At that point, ice and water are in equilibrium: at the freezing point of water. At the freezing point the vapor pressures of ice and water must be equal, and that condition fixes the temperature. (See Fig. 19.1.)

Now let us consider what happens if we add a soluble liquid or solid to the equilibrium mixture of ice and water. Surprisingly, we find that the temperature of the ice and the solution falls as equilibrium is reestablished. This reflects the fact that in the solution the vapor pressure of water at 0°C is less than that of the pure liquid. At equilibrium, the vapor pressures of the solid and liquid phases must be equal. So the vapor pressure of ice at 0°C is higher than that of the solution, and some ice melts. This requires thermal energy, which is drawn from the solution and from the ice, and the temperature falls. The vapor pressures of the ice and the solution both fall, but that of the ice falls faster, until at some temperature T_f, below 0°C, the two vapor pressures become equal and equilibrium is reestablished at the new freezing point. The situation is shown in Figure 19.1.

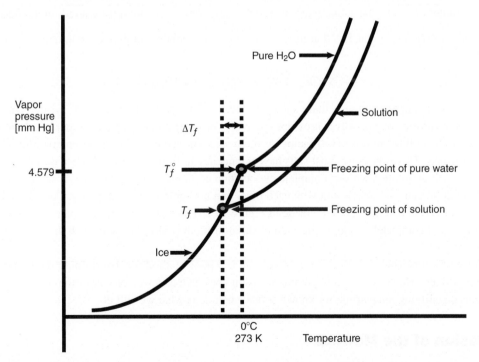

Figure 19.1 Dissolving a solute in a solvent will result in a solution with a lower freezing point than that of the pure solvent.

The change in the freezing point that is observed is called the freezing point depression, ΔT_f, equal to $T_f^\circ - T_f$. It is observed with solutions of any solvent. Freezing point depression is one of the colligative properties of solutions. Others are boiling point elevation (which requires a nonvolatile solute), osmotic pressure, and vapor pressure lowering. The colligative properties of solutions depend on the number of solute particles present in a given amount of solvent but not on the kinds of particles dissolved, be they molecules, atoms, or ions.

When working with colligative properties it is convenient to express the solute concentration in terms of its molality, m, as defined by the equation:

$$\text{molality of } A = m_A = \frac{\text{moles } A \text{ dissolved}}{\text{kg solvent in the solution}} \tag{1}$$

For this unit of concentration, the boiling point elevation, $T_b - T_b^\circ$, or ΔT_b, and the freezing point depression, $T_f^\circ - T_f$, or ΔT_f, in C° at very low concentrations are estimated reasonably well by the equations:

$$\Delta T_b = k_b m \quad \text{and} \quad \Delta T_f = k_f m \tag{2}$$

where k_b and k_f are characteristic of the solvent used. For water, $k_b = 0.52$ and $k_f = 1.86$. For benzene, $k_b = 2.53$ and $k_f = 5.10$. In this experiment we will assume that Equation 2 is valid, even though our solutions are moderately concentrated. We write temperature differences in C°, while temperatures are in °C. Note that a temperature difference of 4 C° is the same as a temperature difference of 4 K, but a temperature of 4°C is very different from a temperature of 4 K!

One of the classic uses of colligative properties was in determining molar masses of unknown substances. With organic molecules, molar masses indicated by freezing point depression agreed with those found by other methods. However, with ionic salts, like NaCl, molar masses suggested by freezing point depression measurements were much lower than the actual formula masses. On the basis of such experiments, Arrhenius suggested that ionic substances exist as ions in aqueous solution, consistent with the observation that such solutions conduct an electric current. Arrhenius's general idea turned out to be correct and is now a basic part of modern chemical theory.

The van't Hoff factor, i, provides insight into how an ionic solute dissociates, and any solid dissolves (becomes solvated by water molecules), in a liquid at a given concentration. The van't Hoff factor is defined as

$$i \equiv \frac{\text{moles of independent particles released into a solution}}{\text{moles of solute that actually dissolve in that solution}} = \frac{\text{actual molar mass}}{\text{apparent (measured) molar mass}} \tag{3}$$

Introductory chemistry teaches that most ionic solids dissociate completely in aqueous solution. For example:

$$NaCl(s) \xrightarrow{H_2O(\ell)} Na^+(aq) + Cl^-(aq)$$

This chemical equation suggests that i should be exactly 2 for sodium chloride. As written above, every mole of sodium chloride that dissolves provides two moles of independent particles (ions) in solution. It may surprise you to learn that the van't Hoff factor for these salts only equals 2.00 in extremely dilute solutions; generally, it is less. This is because most salts have a tendency to form "ion pairs" when dissolved in water: these result in salt dissolution without complete dissociation of all the component ions. At any instant in time, a few anion-cation pairs are interacting with each other and not moving independently. Consider the simplified drawing of an ion pair of sodium chloride in Figure 19.2. The water solvates the pair of ions simultaneously, but does not completely separate the sodium cation from the chloride anion, such that they continue to move together.

Because of ion pairing, freezing point depression is a poor method for determining the molar mass of unknown ionic salts. However, freezing point depression does give us a way of measuring the van't Hoff factor, and thus quantifying ion pairing, for known solids, and that is what we will do.

Discussion of the Method

In this experiment we will study the freezing point behavior of some aqueous solutions. First you will measure the freezing point of pure water, using a slurry of ice and water. Then you will determine the freezing

Interaction between ions persists

Independent ions

An ion pair of Na$^+$ and Cl$^-$

Fully solvated Na$^+$ and Cl$^-$

Figure 19.2 Ion pairs exist in solution because the solvation shells around dissolved ions do not perfectly shield the charges of the solvated ions, and so solvated ions of opposite charge are still loosely attracted to one another, and do not always move independently.

point of an aqueous solution containing a known mass of an unknown liquid, thus determining the freezing point depression, ΔT_f. This will allow you to calculate the molality of the solution. Finally, you will separate the solution from the ice in the mixture, and weigh the solution, which will contain all of the solute. This information furnishes you with the composition of the solution: it lets you calculate the mass of solute present per kilogram of water. The molar mass follows from Equations 1 and 2.

Experimental Procedure

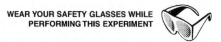

WEAR YOUR SAFETY GLASSES WHILE
PERFORMING THIS EXPERIMENT

Obtain a digital thermometer, an insulated cup, and two unknowns: one liquid and one solid. The actual molar mass of the solid will be furnished to you.

A. Measuring the Freezing Point of Pure Water

You will first need to determine what your thermometer thinks the freezing point of pure water is. Be certain that your insulated cup is clean; then prepare a mixture of deionized water and ice in it by first filling the cup with ice, then adding a bit of water at a time and stirring between each addition. Once the water level reaches the top of the ice, stir well for at least 30 seconds and then record the temperature. (Since your thermometer may not be perfectly calibrated, that temperature may not be exactly 0.0°C; that is the point of this step: it actually serves to calibrate your thermometer.)

B. Finding the Freezing Point of a Solution of Liquid Unknown

Pour the ice water out of your cup and shake out most of the water from it, but do not bother drying the inside. (You do want the *outside* to be dry.) Place the empty cup on a top-loading balance and tare it out. Add 10 ± 2 g of your liquid unknown, recording the exact mass you end up using, to ± 0.01 g. Bring the cup back to your workstation and add ice (for now just ice, no water), so that you end up with an ice layer about one centimeter thick, and stir while monitoring the temperature. Keep stirring until the temperature reading no longer systematically drops; add ice as needed to maintain about a 1-cm layer at the top of the solution. If this final temperature is warmer than −6°C, you are ready to proceed. If not, add water a little at a time and stir, until the equilibrium temperature is warmer than −6°C and a roughly one-centimeter layer of ice remains on top. (Do not obsess about the amount of ice—just make sure that the liquid surface is completely covered with ice, and that you do not have so much ice that it will not all fit on the wire screen.) Now, stir well and record the lowest stable temperature that you observe. Be patient here—make sure that the temperature has stabilized, and that you are stirring when you record the depressed freezing point of the solution. However, once you have confidently determined this temperature, do not wait too long (because the ice will continue to melt, and further dilute the solution!) before separating your solution from the ice by pouring it through a wire screen into a tared beaker on a top-loading balance. Give the screen a good shake, to get as much of the liquid out of the ice as possible, then lift it above the beaker so it is not contributing to the mass measured on the balance. Record the mass of the solution, then pour the solution and the ice back into the cup.

Conduct a second trial with your unknown liquid, at a more dilute solute concentration. This is easily done with the decanted solution and ice you just poured back into the cup. Rinse the beaker with a *small amount* of deionized water and pour the rinsing into the cup, thereby slightly lowering the concentration of the solute in your solution. Add more water as necessary to reduce the freezing point depression to *roughly one-half* of its original value, all the while maintaining a one-centimeter layer of ice in the cup. (If your initial freezing point was −6°C, you would want to aim for about −3°C. Again, do not stress about this, just aim for something in the ballpark. *Note that if you overdilute, there is no way to undo it—you will not be able to make the freezing point of the solution colder again. Just go with what you have: do not add more ice in an effort to cool it down. The ice itself is at 0°C and is warmer than your solution!*) Accurately measure the new equilibrium freezing point for the mixture, then separate the solution from the ice and determine its mass, as you did before.

When you have finished this part of the experiment, dispose of the mixture of ice and solution as directed by your instructor.

C. Finding the Freezing Point of a Solution of a Known Solid

The objective of this part of the experiment will be to experimentally determine the van't Hoff factor for your solid. You will use the same general procedure as you did for the liquid.

On a top-loading balance, add about ten grams of your solid to a thoroughly rinsed, empty cup that is dry on the outside, recording the exact amount used. Return to your workstation with the cup and, while stirring, add a minimum amount of water to dissolve the solid. Stir until you can see that dissolution is complete, adding more water as needed. Then proceed as you did with your unknown liquid, adding ice (and, if necessary, water) to obtain the minimum freezing point possible that is not colder than −6°C. Decant and weigh the solution as you did with the liquid unknown, and record your results.

As with the liquid unknown, return the ice and solution to the cup and take a second measurement, aiming for a freezing point depression about half that observed in your first trial.

If the molar mass you find for your solid is appreciably less than the actual molar mass, you probably have an ionic solid. The ratio of the true molar mass to the value you determine experimentally is the van't Hoff factor, i.

When you are finished with the experiment, pour the solution and the contents of the cup into the sink. Return the digital thermometer and the insulated cup.

Take it Further (Optional): Experimentally determine the largest possible freezing point depression for a solution of sodium chloride, NaCl. Find the composition of that solution. What phases are in equilibrium in the final mixture, at the depressed freezing point?

Name _____ **Section** _____

Experiment 19

Data and Calculations: Molar Mass Determination by Depression of the Freezing Point

A. Measured Freezing Point of Pure Water _____ °C

B. Finding the Freezing Point of a Solution of Liquid Unknown

Unknown # _____

Actual mass of solute used _____ g

Trial I

Freezing point of solution (observed) _____ °C

Mass of solution _____ g

Trial II

Freezing point of solution _____ °C

Mass of solution _____ g

Calculations:

	Trial I	Trial II
Freezing point depression	_____ C°	_____ C°
Molality of unknown solution, m_u	_____ molal	_____ molal
Mass of solution	_____ g	_____ g
Mass of solute	_____ g	_____ g
Mass of solvent (water)	_____ g	_____ g
Moles of solute	_____ mol	_____ mol
Molar mass of unknown	_____ g/mol	_____ g/mol

(continued on following page)

C. Finding the Freezing Point of a Solution of a Known Solid

Solid # _____

Mass of solid used _____ g

Trial I

Freezing point of solution _____ °C

Mass of solution _____ g

Trial II

Freezing point of solution _____ °C

Mass of solution _____ g

Calculations:

	Trial I	**Trial II**
Freezing point depression	_____ C°	_____ C°
Molality of solid solution, m_s	_____ molal	_____ molal
Mass of solution	_____ g	_____ g
Mass of solid solute used	_____ g	_____ g
Mass of solvent (water)	_____ g	_____ g
(Apparent) moles of solute	_____ mol	_____ mol
(Apparent) molar mass of unknown	_____ g/mol	_____ g/mol
Value of i, the van't Hoff factor	_____	_____

Experiment 19

Advance Study Assignment: Determination of Molar Mass by Depression
of the Freezing Point

1. A student determined the molar mass of an unknown non-dissociating liquid by the method described in this experiment. She found that the equilibrium temperature of a mixture of ice and pure water was indicated to be −0.1°C on her thermometer. When she added 9.9 g of her sample to the mixture, the temperature, after thorough stirring, fell to −3.7°C. She then poured off the solution through a screen into a beaker. The mass of the solution was 84.2 g.

   a.   What was the freezing point depression?

                                                                                  _____ C°

   b.   What was the molality of the unknown liquid?

                                                                                  _____ m

   c.   What mass of unknown liquid was in the decanted solution?

                                                                                  _____ g

   d.   What mass of water was in the decanted solution?

                                                                                  _____ g

   e.   How much unknown liquid would there be in a solution containing 1 kg of water, with the student's unknown liquid at the same concentration as she had in her experiment?

                                                                    _____ g unknown liquid

   f.   Based on these data, what value did she calculate for the molar mass of her unknown liquid, assuming she carried out the calculation correctly?

                                                                                   _____ g/mol

Experiment 20

Rates of Chemical Reactions, I. The Iodination of Acetone

The rate at which a chemical reaction occurs depends on several factors: the nature of the reaction, the concentrations of the reactants, the temperature, and the presence of possible catalysts. All of these factors can markedly influence the observed rate of reaction.

Some reactions at a given temperature are very slow indeed: the oxidation of gaseous hydrogen or wood at room temperature would not proceed appreciably in a century. Other reactions are essentially instantaneous: the precipitation of silver chloride when solutions containing silver ions and chloride ions are mixed and the formation of water when acidic and basic solutions are mixed are examples of extremely rapid reactions. In this experiment we will study a reaction that, in the vicinity of room temperature, proceeds at a moderate, relatively easily measured rate.

For a given reaction, the rate typically increases with an increase in the concentration of any reactant. The relation between rate and concentration is a remarkably simple one in many cases, and for the balanced chemical reaction

$$a \, A + b \, B \rightarrow c \, C$$

in which A, B, and C are chemical species and a, b, and c are stoichiometric coefficients. The rate can usually be expressed by the equation

$$\text{rate} = k[A]^m[B]^n \tag{1}$$

where m and n are generally, but not always, integers: specifically 0, 1, 2, or possibly 3; [A] and [B] are the concentrations of A and B (ordinarily in moles per liter); and k is a constant, called the rate constant of the reaction, which makes the relation quantitatively correct. The numbers m and n are called the orders of the reaction with respect to A and B. If m is 1, the reaction is said to be first order with respect to the reactant A. If n is 2, the reaction is second order with respect to reactant B. The overall order is the sum of m and n. In this example, the reaction would be third order overall. Note that the orders of the reaction do not bear a necessary relationship to the stoichiometric coefficients (a, b, and c) of the overall balanced reaction.

The rate of a reaction is also significantly dependent on the temperature at which the reaction occurs. An increase in temperature increases the rate; an often-cited general rule being that a 10 C° rise in temperature will double the rate. This rule is only approximately correct; nevertheless, it is clear that a rise in temperature on the order of 100 C° could change the rate of a reaction very appreciably.

As with concentration, there is a quantitative relation between reaction rate and temperature, but here the relation is somewhat more complicated. This relation is based on the idea that to react, the reactant species must have a certain minimum amount of energy present at the time the reactants collide in the reaction step; this amount of energy, which is typically furnished by the kinetic energy of motion of the species present, is called the activation energy for the reaction. The equation relating the rate constant k to the absolute temperature T and the activation energy E_a is called the Arrhenius equation:

$$\ln k = \frac{-E_a}{RT} + \text{constant} \tag{2}$$

where $\ln k$ is the natural logarithm of k and R is the gas constant (8.31 joules/mole K for E_a in joules per mole). This equation is identical in form to Equation 1 in Experiment 15. By measuring k at different temperatures we can determine graphically the activation energy for a reaction.

In this experiment we will study the kinetics of the reaction between iodine and acetone:

$$CH_3 - \overset{\overset{\displaystyle O}{\|}}{C} - CH_3(aq) + I_2(aq) \rightarrow CH_3 - \overset{\overset{\displaystyle O}{\|}}{C} - CH_2I(aq) + H^+(aq) + I^-(aq)$$

The rate of this reaction is found to depend on the concentration of hydrogen ion in the solution as well as (presumably) on the concentrations of the two reactants. (This is not because H^+ is a product of the reaction, but rather because the mechanism by which it actually occurs involves hydrogen ion as a reactant in the slow, or rate-determining, step.) By Equation 1, the rate law for this reaction is

$$\text{rate} = k[\text{acetone}]^m[I_2]^n[H^+]^p \tag{3}$$

where m, n, and p are the orders of the reaction with respect to acetone, iodine, and hydrogen ion, respectively, and k is the rate constant for the reaction.

The rate of this reaction can be expressed as the (small) change in the concentration of I_2, $\Delta[I_2]$, that occurs, divided by the time interval Δt required for the change:

$$\text{rate} = \frac{-\Delta[I_2]}{\Delta t} \tag{4}$$

The minus sign is to make the rate positive (as $\Delta[I_2]$ is negative). Ordinarily, since rate varies with the concentrations of the reactants according to Equation 3, in a rate study it would be necessary to measure, directly or indirectly, the concentration of each reactant as a function of time; the rate would typically vary markedly with time, decreasing to very low values as the concentration of at least one reactant becomes very low. This makes reaction rate studies relatively difficult to carry out and introduces mathematical complexities that are difficult for beginning students to understand.

The iodination of acetone is a rather atypical reaction, in that it can be easily investigated experimentally. First of all, iodine has color, so that we can readily follow changes in iodine concentration visually. A second and very important characteristic of this reaction is that it turns out to be zero order in I_2 concentration. This means (see Equation 3) that the rate of the reaction does not depend on $[I_2]$ at all; $[I_2]^0 = 1$, no matter what the value of $[I_2]$ is, as long as it is not itself zero.

Because the rate of the reaction does not depend on $[I_2]$, we can study the rate by simply making I_2 the limiting reagent present in a large excess of acetone and H^+ ion. We then measure the time required for a known initial concentration of I_2 to be used up completely. If both acetone and H^+ are present at much higher concentrations than that of I_2, their concentrations will not change appreciably during the course of the reaction, and the rate will remain, by Equation 3, effectively constant until all the iodine is gone, at which time the reaction will stop. Under such circumstances, if it takes t seconds for the color of a solution having an initial concentration of I_2 equal to $[I_2]_0$ to disappear, the rate of the reaction, by Equation 4, would be

$$\text{rate} = \frac{-\Delta[I_2]}{\Delta t} = \frac{[I_2]_0}{t} \tag{5}$$

Although the rate of the reaction is constant during its course under the conditions we have set up, we can vary it by changing the initial concentrations of acetone and H^+ ion. For example, should we double the initial concentration of acetone over that in Mixture 1, keeping $[H^+]$ and $[I_2]$ at the same values they had previously, then the rate of Mixture 2 would, according to Equation 3, differ from that in Mixture 1:

$$\text{rate } 2 = k(2[A])^m[I_2]^0[H^+]^p \tag{6a}$$

$$\text{rate } 1 = k[A]^m[I_2]^0[H^+]^p \tag{6b}$$

Dividing the first equation by the second, we see that the k's cancel, as do the terms in the iodine and hydrogen ion concentrations, since they have the same values in both reactions, and we obtain simply

$$\frac{\text{rate } 2}{\text{rate } 1} = \frac{(2[A])^m}{[A]^m} = \left(\frac{2[A]}{[A]}\right)^m = 2^m \tag{6}$$

Having measured both rate 2 and rate 1 by Equation 5, we can find their ratio, which must be equal to 2^m. We can then solve for m either by inspection or using logarithms and so find the order of the reaction with respect to acetone.

By a similar procedure we can measure the order of the reaction with respect to H^+ ion concentration and also confirm that the reaction is zero order with respect to I_2. Having found the order with respect to each reactant, we can then evaluate k, the rate constant for the reaction.

The determination of the orders m and p, the confirmation of the fact that n, the order with respect to I_2, equals zero, and the evaluation of the rate constant k for the reaction at room temperature comprise your assignment in this experiment. You will be furnished with standard solutions of acetone, iodine, and hydrogen ion, and with the composition of one mixture that will give a reasonable rate. The rest of the planning and the execution of the experiment will be your responsibility.

An optional part of the experiment is to study the rate of this reaction at different temperatures to find its activation energy. The general procedure here would be to study the rate of reaction in one of the mixtures at room temperature and at two other temperatures, one above and one below room temperature. Knowing the rates, and hence the k's, at the three temperatures, you can then find E_a, the activation energy for the reaction, by plotting $\ln k$ vs $1/T$. The slope of the resultant straight line, by Equation 2, must be $-E_a/R$.

Experimental Procedure

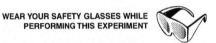

WEAR YOUR SAFETY GLASSES WHILE
PERFORMING THIS EXPERIMENT

Select two medium (18×150 mm) test tubes; when filled with deionized water, they should appear to have the same color when viewed from above the opening (down the tube) against a white background.

Draw 50 mL of each of the following solutions into clean, dry, 100-mL beakers, one solution to a beaker: 4.0 M acetone, 1.0 M HCl, and 0.0050 M I_2. Cover each beaker with a watch glass.

With your graduated cylinder, measure out 10. mL of the 4 M acetone solution and pour it into a clean 125-mL Erlenmeyer flask. Then measure out 10. mL of 1.0 M HCl and add that to the acetone in the flask. Add 20. mL of deionized H_2O to the flask. Drain the graduated cylinder, shaking out any excess water, and then use the cylinder to measure out 10. mL of 0.0050 M I_2 solution. Be careful not to spill the iodine solution on your hands or clothes.

Noting the time to the nearest second, or starting a stopwatch simultaneously, pour the iodine solution into the Erlenmeyer flask and quickly swirl the flask to mix the reagents thoroughly. The reaction mixture will appear yellow because of the presence of the iodine, and the color will fade slowly as the iodine reacts with the acetone. Fill one of the test tubes three-quarters full with the reaction mixture, and fill the other test tube to the same depth with deionized water. Look down the test tubes toward a well-lit piece of white paper, and note the time the color of the iodine disappears. Measure the temperature of the reaction mixture in the test tube.

Repeat the experiment, using as a color reference the reacted solution instead of deionized water. The amount of time required in the two runs should agree within about twenty seconds.

The rate of the reaction equals the initial concentration of I_2 *in the reaction mixture* divided by the elapsed time. Since the reaction is zero order in I_2, and since both acetone and H^+ ion are present in great excess, the concentrations of both acetone and H^+ remain essentially at their initial values in the reaction mixture, and the reaction rate is constant, throughout the reaction.

Having found the reaction rate for one composition of the system, think for a moment about what changes in composition you might make to decrease the time and hence increase the rate of reaction. In particular, how could you change the composition in such a way as to allow you to determine how the rate depends upon acetone concentration? If it is not clear how to proceed, reread the discussion preceding Equation 6. In your new mixture you should keep the total volume at 50. mL, and be sure that the concentrations of H^+ and I_2 are the *same* as in the first experiment. Carry out the reaction twice with your new mixture; the times should not differ from one another by more than about fifteen seconds. The temperature should be kept within about one degree of that in the initial run. Calculate the rate of the reaction. Compare it with that for the first mixture, and then calculate the order of the reaction with respect to acetone, using a relation similar to Equation 6. First, write an equation like 6a for the second reaction mixture, substituting in the values for the rate as obtained by Equation 5 and the initial concentrations of acetone, I_2, and H^+ in the reaction mixture. Then write an equation like 6b for the first reaction mixture, using the observed rate and the initial concentrations in that mixture. Obtain an equation like 6 by dividing Equation 6a by Equation 6b. Solve Equation 6 for the order m of the reaction with respect to acetone.

Again change the composition of the initial reaction mixture so that this time a measurement of the reaction rate will give you information about the order of the reaction with respect to H^+. Repeat the experiment with this mixture to establish the time of reaction to within 15 seconds, again making sure that the temperature is within about one degree of that observed previously. From the rate you determine for this mixture find p, the order of the reaction with respect to H^+.

Finally, change the reaction mixture composition in such a way as to allow you to show that the order of the reaction with respect to I_2 is zero. Measure the rate of the reaction twice, and calculate n, the order with respect to I_2.

Having found the order of the reaction for each species on which the rate depends, evaluate k, the rate constant for the reaction, from the rate and concentration data in each of the mixtures you studied. If the temperatures at which the reactions were run are all equal to within one or two degrees, k should be about the same for each mixture. Calculate the mean value of k and the standard deviation for the set of values. (See Appendix IX.)

Optional   As a final reaction, make up a mixture using reactant volumes that you did not use in any previous experiments. Using Equation 3, the values of concentrations in the mixtures, the orders, and the rate constant you calculated from your experimental data, predict how long it will take for the I_2 color to disappear from your mixture. Measure the time for the reaction and compare it with your prediction. If time permits, select one of the reaction mixtures you have already used that gave a convenient time, and use that mixture to measure the rate of reaction at about 10°C and at about 40°C. From the two rates you find, plus the rate at room temperature, calculate the activation energy for the reaction, using Equation 2. ■

DISPOSAL OF REACTION PRODUCTS.   Dispose of your solutions from this experiment as directed by your instructor.

Experiment 20

Data and Calculations: The Iodination of Acetone

A. Reaction Rate Data

Mixture	Volume in mL 4.0 M acetone	Volume in mL 1.0 M HCl	Volume in mL 0.0050 M I_2	Volume in mL H_2O	Time for reaction in sec 1st run	Time for reaction in sec 2nd run	Temp. in °C
I	10.	10.	10.	20.	_____	_____	_____
II	_____	_____	_____	_____	_____	_____	_____
III	_____	_____	_____	_____	_____	_____	_____
IV	_____	_____	_____	_____	_____	_____	_____

B. Determination of Reaction Orders with Respect to Acetone, H^+ Ion, and I_2

$$\text{rate} = k\,[\text{acetone}]^m\,[I_2]^n\,[H^+]^p \tag{3}$$

Calculate the initial concentrations of acetone, H^+ ion, and I_2 in each of the mixtures you studied. Use Equation 5 to find the rate of each reaction.

Mixture	[acetone]	$[H^+]$	$[I_2]_0$	Rate $= \dfrac{[I_2]_0}{\text{avg. time}}$
I	0.80 M	0.20 M	0.0010 M	_____
II	_____	_____	_____	_____
III	_____	_____	_____	_____
IV	_____	_____	_____	_____

Substituting the initial concentrations and the rate from this table, write Equation 3 as it would apply to Reaction Mixture II:

Rate II =

Now write Equation 3 for Reaction Mixture I, substituting concentrations and the calculated rate from the table:

Rate I =

(continued on following page)

Divide the equation for Mixture II by the equation for Mixture I; the resulting equation should have the ratio of Rate II to Rate I on the left side, and a ratio of acetone concentrations raised to the power m on the right. It should be similar in appearance to Equation 6. Write the resulting equation here:

$$\frac{\text{Rate II}}{\text{Rate I}} =$$

The only unknown in the equation is m. Solve for m.　　　　　　　　　　　　　$m =$ _____

Now write Equation 3 as it would apply to Reaction Mixture III and as it would apply to Reaction Mixture IV:

Rate III =

Rate IV =

Using the ratios of the rates of Mixtures III and IV to those of Mixtures II or I, find the orders of the reaction with respect to H^+ ion and I_2:

$$\frac{\text{Rate III}}{\text{Rate}} =$$　　　　　　　　　　　　　　　　　　$p =$ _____

$$\frac{\text{Rate IV}}{\text{Rate}} =$$　　　　　　　　　　　　　　　　　　$n =$ _____

C. Determination of the Rate Constant, *k*

Given the values of m, p, and n as determined in Part B, calculate the rate constant k for each mixture by simply substituting those orders, the initial concentrations, and the observed rate from the table into Equation 3.

Mixture	I	II	III	IV	Mean	Standard deviation
k	_____	_____	_____	_____	_____	_____

Optional **D. Prediction of Reaction Rate**

Reaction mixture

Volume in mL	Volume in mL	Volume in mL	Volume in mL
4.0 M acetone _____	1.0 M HCl _____	0.0050 M I_2 _____	H_2O _____

Initial concentrations

[acetone] _____ M $[H^+]$ _____ M $[I_2]_0$ _____ M

Predicted rate _____ (Eq. 3)

Predicted time for reaction _____ sec (Eq. 5)

Observed time for reaction _____ sec ▪

Optional **E. Determination of Energy of Activation**

Reaction mixture used _____ (same for all temperatures)

Time for reaction at about 10°C _____ sec   Temperature _____ °C; _____ K

Time for reaction at about 40°C _____ sec   Temperature _____ °C; _____ K

Time for reaction at room temp _____ sec   Temperature _____ °C; _____ K

Calculate the rate constant at each temperature from your data, following the procedure in Part C.

	Rate	**k**	**ln k**	$\dfrac{1}{T[K]}$
≈10°C	_____	_____	_____	_____
≈40°C	_____	_____	_____	_____
Room temp	_____	_____	_____	_____

Plot ln k vs $1/T$ using Excel or the graph paper on the following page. Find the slope of the best straight line through the points. (If you need to, see Appendix V.)

Slope = _____ K

By Equation 2: E_a [J/mol] $= -8.31 \times$ slope

E_a _____ joules/mol

(continued on following page)

Experiment 20

Data and Calculations: The Iodination of Acetone

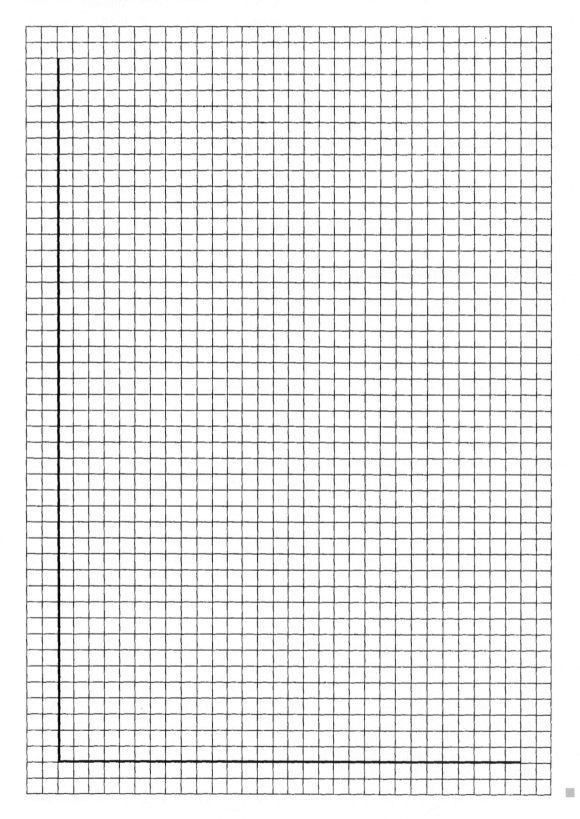

Experiment 20

Advance Study Assignment: The Iodination of Acetone

1. In a reaction involving the iodination of acetone, the following volumes were used to make up the reaction mixture:

   10. mL 4.0 M acetone + 10. mL 1.0 M HCl + 10. mL 0.0050 M I_2 + 20. mL H_2O

   a. How many moles of acetone were in the reaction mixture? Recall that, for a component A, moles of A = $[A] \times V$, where $[A]$ is the molarity of A and V is the volume in liters of the solution of A that was used.

   _____ moles acetone

   b. What was the molarity of acetone in the reaction mixture? The volume of the mixture was 50. mL, or 0.050 L, and the number of moles of acetone was found in Part (a). Again,

   $$[A] = \frac{\text{moles of A}}{V \text{ of soln. in liters}}$$

   _____ M acetone

   c. How could you double the molarity of the acetone in the reaction mixture, keeping the total volume at 50. mL and keeping the same concentrations of H^+ ion and I_2 as in the original mixture?

2. Using the reaction mixture in Problem 1, a student found that it took 510 seconds for the color of the I_2 to disappear.

   a. What was the rate of the reaction? (Hint: First find the initial concentration of I_2 in the reaction mixture, $[I_2]_0$. Then use Equation 5.)

   rate = _____ M/sec

   b. Given the rate from Part (a), and the initial concentrations of acetone, H^+ ion, and I_2 in the reaction mixture (you have yet to calculate $[H^+]$; do so now), write Equation 3 as it would apply to the mixture.

   rate =

   c. What are the unknowns that remain in the equation in Part (b)?

   _____ _____ _____ _____

(continued on following page)

3. A second reaction mixture was made up in the following way:

    10. mL 4.0 M acetone + 20. mL 1.0 M HCl + 10. mL 0.0050 M I_2 + 10. mL H_2O

    a. What were the initial concentrations of acetone, H^+ ion, and I_2 in the reaction mixture?

        [acetone] _____ M; $[H^+]$ _____ M; $[I_2]_0$ _____ M

    b. It took 127 seconds for the I_2 color to disappear from the reaction mixture when it occurred at the same temperature as the reaction in Problem 2. What was the rate of the reaction?

        rate = _____ M/sec

        Write Equation 3 as it would apply to the second reaction mixture:

        rate =

    c. Divide the equation in Part (b) by the equation in Problem 2(b). The resulting equation should have the ratio of the two rates on the left side and a ratio of H^+ concentrations raised to the p power on the right. Write the resulting equation and solve for the value of p, the order of the reaction with respect to H^+. (Round off the value of p to the nearest integer.)

        p = _____

4. A third reaction mixture was made up in the following way:

    10. mL 4.0 M acetone + 20. mL 1.0 M HCl + 5.0 mL 0.0050 M I_2 + 15. mL H_2O

    If the reaction is zero order in I_2, how long would it take for the I_2 color to disappear at the temperature of the reaction mixture in Problem 3?

        _____ seconds

Experiment 21

Rates of Chemical Reactions,
II. A Clock Reaction

In the previous experiment we discussed the factors that influence the rate of a chemical reaction and presented the terminology used in quantitative relations in studies of the kinetics of chemical reactions. That material is also pertinent to this experiment and should be studied before you proceed further.

This experiment involves the study of the rate properties, or chemical kinetics, of the following reaction between iodide ion and bromate ion under acidic conditions:

$$6\,I^-(aq) + BrO_3^-(aq) + 6\,H^+(aq) \rightarrow 3\,I_2(aq) + Br^-(aq) + 3\,H_2O(\ell) \tag{1}$$

This reaction proceeds reasonably slowly at room temperature, its rate depending on the concentrations of the I^-, BrO_3^-, and H^+ ions according to the rate law discussed in the previous experiment. For this reaction the rate law takes the form

$$\text{rate} = k[I^-]^m[BrO_3^-]^n[H^+]^p \tag{2}$$

One of the main purposes of this experiment will be to evaluate the rate constant k and the reaction orders m, n, and p for this reaction. We will also investigate the manner in which the reaction rate depends on temperature, and will evaluate the activation energy of the reaction, E_a.

Our method for measuring the rate of the reaction involves what is frequently called a clock reaction. In addition to Reaction 1, whose kinetics we will study, the following reaction will also be made to occur simultaneously in the reaction flask:

$$I_2(aq) + 2\,S_2O_3^{2-}(aq) \rightarrow 2\,I^-(aq) + S_4O_6^{2-}(aq) \tag{3}$$

Relative to Reaction 1, this reaction is essentially instantaneous. The I_2 produced in (1) reacts completely with the thiosulfate ion, $S_2O_3^{2-}$, present in the solution, so that until all the thiosulfate ion has reacted, the concentration of I_2 remains effectively zero. As soon as all the $S_2O_3^{2-}$ is consumed, however, the I_2 produced by (1) begins to accumulate in the solution. The presence of I_2 is made strikingly apparent by a starch indicator that is added to the reaction mixture, since I_2 even in small concentrations reacts with starch in solution to produce a deep blue color.

By carrying out Reaction 1 in the presence of $S_2O_3^{2-}$ and a starch indicator, we introduce a "clock" into the system. Our clock tells us when a given amount of BrO_3^- ion has reacted (1/6 mole BrO_3^- per mole $S_2O_3^{2-}$), which is just what we need to know, since the rate of reaction can be expressed in terms of the time it takes for a particular amount of BrO_3^- to be used up. In all our reactions, the amount of BrO_3^- that reacts in the time we measure will be constant and small compared to the amounts of any of the reactants, including bromate ion itself. This means that the concentrations of all reactants will be essentially constant in Equation 2, and hence so will the rate during each trial.

In our experiment we will carry out the reaction between BrO_3^-, I^-, and H^+ ions under different concentration conditions. Measured amounts of each of these ions in aqueous solution will be mixed in the presence of a constant, small amount of $S_2O_3^{2-}$. The time it takes for each mixture to turn blue will be measured. The time before the color change occurs in each reaction mixture will be inversely proportional to the rate of Reaction 1 in that mixture. By changing the concentration of one reactant and keeping the other concentrations constant, we can investigate how the rate of the reaction varies with the concentration of a particular reactant. Once we know the order of Reaction 1 in each reactant we can determine the rate constant for the reaction.

In the last part of this experiment we will investigate how the rate of the reaction depends on temperature. You will recall that in general the rate increases sharply with temperature. By measuring how the rate varies with temperature we can determine the activation energy, E_a, for the reaction by making use of the Arrhenius equation:

$$\ln k = -\frac{E_a}{RT} + \text{constant} \tag{4}$$

In this equation, k is the rate constant at the Kelvin temperature T, E_a is the activation energy, and R is the gas constant. By plotting $\ln k$ against $1/T$ we should obtain, by Equation 4, a straight line whose slope equals $-E_a/R$. From the slope of that line we can easily calculate the activation energy.

WEAR YOUR SAFETY GLASSES WHILE PERFORMING THIS EXPERIMENT

Experimental Procedure

A. Dependence of Reaction Rate on Concentration

In Table 21.1 we have summarized the reagent volumes to be used in carrying out the several trials whose rates we need to know to find the general rate law for Reaction 1. First, measure out about one-hundred milliliters of each of the listed reagents (except H_2O) into clean, labeled flasks or beakers. Use these reagents in preparing your reaction mixtures.

Table 21.1

Reaction Mixtures at Room Temperature [Reagent volumes in mL]					
	Reaction Flask I (250 mL)			Reaction Flask II (125 mL)	
Reaction Mixture	0.010 M KI	0.0010 M $Na_2S_2O_3$	H_2O	0.040 M $KBrO_3$	0.10 M HCl
1	10.0	10.0	10.0	10.0	10.0
2	20.0	10.0	0	10.0	10.0
3	10.0	10.0	0	20.0	10.0
4	10.0	10.0	0	10.0	20.0
5	8.0	10.0	12.0	5.0	15.0

The actual procedure for each reaction mixture will be much the same, and we will describe it now for the first reaction mixture, Reaction Mixture 1.

Since there are several reagents to mix, and since we do not want the reaction to start until we are ready, we will put some of the reagents into one flask and the rest into another, selecting them so that no reaction occurs until the contents of the two flasks are mixed. Using a 10-mL graduated cylinder to measure volumes, measure out 10.0 mL of 0.010 M KI, 10.0 mL of 0.0010 M $Na_2S_2O_3$, and 10.0 mL of deionized water into a 250-mL Erlenmeyer flask (Reaction Flask I). When measuring out reagents, rinse the graduated cylinder with deionized water after you have added the reagents to Reaction Flask I, and *before* you measure out the reagents for Reaction Flask II. Having rinsed out the graduated cylinder, measure out 10.0 mL of 0.040 M $KBrO_3$ and 10.0 mL of 0.10 M HCl into a 125-mL Erlenmeyer flask (Reaction Flask II). To Flask II add three or four drops of starch indicator solution.

Pour the contents of Reaction Flask II into Reaction Flask I and swirl the solutions to mix them thoroughly. Note the time or start a stopwatch at the instant the solutions are mixed. Continue swirling the solution. It should turn blue in less than 2 minutes. Note the time or stop the stopwatch at the instant that the blue color appears. Record the temperature of the blue solution to the nearest 0.2°C.

Repeat this procedure with the other mixtures in Table 21.1. *Do not forget to add the indicator* before mixing the solutions in the two flasks. The reaction flasks should be rinsed with deionized water between runs. When measuring out the reagents, be certain to rinse the graduated cylinder with deionized water after you have added the reagents to Reaction Flask I and *before* you measure out the reagents for Reaction Flask II. Try to keep the temperature just about the same in all the runs. Repeat any experiments that did not appear to proceed properly or for which you did not get a reliable time measurement.

B. Dependence of Reaction Rate on Temperature

In this part of the experiment, the reaction will be carried out at several different temperatures, using Reaction Mixture 1 in all cases. We will target temperatures of 20°C, 40°C, 10°C, and 0°C.

We will take the 20°C time to be that for Reaction Mixture 1, as determined at room temperature. To determine the time at elevated temperature, proceed as follows. Make up Reaction Mixture 1 as you did in Part A, including the indicator. However, instead of mixing the solutions in the two flasks at room temperature, put the flasks into water at 40°C (this number has one significant figure, so it indicates a temperature between 30°C and 50°C), drawn from the hot-water tap, into one or more large beakers. Check to see that the water is indeed in the required temperature range, and leave the flasks in the water for several minutes to bring them to that temperature. Be prepared for a much shorter reaction time when you mix the two solutions and continue swirling the reaction flask in the warm water. When the color change occurs, record the time, and also the temperature of the solution in the flask.

Repeat the experiment at 10 ± 5°C, cooling all the reactants in water at that temperature before starting the reaction. Record the time required for the color to change and the final temperature of the reaction mixture. Repeat once again at 0°C (between -1°C and 1°C), this time using an ice-water bath to cool the reactants.

Optional ## C. Dependence of the Reaction Rate on the Presence of Catalyst

Some ions have a pronounced catalytic effect on the rates of many reactions in water solution. Observe the effect on this reaction by once again making up Reaction Mixture 1. Before mixing, add 1 drop of 0.5 M $(NH_4)_2MoO_4$, ammonium molybdate, and a few drops of starch indicator to Reaction Flask II. Swirl the flask to mix the catalyst in thoroughly. Then combine the solutions, noting the time required for the color to change. ■

DISPOSAL OF REACTION PRODUCTS. The reaction products in this experiment are very dilute and may be poured into the sink as you complete each part of the experiment if so directed by your instructor.

Experiment 21

Data and Calculations: Rates of Chemical Reactions,
II. A Clock Reaction

A. Dependence of Reaction Rate on Concentration

Reaction: $6\,I^-(aq) + BrO_3^-(aq) + 6\,H^+(aq) \rightarrow 3\,I_2(aq) + Br^-(aq) + 3\,H_2O(\ell)$ **(1)**

$$\text{rate} = k[I^-]^m[BrO_3^-]^n[H^+]^p = -\frac{\Delta[BrO_3^-]}{t} \quad \textbf{(2)}$$

In all the reaction mixtures used in this experiment, the color change occurred when a constant, predetermined number of moles of BrO_3^- had been consumed in the reaction. The color "clock" allows you to measure the time required for this fixed quantity of BrO_3^- to react. The rate of each reaction is determined by the time t required for the color to change; since in Equation 2 the change in concentration of BrO_3^- ion, $\Delta[BrO_3^-]$, is the same in each mixture, the relative rate of each reaction is inversely proportional to the time t. We are mainly concerned with relative rather than absolute rates, so we will use relative rates equal to $(1000\ \text{sec})/t$ because they are easy to compare. Fill in the following table, first calculating the relative reaction rate for each mixture.

Reaction Mixture	Time (t) [sec] for color to change	Relative rate of reaction $(1000\ \text{sec})/t$	Reactant concentrations in reacting mixture [M]			Temp. [°C]
			$[I^-]$	$[BrO_3^-]$	$[H^+]$	
1	_____	_____	0.0020	_____	_____	_____
2	_____	_____	_____	_____	_____	_____
3	_____	_____	_____	_____	_____	_____
4	_____	_____	_____	_____	_____	_____
5	_____	_____	_____	_____	_____	_____

The reactant concentrations in the reacting mixtures are *not* those of the stock solutions, since the reagents were diluted by the other solutions. The final volume of the reaction mixture is 50.0 mL in all cases. Since the number of moles of reactant does not change on dilution we can say, for example, for I^- ion, that

$$\text{moles of } I^- = [I^-]_{\text{stock}} \times V_{\text{stock}} = [I^-]_{\text{mixture}} \times V_{\text{mixture}}$$

For Reaction Mixture 1,

$$[I^-]_{\text{stock}} = 0.010\ \text{M}, \quad V_{\text{stock}} = 10.0\ \text{mL}, \quad V_{\text{mixture}} = 50.0\ \text{mL}$$

Therefore,

$$[I^-]_{\text{mixture}} = \frac{0.010\ \text{M} \times 10.0\ \text{mL}}{50.0\ \text{mL}} = 0.0020\ \text{M}$$

Calculate the rest of the concentrations in the table using the same approach.

Determination of the Orders of the Reaction

Given the data in the table, the problem is to find the order for each reactant and the rate constant for the reaction. Since we are dealing with relative rates, we can modify Equation 2 to read as follows:

$$\text{relative rate} = k'[I^-]^m[BrO_3^-]^n[H^+]^p \quad \textbf{(5)}$$

(continued on following page)

We need to determine the relative rate constant k' and the orders m, n, and p in such a way as to be consistent with the data in the table. The solution to this problem is simple, once you make a few observations on the reaction mixtures. Each mixture (2 to 4) differs from Reaction Mixture 1 in the concentration of only one species (see table). This means that for any pair of mixtures that includes Reaction Mixture 1, there is only one concentration that changes. From the ratio of the relative rates for such a pair of mixtures we can find the order for the reactant whose concentration was changed. Proceed as follows:

Write Equation 5 below for Reaction Mixtures 1 and 2, substituting the relative rates and the concentrations of I^-, BrO_3^-, and H^+ ions from the table you have just completed.

Relative Rate 1 = _____ = k'[_____]^m[_____]^n[_____]^p

Relative Rate 2 = _____ = k'[_____]^m[_____]^n[_____]^p

Divide the first equation by the second, noting that nearly all the terms cancel out. The result is simply

$$\frac{\text{Relative Rate 1}}{\text{Relative Rate 2}} =$$

If you have done this properly, you will have an equation involving only m as an unknown. Solve this equation for m, the order of the reaction with respect to I^- ion.

$m =$ _____ (nearest integer)

Applying the same approach to Reaction Mixtures 1 and 3, find the value of n, the order of the reaction with respect to BrO_3^- ion.

Relative Rate 1 = _____ = k'[_____]^m[_____]^n[_____]^p

Relative Rate 3 = _____ = k'[_____]^m[_____]^n[_____]^p

Dividing one equation by the other:

$$=$$

$n =$ _____

Now that you have the idea, apply the method once again, this time to Reaction Mixtures 1 and 4, and find p, the order with respect to H^+ ion.

Relative Rate 4 = _____ = k'[_____]^m[_____]^n[_____]^p

Dividing the equation for Relative Rate 1 by that for Relative Rate 4, we get

$$=$$

$p =$ _____

Having found m, n, and p (nearest integers), the relative rate constant, k', can be calculated by substitution of m, n, p, and the known rates and reactant concentrations into Equation 5. Evaluate k' for Reaction Mixtures 1 to 4.

Mixture	1	2	3	4
k'	_____	_____	_____	_____

k'_{mean} _____ Standard deviation in k' _____ (See Appendix IX)

Why should k' have nearly the same value for each of the above reactions?

Using k'_{mean} in Equation 5, predict the relative rate and time, t_{pred}, for Reaction Mixture 5. Use the concentrations in the table.

Relative rate$_{pred}$ _____ t_{pred} _____ t_{obs} _____

B. Effect of Temperature on Reaction Rate: The Activation Energy

To find the activation energy for the reaction it will be helpful to complete the table in this section. The dependence of the rate constant, k', for a reaction is given by Equation 4:

$$\ln k' = -\frac{E_a}{RT} + \text{constant} \tag{4}$$

Since the reactions at the different temperatures all involve the same reactant concentrations, the rate constants, k', for two different mixtures will have the same ratio as the reaction rates themselves for the two mixtures. This means that in the calculation of E_a, we can use the observed relative rates instead of rate constants. Proceeding as before, calculate the relative rates of reaction in each of the mixtures and enter these values in (c). Take the natural logarithm of the relative rate for each mixture and enter these values in (d). To set up the terms in $1/T$, fill in (b), (e), and (f) in the table.

	Approximate temperature in °C			
	20	**40**	**10**	**0**
(a) Time, t, in seconds, for color to appear	_____	_____	_____	_____
(b) Temperature of the reaction mixture in °C	_____	_____	_____	_____
(c) Relative rate = $(1000\ \text{sec})/t$	_____	_____	_____	_____
(d) ln of relative rate	_____	_____	_____	_____
(e) Temperature T in K	_____	_____	_____	_____
(f) $1/T\ [\text{K}^{-1}]$	_____	_____	_____	_____

To evaluate E_a, make a graph of (ln relative rate) vs $1/T$, using a spreadsheet or the graph paper provided. (See Appendix V.)
Find the slope of the line obtained by drawing the best straight line through the experimental points.

Slope = _____ K

The slope of the line equals $-E_a/R$, where $R = (8.31\ \text{joules}/\text{mole}\,K)$ if E_a is to be in joules per mole. Calculate the activation energy, E_a, for the reaction.

E_a = _____ joules/mole

Optional   C. Effect of a Catalyst on Reaction Rate

	Reaction 1	Catalyzed Reaction 1
Time for color to appear [seconds]	_____	_____

Would you expect the activation energy, E_a, for the catalyzed reaction to be greater than, less than, or equal to the activation energy for the uncatalyzed reaction? Why? ■

Experiment 21

Data and Calculations: Rates of Chemical Reactions,
II. A Clock Reaction

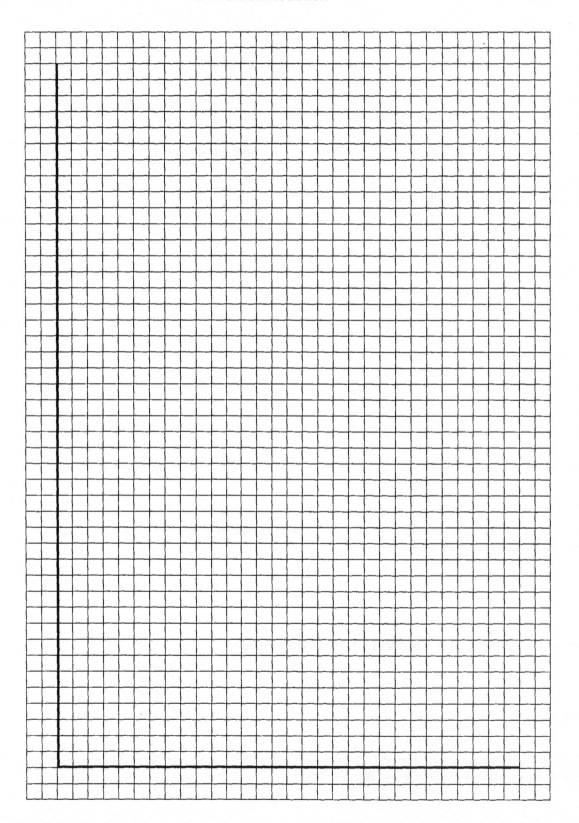

Experiment 21

Advance Study Assignment: Rates of Chemical Reactions,
II. A Clock Reaction

1. A student studied the clock reaction described in this experiment. They set up Reaction Mixture 4 by mixing 10.0 mL of 0.010 M KI, 10.0 mL of 0.0010 M $Na_2S_2O_3$, 10.0 mL of 0.040 M $KBrO_3$, and 20.0 mL of 0.10 M HCl using the procedure given. It took 22 sec for the color to turn blue.

   a. They found the concentrations of each reactant in the reacting mixture by realizing that the number of moles of each reactant did not change when that reactant was mixed with the others, but that its concentration did. For any reactant A,

   $$\text{moles } A = M_{A \text{ stock}} \times V_{\text{stock}} = M_{A \text{ mixture}} \times V_{\text{mixture}}$$

   The volume of the mixture was 50.0 mL. Revising the above equation, they obtained

   $$M_{A \text{ mixture}} = M_{A \text{ stock}} \times \frac{V_{\text{stock}}[\text{mL}]}{50.0 \text{ mL}}$$

   Find the concentrations of each reactant in the reaction mixture, using the equation above:

   $[I^-] =$ _____ M $[BrO_3^-] =$ _____ M $[H^+] =$ _____ M

   b. What was the relative rate of the reaction, $(1000 \text{ sec})/t$? _____ $1/\text{sec}$

   c. Knowing the relative rate of reaction for Mixture 4 and the concentrations of I^-, BrO_3^-, and H^+ in that mixture, the student was able to set up Equation 5 for the relative rate of the reaction. The only quantities that remained unknown were k', m, n, and p. Set up Equation 5 as they did, presuming they did it properly.

2. For Reaction Mixture 1, the student found that 92 seconds were required. On dividing Equation 5 for Reaction Mixture 1 by Equation 5 for Reaction Mixture 4, and after canceling out the common terms (k' and terms in $[I^-]$ and $[BrO_3^-]$), they got the following equation:

   $$\frac{10.9}{45} = \left(\frac{0.020}{0.040}\right)^p = \left(\frac{1}{2}\right)^p$$

   Recognizing that $10.9/45$ is about equal to one-fourth, they obtained an approximate value for p. What was that value?

   $p =$ _____

(continued on following page)

By taking logarithms of both sides of the equation, the student got an exact value for p. What was that value?

$$p = \underline{\hspace{2cm}}$$

Since orders of reactions are often integers, the student rounded their value of p to the nearest integer and reported that value as the order of the reaction with respect to H^+.

Experiment 22

Properties of Systems in Chemical Equilibrium—Le Châtelier's Principle

When working in the laboratory, we often make observations that at first sight are surprising and hard to explain. We might add a reagent to a solution and obtain a precipitate. Addition of more of that same reagent to the precipitate causes it to dissolve. A violet solution turns yellow on addition of a reagent. Subsequent addition of another reagent brings back first a green solution and then the original violet one. Clearly, chemical reactions are occurring, but how and why they behave as they do is not immediately obvious.

In this experiment we will examine and attempt to explain several observations of the sort we have just described. Central to our explanation will be recognition of the fact that chemical systems tend to exist in a state of equilibrium. If we disturb the equilibrium in one way or another, the reaction may shift to the left or right, producing the kinds of effects we have mentioned. If we can understand the principles governing the equilibrium system, it is often possible to see how we might disturb the system, such as by adding a particular reagent or heat, and so cause it to change in a desired way.

Before proceeding to specific examples, let us examine the situation in a general way, noting the key principle that allows us to make a system in equilibrium behave as we wish. Consider the reaction

$$A(aq) \rightleftharpoons B(aq) + C(aq) \tag{1}$$

where A, B, and C are molecules or ions in solution. If we have a mixture of these species in equilibrium, it turns out that their concentrations are not completely unrelated. Rather, there is a condition that those concentrations must meet, namely that

$$\frac{[B] \times [C]}{[A]} = K_c \tag{2}$$

where K_c is a constant, called the equilibrium constant for the reaction. For a given reaction at any given temperature, K_c has a particular value.

When we say that K_c has a particular value, we mean just that. For example, we might find that, for a given solution in which Reaction 1 can occur, when we substitute the equilibrium values for the molarities of A, B, and C into Equation 2, we get a value of 10 for K_c. Now, suppose we add more of species A to that solution. What will happen? Remember, K_c can't change. If we substitute the new higher molarity of A into Equation 2 we get a value that is smaller than K_c. This means that the system is not in equilibrium, and *must* change in some way to get back to equilibrium. How can it do this? It can do this by shifting to the right, producing more B and C and using up some A. It *must* do this, and *will* (though it may be quickly or slowly), until the molarities of C, B, and A reach values that, on substitution into Equation 2, equal 10. At that point the system is once again in equilibrium. In the new equilibrium state, [B] and [C] are greater than they were initially, and [A] is larger than its initial value but smaller than if there had been no forced shift to the right.

The conclusion you should reach on reading the last paragraph is that *we can always cause a reaction to shift to the right by increasing the concentration of a reactant*. An *increase* in concentration of a *product* will force a *shift* to the *left*. By a similar argument we find that a *decrease* in *reactant* concentration causes a *shift* to the *left*; a *decrease* in *product* concentration produces a *shift* to the *right*. This is all true because K_c does not change (unless you change the temperature). The changes in concentration that we can produce by adding particular reagents may be simply enormous, so the shifts in the equilibrium system may also be enormous. Much of the mystery of chemical behavior disappears once you understand this idea.

Another way we might disturb an equilibrium system is by changing its temperature. When this happens, the value of K_c changes. It turns out that the change in K_c depends upon the enthalpy change, ΔH, for the reaction. If ΔH is positive, greater than zero (an "endothermic reaction"), K_c increases with increasing T. If ΔH is negative (an "exothermic" reaction), K_c decreases with an increase in T. Let us return to our original equilibrium between A, B, and C, where K_c equals 10. Let us assume that ΔH for Reaction 1 is −40 kJ. If we raise the temperature, K_c will go down ($\Delta H < 0$), say to a value of 1. This means that the system will no longer be in equilibrium. Substitution of the initial values of [A], [B], and [C] into Equation 2 produces a value that is too big, 10 instead of 1. How can the system change itself to regain equilibrium? It must shift to the left, lowering [B] and [C] and raising [A]. This will make the expression in Equation 2 smaller. The shift will continue until the concentrations of A, B, and C, on substitution into Equation 2, give the expression a value of 1.

From the discussion in the previous paragraph, you should be able to conclude that an equilibrium system will shift to the left on being heated if the reaction is exothermic ($\Delta H < 0$, K_c decreases with temperature). It will shift to the right if the reaction is endothermic ($\Delta H > 0$, K_c increases with temperature). Again, since we can change temperatures very markedly, we can shift equilibria a long, long way. An endothermic reaction that at 25°C has an equilibrium state that consists mainly of reactants might, at 1000°C, exist almost completely as products.

The effects of concentration and temperature on systems in chemical equilibrium are often summarized by Le Châtelier's principle. The principle states that:

If you change a system in chemical equilibrium, it will react in such a way as to partially counteract the change you made.

If you think about the principle for a while, you will see that it predicts the same kind of behavior as we did by using the properties of K_c. Increasing the concentration of a reactant will, by the principle, cause a change that decreases that concentration; that change must be a shift to the right. Increasing the temperature of a reaction mixture will cause a change that tends to absorb heat; that change must be a shift in the endothermic direction. The principle is an interesting one, but does require more careful reasoning in some cases than the more direct approach we employed. For the most part we will find it more useful to base our arguments on the properties of K_c.

In working with aqueous systems, the most important equilibrium is often that which involves the dissociation of water into H^+ and OH^- ions:

$$H_2O(\ell) \rightleftharpoons H^+(aq) + OH^-(aq) \qquad K_c = [H^+][OH^-] = 1 \times 10^{-14} \tag{3}$$

In this reaction the concentration of water is very high and is essentially constant at 55 M; it is incorporated into K_c. The value of K_c is very small, which means that in *any* aqueous system the product of [H^+] and [OH^-] must be very small. In pure water at 25°C, [H^+] equals [OH^-] equals 1×10^{-7} M.

Although the product $[H^+] \times [OH^-]$ is small, this does not mean that both concentrations are necessarily small. If, for example, we dissolve HCl in water, the HCl in the solution will dissociate completely to H^+ and Cl^- ions; in 1.0 M HCl, [H^+] will become 1.0 M, and there is nothing that Reaction 3 can do about changing that concentration appreciably. Rather, Reaction 3 must occur in such a direction as to maintain equilibrium. It does this by lowering [OH^-] by reaction to the left; this uses up a little bit of H^+ ion and drives [OH^-] to the value it must have when [H^+] is 1.0 M, namely, 1×10^{-14} M. In 1.0 M HCl, [OH^-] is a factor of 10 million *smaller* than it is in water. This makes the properties of 1.0 M HCl different from those of water, particularly where H^+ and OH^- ions are involved.

If we take 1.0 M HCl and add an aqueous solution of NaOH to it, an interesting situation develops. Like HCl, NaOH dissociates completely in water, so in 1.0 M NaOH, [OH^-] is equal to 1.0 M. If we add 1.0 M NaOH to 1.0 M HCl, we will initially raise [OH^-] ions way above 1×10^{-14} M. However, Reaction 3 cannot be in equilibrium when both [H^+] and [OH^-] are high; reaction must occur to re-establish equilibrium. The added OH^- ions react with H^+ ions to form H_2O, decreasing both concentrations until equilibrium is established. If only a small amount of OH^- ion is added, it will essentially all be used up; [H^+] will remain high, and [OH^-] will still be very small, but somewhat larger than 10^{-14} M. If we add OH^- ion until the amount added equals in moles the amount of H^+ originally present, then Reaction 3 will go to the left until [H^+]

equals [OH⁻] equals 1×10^{-7} M, and both concentrations will be very small. Further addition of OH⁻ ion will raise [OH⁻] to much higher values, easily as high as 1 M. In such a solution, [H⁺] would be very low, 1×10^{-14} M. So, in aqueous solution, depending on the solutes present, we can have [H⁺] and [OH⁻] range from 1 M to 10^{-14} M, or 14 orders of magnitude. This will have a tremendous effect on any *other* equilibrium system in which [H⁺] or [OH⁻] ions are reactants. Similar situations arise in other equilibrium systems in which the concentration of a reactant or product can be changed significantly by adding a particular reagent.

In many equilibrium systems, several equilibria are present simultaneously. For example, in aqueous solution, Reaction 3 must *always* be in equilibrium. There may, in addition, be equilibria between solutes in the aqueous solution. Some examples are those in Reactions 4, 5, 7, 8, 9, and 10 in the Experimental Procedure section. In some of those reactions, H⁺ and OH⁻ ions appear; in others, they do not. In Reaction 4, for example, H⁺ ion is a product. The molarity of H⁺ in Reaction 4 is *not* determined by the indicator (in this case methyl violet, or HMV), since it is only present in a tiny amount. Reaction 4 will have an equilibrium state that is fixed by the state of Reaction 3, which as we have seen depends markedly on the presence of solutes such as HCl or NaOH. Reactions 8 and 9 can similarly be controlled by Reaction 3. Reactions 5 and 7, which do not involve H⁺ or OH⁻ ions, are not dependent on Reaction 3 for their equilibrium state. Reaction 10 is sensitive to NH_3 concentration and can be driven far to the right by addition of a reagent such as 6 M NH_3.

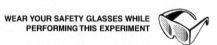

WEAR YOUR SAFETY GLASSES WHILE
PERFORMING THIS EXPERIMENT

Experimental Procedure

In this experiment we will work with several equilibrium systems, each of which is similar to the A-B-C system we discussed. We will alter these systems in various ways, forcing shifts to the right and left by changing concentrations and temperature. You will be asked to interpret your observations in terms of the principles we have presented.

A. Acid-Base Indicators

There is a large group of chemical substances, called acid-base indicators, which change color in solution when [H⁺] changes. A typical substance of this sort is called methyl violet, to which we will give the formula HMV. In solution HMV dissociates as follows:

$$HMV(aq) \rightleftharpoons H^+(aq) + MV^-(aq) \qquad (4)$$
$$\text{yellow} \qquad\qquad\qquad \text{violet}$$

HMV has an intense yellow color, while the anion MV⁻ is violet. The color of the indicator in solution depends very strongly on [H⁺].

Step 1 Add about five milliliters of deionized water to a medium (18×150 mm) test tube. Add a few drops of methyl violet (HMV) indicator. Report the color of the solution on the report page.

Step 2 How could you force the equilibrium system to go to the other form (color)? Select a reagent that should do this and add it to the solution, drop by drop, until the color change is complete. If your reagent works, write its formula on the report page. If it doesn't, try another until you find one that does. Work with 6 M reagents if they are available.

Step 3 Equilibrium systems are reversible. That is, the reaction can be driven to the left and right many times by changing the conditions in the system. How can you force the system in Step 2 to revert to its original color? Select a reagent that should do this and add it drop by drop until the color has become the original one. Again, if your first choice was incorrect, try another reagent. On the report page write the formula of the reagent that was effective. Answer all the questions for Part A before going on to Part B.

B. Solubility Equilibrium; Finding a Value for K_{sp}

Many ionic substances have limited solubility in water. A typical example is lead(II) chloride ($PbCl_2$), which dissolves to a limited extent in water, according to the reaction

$$PbCl_2(s) \rightleftharpoons Pb^{2+}(aq) + 2\ Cl^-(aq) \tag{5}$$

The equilibrium constant for this reaction takes the form

$$K_c = [Pb^{2+}] \times [Cl^-]^2 = K_{sp} \tag{6}$$

The $PbCl_2$ does not enter into the expression because it is a solid, and so has a constant effect on the system, independent of its amount. The equilibrium constant for a solubility equilibrium is called the solubility product, and is given the symbol K_{sp}.

For the equilibrium in Reaction 5 to exist, there *must* be some solid $PbCl_2$ present in the system. If there is no solid, there is no equilibrium; Equation 6 is not obeyed, and $[Pb^{2+}] \times [Cl^-]^2$ must be *less* than the value of K_{sp}. If the solid is present, even in a tiny amount, then the values of $[Pb^{2+}]$ and $[Cl^-]$ are subject to Equation 6.

Step 1 Set up a hot-water bath, using a 400-mL beaker half full of water. Start heating the water while proceeding with Step 2.

Step 2 To a medium (18×150 mm) test tube add 5 mL of 0.3 M $Pb(NO_3)_2$. In this solution $[Pb^{2+}]$ equals 0.3 M. Add 5 mL of 0.3 M HCl to a 10-mL graduated cylinder. In this solution $[Cl^-]$ equals 0.3 M. Add 1.0 mL of the HCl solution to the $Pb(NO_3)_2$ solution. Stir, and wait for about fifteen seconds. What happens? Record your result.

Step 3 To the $Pb(NO_3)_2$ solution add the HCl in 1.0-mL increments until a noticeable amount of white solid $PbCl_2$ is present after stirring. Record the total volume of HCl added at that point.

Step 4 Put the test tube with the precipitate of $PbCl_2$ into the hot-water bath. Stir for a few moments. What happens? Record your observations. Cool the test tube under the cold-water tap. What happens?

Step 5 Rinse out your graduated cylinder, then add about five milliliters of deionized water to the cylinder. Add more water, in 1.0-mL increments, to the mixture in the test tube, stirring well after each addition. When the precipitate completely dissolves, record the total amount of water you added to it in this step. Answer the questions and do the calculations in Part B before proceeding.

C. Complex Ion Equilibria

Many metallic ions in solution exist not as simple ions but, rather, as complex ions incorporating other ions or molecules, called ligands. For example, the Co^{2+} ion in solution exists as the pink $Co(H_2O)_6^{2+}$ complex ion, and Cu^{2+} as the blue $Cu(H_2O)_4^{2+}$ complex ion. In both of these ions the ligands are H_2O molecules. Complex ions are reasonably stable but may be converted to other complex ions on addition of ligands that form more stable complexes than the original ones. Among the common ligands that may form complex species, OH^-, NH_3, and Cl^- are important.

An interesting Co(II) complex is the $CoCl_4^{2-}$ ion, which is blue. This ion is stable in concentrated Cl^- solutions. Depending upon conditions, Co(II) in solution may exist as either $Co(H_2O)_6^{2+}$ or as $CoCl_4^{2-}$. The principles of chemical equilibrium can be used to predict which ion will be present:

$$Co(H_2O)_6^{2+}(aq) + 4\ Cl^-(aq) \rightleftharpoons CoCl_4^{2-}(aq) + 6\ H_2O(\ell) \tag{7}$$

Step 1 Put a few small crystals (about one-tenth of a gram) of $CoCl_2 \cdot 6H_2O$ in a medium test tube. Add 2 mL of 12 M HCl. **CAUTION:** **This is a concentrated solution of an acidic gas; avoid contact with it and its fumes. This is a concentrated strong acid with a choking odor.** Stir to dissolve the crystals. Record the color of the solution.

Step 2 Add 2-mL portions of deionized water, stirring after each dilution, until no further color change occurs. Record the new color.

Step 3 Place the test tube into the hot-water bath and note any change in color. Cool the tube under the water tap and report your observations. Complete the questions in Part C before continuing.

D. Dissolving Insoluble Solids

We saw in Part B that we can dissolve more $PbCl_2$ by either heating its saturated solution or by simply adding water. In most cases these procedures won't work very well on other solids because they are typically much less soluble than $PbCl_2$.

There are, however, some very powerful methods for dissolving solids whose effectiveness depends upon the principles of equilibrium. As an example of an insoluble substance, we might consider $Zn(OH)_2$:

$$Zn(OH)_2(s) \rightleftharpoons Zn^{2+}(aq) + 2\ OH^-(aq) \qquad K_{sp} = 5 \times 10^{-17} = [Zn^{2+}] \times [OH^-]^2 \qquad \textbf{(8)}$$

The equilibrium constant for Reaction 8 is very small, which tells us that the reaction does not go very far to the right or, equivalently, that $Zn(OH)_2$ is almost completely insoluble in water. Adding a few drops of a solution containing OH^- ion to one containing Zn^{2+} ion will cause precipitation of $Zn(OH)_2$.

At first sight you might well wonder how you could possibly dissolve, say, 1 mole of $Zn(OH)_2$ in an aqueous solution. If, however, you examine Equation 8, you can see, from the equation for K_{sp}, that in the saturated solution $[Zn^{2+}] \times [OH^-]^2$ must equal 5×10^{-17}. If, by some means, we can lower that product to a value *below* 5×10^{-17}, then $Zn(OH)_2$ will dissolve, until the product becomes equal to K_{sp}, where equilibrium will again exist. To lower the product, we need to lower the concentration of either Zn^{2+} or OH^- drastically. This turns out to be easy to do. To lower $[OH^-]$ we can add H^+ ions from an acid. If we do that, we drive Reaction 3 to the left, making $[OH^-]$ very small—small enough to dissolve substantial amounts of $Zn(OH)_2$.

To lower $[Zn^{2+}]$ we can take advantage of the fact that zinc(II) forms stable complex ions with both OH^- and NH_3:

$$Zn^{2+}(aq) + 4\ OH^-(aq) \rightleftharpoons Zn(OH)_4^{2-}(aq) \qquad K_1 = 3 \times 10^{15} \qquad \textbf{(9)}$$

$$Zn^{2+}(aq) + 4\ NH_3(aq) \rightleftharpoons Zn(NH_3)_4^{2+}(aq) \qquad K_2 = 1 \times 10^9 \qquad \textbf{(10)}$$

In high concentrations of OH^- ion, Reaction 9 is driven strongly to the right, making $[Zn^{2+}]$ very low. The same thing would happen in solutions containing high concentrations of NH_3. In both media we would therefore expect that $Zn(OH)_2$ might dissolve, since if $[Zn^{2+}]$ is very low, Reaction 8 must go to the right.

Step 1 To each of three small (13×100 mm) test tubes add about two milliliters (a depth of about two centimeters in the tube) 0.10 M $Zn(NO_3)_2$. In this solution $[Zn^{2+}]$ equals 0.10 M. To each test tube add 1 drop 6 M NaOH and stir. Report your observations.

Step 2 To the first tube add 6 M HCl drop by drop, with stirring. To the second add 6 M NaOH, again drop by drop. To the third add 6 M NH_3. Note what happens in each case.

Step 3 Repeat Steps 1 and 2, this time using a solution of 0.10 M $Mg(NO_3)_2$. Record your observations. Answer the questions in Part D.

DISPOSAL OF REACTION PRODUCTS. The residues from Parts B and C should be poured in the waste container. Those from Parts A and D may be poured down the sink.

Experiment 22

Observations and Analysis: Properties of Systems in Chemical Equilibrium

A. Acid-Base Indicators

1. Color of methyl violet in water _____

2. Reagent causing color change _____

3. Reagent causing shift back _____. Explain, by considering how changes in $[H^+]$ will cause Reaction 4 to shift to right and left, why the reagents in Steps 2 and 3 caused the solution to change color. Note that Reactions 3 and 4 must both come to equilibrium after a reagent is added.

B. Solubility Equilibrium; Finding a Value for K_{sp}

2. Vol. 0.3 M $Pb(NO_3)_2$ = 5 mL   moles Pb^{2+} = $[M \times V(\text{stock})]$ = 1.5×10^{-3} moles

   Observations:

3. Vol. 0.3 M HCl used _____ mL; moles Cl^- added _____ moles

4. Observations: in hot water _____ in cold water _____

5. Volume of H_2O added to dissolve $PbCl_2$ _____ mL

   Total volume of solution _____ mL

   a. Explain why $PbCl_2$ did not precipitate immediately on addition of HCl. (What condition must be met by $[Pb^{2+}]$ and $[Cl^-]$ if $PbCl_2$ is to form?)

(continued on following page)

b. Explain your observations in Step 4. (In which direction did Reaction 5 shift when heated? What must have happened to the value of K_{sp} in the hot solution? What does this tell you about the sign of ΔH in Reaction 5?)

c. Explain why the $PbCl_2$ dissolved when water was added in Step 5. (What was the effect of the added water on $[Pb^{2+}]$ and $[Cl^-]$? In what direction would such a change drive Reaction 5?)

d. Given the numbers of moles of Pb^{2+} and Cl^- in the final solution in Step 5, and the volume of that solution, calculate $[Pb^{2+}]$ and $[Cl^-]$ in that solution.

$[Pb^{2+}]$ _____ M; $[Cl^-]$ _____ M

Noting that the molarities just calculated are essentially those in equilibrium with solid $PbCl_2$, calculate $[Pb^{2+}] \times [Cl^-]^2$. This is equal to K_{sp} for $PbCl_2$.

$K_{sp} =$ _____

C. Complex Ion Equilibria

1. Color of $CoCl_2 \cdot 6H_2O$ _____

    Color in solution in 12 M HCl   _____

2. Color in diluted solution   _____

3. Color of hot solution   _____

    Color of cooled solution   _____

    Formula of Co(II) complex ion present in solution in

    a. 12 M HCl   _____

b.   diluted solution   _____

c.   hot solution   _____

d.   cooled solution   _____

Explain the color change that occurred when

a.   water was added in Step 2. (Consider how a change in $[Cl^-]$ and $[H_2O]$ will shift Reaction 7.)

b.   the diluted solution was heated. (How would Reaction 7 shift if K_c went up? How did increasing the temperature affect the value of K_c? What is the sign of ΔH in Reaction 7?)

D. Dissolving Insoluble Solids

1.   Observations on addition of 1 drop of 6 M NaOH to $Zn(NO_3)_2$ solution:

2.   Effect on solubility of $Zn(OH)_2$:

a.   of added HCl solution

b.   of added NaOH solution

c.   of added NH_3 solution

3.   Observations on addition of 1 drop of 6 M NaOH to $Mg(NO_3)_2$ solution:

Effect on solubility of $Mg(OH)_2$:

a.   of added HCl solution

b.   of added NaOH solution

c.   of added NH_3 solution

(continued on following page)

Explain your observations in Step 1. (Consider Reaction 8; how is it affected by addition of OH^- ion?)

In Step 2(a), how does an increase in $[H^+]$ affect Reaction 3? (What does that do to Reaction 8?) Explain your observations in Step 2(a).

In Step 2(b), how does an increase in $[OH^-]$ affect Reaction 9? What does that do to Reaction 8? Explain your observations in Step 2(b).

In Step 2(c), how does an increase in $[NH_3]$ affect Reaction 10? What does that do to Reaction 8? Explain your observations in Step 2(c).

In Step 3, you probably found that $Mg(OH)_2$ was similar in some ways in its behavior to that of $Zn(OH)_2$, but different in others.

  a. How was it similar? Explain that similarity. (In particular, why would any insoluble hydroxide tend to dissolve in acidic solution?)

  b. How was it different? Explain that difference. (In particular, does Mg^{2+} appear to form complex ions with OH^- and NH_3? What would we observe if it did? If it did not?)

Experiment 22

Advance Study Assignment: Properties of Systems in Chemical Equilibrium

1. Methyl orange, HMO, is a common acid-base indicator. In solution it ionizes according to the equation:

$$HMO(aq) \rightleftharpoons H^+(aq) + MO^-(aq)$$
$$\text{red} \qquad\qquad\qquad \text{yellow}$$

   If methyl orange is added to deionized water, the solution turns yellow. If one or two drops of 6 M HCl are added to the yellow solution, it turns red. If a few more drops of 6 M NaOH are added to the solution, the color reverts to yellow.

   a. Why does adding 6 M HCl to the yellow solution of methyl orange tend to cause the color to change to red? (Note that in solution HCl exists as H^+ and Cl^- ions.)

   b. Why does adding 6 M NaOH to the red solution tend to make it turn back to yellow? Note that in solution NaOH exists as Na^+ and OH^- ions. (Hint: How does increasing $[OH^-]$ shift Reaction 3 in the discussion section? How would the resulting change in $[H^+]$ affect the dissociation reaction of HMO? Explain these as part of your answer.)

(continued on following page)

2. Calcium hydroxide is only slightly soluble in water. The reaction by which it goes into solution is:

$$Ca(OH)_2(s) \rightleftharpoons Ca^{2+}(aq) + 2\,OH^-(aq)$$

a. Formulate the expression for the equilibrium constant, K_{sp}, for the above reaction.

b. It is possible to dissolve significant amounts of $Ca(OH)_2$ in solutions in which the concentration of either Ca^{2+} or OH^- is kept very small. Explain, using K_{sp}, why this is the case.

c. Explain why $Ca(OH)_2$ might have very appreciable solubility in 1.0 M HCl. [Hint: Consider the effect of Reaction 3 on the $Ca(OH)_2$ dissolution reaction.]

Experiment 23

Determination of the Equilibrium Constant for a Chemical Reaction

When chemical substances react, the reaction typically does not go to completion. Rather, the system goes to some intermediate state in which both the reactants and products have concentrations that do not change with time. Such a system is said to be in chemical equilibrium. When in equilibrium at a particular temperature, a reaction mixture obeys the Law of Chemical Equilibrium, which imposes a mathematical condition on the concentrations of reactants and products, requiring that some ratio of them be equal to the equilibrium constant K_c for the reaction.

In this experiment we will study the equilibrium properties of the reaction between the iron(III) ion, Fe^{3+}, and the thiocyanate ion, SCN^-:

$$Fe^{3+}(aq) + SCN^-(aq) \rightleftharpoons FeSCN^{2+}(aq) \tag{1}$$

When solutions containing Fe^{3+} and SCN^- are mixed, Reaction 1 occurs to some extent, forming the $FeSCN^{2+}$ complex ion, which has a deep red color. As a result of the reaction, the equilibrium amounts of Fe^{3+} and SCN^- will be less than they would have been if no reaction had occurred; for every mole of $FeSCN^{2+}$ that is formed, 1 mole of Fe^{3+} and 1 mole of SCN^- will react. According to the Law of Chemical Equilibrium, the equilibrium constant expression K_c for Reaction 1 is formulated as follows:

$$\frac{[FeSCN^{2+}]}{[Fe^{3+}][SCN^-]} = K_c \tag{2}$$

The value of K_c in Equation 2 is constant at a given temperature. This means that mixtures containing Fe^{3+} and SCN^- will react until Equation 2 is satisfied, so that the same value of the K_c will be obtained no matter what initial amounts of Fe^{3+} and SCN^- were used. Our purpose in this experiment will be to find K_c for this reaction for several mixtures made up in different ways, and to show that K_c indeed has the same value in each of the mixtures. This reaction is a particularly good one to study because K_c is of a convenient magnitude and the color of the $FeSCN^{2+}$ ion makes for an easy analysis of the equilibrium mixture.

The mixtures will be prepared by combining solutions containing known concentrations of iron(III) nitrate, $Fe(NO_3)_3$, and potassium thiocyanate, KSCN. The color of the $FeSCN^{2+}$ ion formed will allow us to determine its equilibrium concentration. Knowing the initial composition of a mixture and the equilibrium concentration of $FeSCN^{2+}$, we can calculate the equilibrium concentrations of the rest of the pertinent species and then determine K_c.

Since the calculations required in this experiment may not be apparent, we will go through a step-by-step procedure by which they can be made. As a specific example, let us assume that we prepare a mixture by mixing 10.0 mL of 2.00×10^{-3} M $Fe(NO_3)_3$ with 10.0 mL of 2.00×10^{-3} M KSCN. As a result of Reaction 1, some red $FeSCN^{2+}$ ion is formed. By the method of analysis described later, its concentration at equilibrium is found to be 1.50×10^{-4} M. Our problem is to find K_c for the reaction from this information. To do this we first need to find the initial number of moles of each reactant in the mixture. Second, we determine the number of moles of product that were formed at equilibrium. Since the product was formed from the reactants, we can calculate the amount of each reactant that was used up. In the third step we find the number of moles of each reactant remaining in the equilibrium mixture. Fourth, we determine the concentration of each reactant. Finally, in the fifth step, we evaluate K_c for the reaction.

Step 1 Finding the Initial Number of Moles of Each Reactant. From the volumes and concentrations of the reagent solutions that were mixed, the numbers of moles of each reactant species initially present can be calculated. By the definition of the molarity, M_A, of a species A,

$$M_A = \frac{\text{moles } A}{\text{liters of solution, } V} \quad \text{so} \quad \text{moles } A = M_A \times V \tag{3}$$

Using Equation 3, we find the initial number of moles of Fe^{3+} and SCN^-. For each reagent solution the volume used was 10.0 mL, or 0.0100 L. The molarity of each of the solutions was 2.00×10^{-3} M, so $M_{Fe^{3+}} = 2.00 \times 10^{-3}$ M and $M_{SCN^-} = 2.00 \times 10^{-3}$ M. Therefore,

initial moles $Fe^{3+} = M_{Fe^{3+}} \times V = 2.00 \times 10^{-3}$ M $\times 0.0100$ L $= 20.0 \times 10^{-6}$ moles

initial moles $SCN^- = M_{SCN^-} \times V = 2.00 \times 10^{-3}$ M $\times 0.0100$ L $= 20.0 \times 10^{-6}$ moles

Step 2 Finding the Number of Moles of Product Formed. The concentration of $FeSCN^{2+}$ was found to be 1.50×10^{-4} M at equilibrium. The volume of the mixture at equilibrium is the *sum* of the two volumes that were mixed, which is 20.0 mL, or 0.0200 L. So, by Equation 3,

moles $FeSCN^{2+} = M_{FeSCN^{2+}} \times V = 1.50 \times 10^{-4}$ M $\times 0.0200$ L $= 3.00 \times 10^{-6}$ moles

The number of moles of Fe^{3+} and SCN^- that were *used up* in producing the $FeSCN^{2+}$ must also both be equal to 3.00×10^{-6} moles since, by Equation 1, it takes *one mole* Fe^{3+} *and one mole* SCN^- to make each mole of $FeSCN^{2+}$.

Step 3 Finding the Number of Moles of Each Reactant Present at Equilibrium. In Step 1 we determined that initially we had 20.0×10^{-6} moles Fe^{3+} and 20.0×10^{-6} moles SCN^- present. In Step 2 we found that in the reaction 3.00×10^{-6} moles of Fe^{3+} and 3.00×10^{-6} moles of SCN^- were used up. The number of moles present at equilibrium must equal the number we started with minus the number that reacted. Therefore, *at equilibrium,*

moles at equilibrium = initial moles − moles used up (4)

equilibrium moles $Fe^{3+} = 20.0 \times 10^{-6} - 3.00 \times 10^{-6} = 17.0 \times 10^{-6}$ moles

equilibrium moles $SCN^- = 20.0 \times 10^{-6} - 3.00 \times 10^{-6} = 17.0 \times 10^{-6}$ moles

Step 4 Find the Concentrations of All Species at Equilibrium. Experimentally, we obtained the equilibrium concentration of $FeSCN^{2+}$ directly. $[FeSCN^{2+}] = 1.50 \times 10^{-4}$ M. The concentrations of Fe^{3+} and SCN^- follow from Equation 3. The number of moles of each of these species at equilibrium was obtained in Step 3. The volume of the mixture being studied was 20.0 mL, or 0.0200 L. So, *at equilibrium,*

$$[Fe^{3+}] = M_{Fe^{3+}} = \frac{\text{moles } Fe^{3+}}{\text{volume of solution}} = \frac{17.0 \times 10^{-6}\,\text{moles}}{0.0200\,\text{L}} = 8.50 \times 10^{-4}\,\text{M}$$

$$[SCN^-] = M_{SCN^-} = \frac{\text{moles } SCN^-}{\text{volume of solution}} = \frac{17.0 \times 10^{-6}\,\text{moles}}{0.0200\,\text{L}} = 8.50 \times 10^{-4}\,\text{M}$$

Step 5 Finding the Value of K_c for the Reaction. Once the equilibrium concentrations of all the reactants and products are known, we merely need to substitute into Equation 2 to determine K_c:

$$K_c = \frac{[FeSCN^{2+}]}{[Fe^{3+}][SCN^-]} = \frac{1.50 \times 10^{-4}}{(8.50 \times 10^{-4}) \times (8.50 \times 10^{-4})} = 208$$

In this experiment you will obtain data similar to that shown in this example. The calculations involved in processing that data are completely analogous to those we have made. (Actually, your results will differ from the ones we obtained, since the data in our example were obtained at a different temperature and so relate to a different value of K_c.)

In carrying out this analysis we made the assumption that the reaction which occurred was given by Equation 1. There is no inherent reason why the reaction might not have been

$$Fe^{3+}(aq) + 2\ SCN^-(aq) \rightleftharpoons Fe(SCN)_2^+(aq) \qquad (5)$$

If you are interested in matters of this sort, you might ask how we know whether we are actually observing Reaction 1 or Reaction 5. The line of reasoning is as follows. If Reaction 1 is occurring, K_c for that reaction as we calculate it should remain constant with different reagent mixtures. If, however, Reaction 5 is going on, K_c as calculated for that reaction should remain constant. In the optional section of the report page, we will assume that Reaction 5 occurs and make the analysis of K_c on that basis. The results of the two sets of calculations should make it clear that Reaction 1 is the one that we are studying.

Two analytical methods can be used to determine $[FeSCN^{2+}]$ in the equilibrium mixtures. The more precise method uses a spectrophotometer, which measures the amount of light absorbed by the red complex at 447 nm, the wavelength at which the complex most strongly absorbs. The absorbance, A, of the complex is proportional to its concentration, M, and can be measured directly on the spectrophotometer:

$$A = kM \qquad (6)$$

Your instructor will show you how to operate the spectrophotometer, if one is available in your laboratory, and may provide you with a calibration curve or equation from which you can find $[FeSCN^{2+}]$ once you have determined the absorbance of your solutions or, optionally, have you prepare your own calibration curve (which will provide more accurate results). See Appendix IV for information about spectrophotometers.

In the other analytical method a solution of known concentration of $FeSCN^{2+}$ is prepared. The $[FeSCN^{2+}]$ concentrations in the solutions being studied are found by comparing the color intensities of these solutions with that of the known. The method involves matching the color intensity of a given depth of unknown solution with that for an adjusted depth of known solution. The actual procedure and method of calculation are discussed in the Experimental Procedure section.

In preparing the mixtures in this experiment we will maintain the concentration of H^+ ion at 0.5 M. The hydrogen ion does not participate directly in the reaction, but its concentration does influence the concentrations of both Fe^{3+} and SCN^- through other (acid-base) reactions; thus, maintaining a constant H^+ concentration is important.

WEAR YOUR SAFETY GLASSES WHILE PERFORMING THIS EXPERIMENT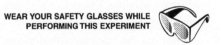

Experimental Procedure

Label five medium (18×150 mm) test tubes 1 to 5, with labels or by noting their positions on your test tube rack. Pour about thirty milliliters of 2.00×10^{-3} M $Fe(NO_3)_3$ in 1.0 M HNO_3 into a dry 100-mL beaker. Pipet 5.00 mL of that solution into each test tube. Then add about twenty milliliters of 2.00×10^{-3} M KSCN to another dry 100-mL beaker. Pipet 1.00, 2.00, 3.00, 4.00, and 5.00 mL from the KSCN beaker into each of the corresponding test tubes labeled 1 to 5. Then pipet the proper number of milliliters of water into each test tube to bring the total volume in each tube to 10.00 mL. The volumes of reagents to be added to each tube are summarized in Table 23.1, which you should complete by filling in the required volumes of water. See Appendix IV for a discussion of the use of pipets.

Mix each solution thoroughly with a glass stirring rod. Be sure to dry the stirring rod after mixing each solution.

Table 23.1

Reagent Solution and Dilution Volumes to be Used in Preparing Samples					
	Test tube number				
	1	**2**	**3**	**4**	**5**
Volume $Fe(NO_3)_3$ solution [mL]	5.00	5.00	5.00	5.00	5.00
Volume KSCN solution [mL]	1.00	2.00	3.00	4.00	5.00
Volume H_2O [mL]	____	____	____	____	____

Method I. Analysis by Spectrophotometric Measurement

Place a portion of the mixture in Tube 1 in a spectrophotometer cell, as demonstrated by your instructor, and measure the absorbance of the solution at 447 nm. Determine the concentration of $FeSCN^{2+}$ from the calibration curve provided for each instrument or from the equation furnished to you, or that you prepare yourself by following the optional instructions, below. Record the value on the report page. Repeat the measurement using the mixtures in each of the other test tubes. For a discussion of how absorbance and concentration are related, see Appendix IV.

Optional   Prepare a calibration curve as follows: In a 100-mL volumetric flask, combine 5.00 mL of 2.00×10^{-3} M KSCN with 50. mL of a special calibration solution, 0.200 M $Fe(NO_3)_3$ in 1.0 M HNO_3, then dilute to the mark on the volumetric flask with deionized water. Repeat this procedure using 10.0 mL, and then 15.0 mL, of 2.00×10^{-3} M KSCN. This will give three solutions, which may be assumed to be 1.00×10^{-4}, 2.00×10^{-4}, and 3.00×10^{-4} M in $FeSCN^{2+}$, respectively, because $[Fe^{3+}] \gg [SCN^-]$ and Reaction 1 is driven strongly to the right. Measure the absorbances of these three solutions at 447 mm, using deionized water as the reference solution. Plot absorbance vs $[FeSCN^{2+}]$ and draw the best straight line possible that passes through the origin.

Method II. Analysis by Comparison with a Standard

Prepare a solution of known $FeSCN^{2+}$ concentration by pipetting 10.00 mL of a special calibration solution, 0.200 M $Fe(NO_3)_3$ in 1.0 M HNO_3, into a test tube and adding 2.00 mL of 2.00×10^{-3} M KSCN and 8.00 mL of water. Mix the solution thoroughly with a stirring rod.

Since in this solution $[Fe^{3+}] \gg [SCN^-]$, Reaction 1 is driven strongly to the right. You can assume without serious error that essentially all the SCN^- added is converted to $FeSCN^{2+}$. Assuming that this is the case, calculate $[FeSCN^{2+}]$ in the standard solution and record the value on the report page.

The $[FeSCN^{2+}]$ in the unknown mixture in test tubes 1 to 5 can be found by comparing the intensity of the red color in these mixtures with that of the standard solution. This can be done by placing the test tube containing Mixture 1 next to a test tube containing the standard. Look down both test tubes toward a well-illuminated piece of white paper on the laboratory bench.

Pour out the standard solution into a dry, clean beaker until the color intensity you see when looking down the tube containing the standard matches that which you see when looking down the tube containing the unknown. When the colors match, the following relation is valid:

$$[FeSCN^{2+}]_{unknown} \times \text{depth of unknown solution} = [FeSCN^{2+}] \times \text{depth of standard solution} \tag{7}$$

Measure the depths of the matching solutions with a ruler and record them. Repeat the measurement for Mixtures 2 through 5, recording the depth of each unknown and that of the standard solution which matches it in intensity.

DISPOSAL OF REACTION PRODUCTS. Dispose of your solutions from this experiment as directed by your instructor.

Experiment 23

Data and Calculations: Determination of the Equilibrium Constant for a Chemical Reaction

Mixture	Volume in mL, 2.00×10^{-3} M Fe(NO$_3$)$_3$	Volume in mL, 2.00×10^{-3} M KSCN	Volume in mL, Water	Method I Absorbance	Method II Depth in mm		[FeSCN]$^{2+}$
					Standard	Unknown	
1	5.00	1.00	_____	_____	_____	_____	_____ $\times 10^{-4}$ M
2	5.00	2.00	_____	_____	_____	_____	_____ $\times 10^{-4}$ M
3	5.00	3.00	_____	_____	_____	_____	_____ $\times 10^{-4}$ M
4	5.00	4.00	_____	_____	_____	_____	_____ $\times 10^{-4}$ M
5	5.00	5.00	_____	_____	_____	_____	_____ $\times 10^{-4}$ M

If Method II was used, [FeSCN^{2+}]$_{standard}$ = _____ $\times 10^{-4}$ M; [FeSCN^{2+}] in Mixtures 1 to 5 is found by Equation 7.

Processing the Data

A. Calculation of K_c assuming the reaction:

$$Fe^{3+}(aq) + SCN^-(aq) \rightleftharpoons Fe(SCN)^{2+}(aq) \tag{1}$$

This calculation is most easily done by following Steps 1 through 5 in the discussion. Results are to be entered in the table on the following page. If you are using Excel, set up the table as we have, and follow the directions below.

Step 1 Find the initial number of moles of Fe^{3+} and SCN$^-$ in the mixtures in test tubes 1 through 5. Use Equation 3 and enter the values in the first two columns of the table on the following page.

Step 2 Enter the experimentally determined value of [FeSCN^{2+}] at equilibrium for each of the mixtures in the next to last column of the table on the following page. Use Equation 3 to find the number of moles of FeSCN^{2+} in each of the mixtures, and enter the values in the fifth column of the table. Note that this is also the number of moles of Fe^{3+} and SCN$^-$ that were used up in the reaction.

(continued on following page)

Step 3 From the number of moles of Fe^{3+} and SCN^- initially present in each mixture, and the number of moles of Fe^{3+} and SCN^- used up in forming $FeSCN^{2+}$, calculate the number of moles of Fe^{3+} and SCN^- that remain in each mixture at equilibrium. Use Equation 4. Enter the results in the third and fourth columns of the table below.

Step 4 Use Equation 3 and the results of Step 3 to find the concentrations of all of the species at equilibrium. The volume of the mixture is 10.00 mL, or 0.0100 liter, in all cases. Enter the values in the sixth and seventh columns of the table.

Step 5 Calculate K_c for the reaction for each of the mixtures by substituting values for the equilibrium concentrations of Fe^{3+}, SCN^-, and $FeSCN^{2+}$ in Equation 2. Record these results in the last column of the table.

Step 6 Calculate and record the mean value and the standard deviation for K_c. (See Appendix IX.)

Mixture	Initial moles		Equilibrium moles			Equilibrium concentrations			K_c
	Fe^{3+}	SCN^-	Fe^{3+}	SCN^-	$FeSCN^{2+}$	$[Fe^{3+}]$	$[SCN^-]$	$[FeSCN^{2+}]$	
1	___ $\times 10^{-6}$	___ $\times 10^{-6}$	___ $\times 10^{-6}$	___ $\times 10^{-6}$	___ $\times 10^{-6}$	___ $\times 10^{-4}$ M	___ $\times 10^{-4}$ M	___ $\times 10^{-4}$ M	___
2	___	___	___	___	___	___	___	___	___
3	___	___	___	___	___	___	___	___	___
4	___	___	___	___	___	___	___	___	___
5	___	___	___	___	___	___	___	___	___

Mean value of K_c _____   Standard deviation _____

On the basis of the results of Part A, what can you conclude about the validity of the equilibrium concept, as exemplified by Equation 2?

(continued on following page)

B. [Optional] In calculating K_c in Part A, we assume, correctly, that the formula of the complex ion is $FeSCN^{2+}$. It is by no means obvious that this is the case and one might have assumed, for instance, that $Fe(SCN)_2^+$ was the species formed. The reaction would then be

$$Fe^{3+}(aq) + 2\ SCN^-(aq) \rightleftharpoons Fe(SCN)_2^+(aq) \tag{5}$$

If we analyze the equilibrium system we have studied, assuming that Reaction 5 occurs rather than Reaction 1, we would presumably obtain nonconstant values of K_c. Using the same kind of procedure as in Part A, calculate K_c for Mixtures 1, 3, and 5 based on the assumption that $Fe(SCN)_2^+$ is the formula of the complex ion formed by the reaction between Fe^{3+} and SCN^-. As a result of the procedure used for calibrating the system by Method I or Method II, $[Fe(SCN)_2^+]$ will equal *one-half* the $[FeSCN^{2+}]$ obtained for each solution in Part A. Note that *two* moles of SCN_2 are needed to form *one* mole of $Fe(SCN)_2^+$. This changes the expression for K_c. Also, in calculating the equilibrium number of moles of SCN^- you will need to subtract ($2 \times$ number of moles $Fe(SCN)_2^+$) from the initial number of moles of SCN^-.

Mixture	Initial moles		Equilibrium moles			Equilibrium concentrations			K_c
	Fe^{3+}	SCN^-	Fe^{3+}	SCN^-	$Fe(SCN)_2^+$	$[Fe^{3+}]$	$[SCN^-]$	$[Fe(SCN)_2^+]$	
1									
3									
5									

What do you conclude about the formula of iron(III) thiocyanate complex ion, based on what you have calculated in Part B?

Experiment 23

Advance Study Assignment: Determination of the Equilibrium Constant
for a Chemical Reaction

1. A student mixes 5.00 mL of 2.00×10^{-3} M $Fe(NO_3)_3$ in 1.0 M HNO_3 with 4.00 mL of 2.00×10^{-3} M KSCN and 1.00 mL of water. He finds that in the equilibrium mixture the concentration of $FeSCN^{2+}$ is 7.0×10^{-5} M. Find K_c for the reaction $Fe^{3+}(aq) + SCN^-(aq) \rightleftharpoons Fe(SCN)^{2+}(aq)$.

Step 1 Calculate the number of moles Fe^{3+} and SCN^- initially present. (Use Eq. 3.)

_____ moles Fe^{3+}; _____ moles SCN^-

Step 2 What is the volume of the equilibrium mixture? How many moles of $FeSCN^{2+}$ are in the mixture at equilibrium? (Use Eq. 3.)

_____ mL; _____ moles $FeSCN^{2+}$

How many moles of Fe^{3+} and SCN^- are used up in making the $FeSCN^{2+}$?

_____ moles Fe^{3+}; _____ moles SCN^-

Step 3 How many moles of Fe^{3+} and SCN^- remain in the solution at equilibrium? (Use Eq. 4 and the results of Steps 1 and 2.)

_____ moles Fe^{3+}; _____ moles SCN^-

Step 4 What are the concentrations of Fe^{3+}, SCN^-, and $FeSCN^{2+}$ at equilibrium? (Use Eq. 3 and the results of Steps 2 and 3.)

$[Fe^{3+}] =$ _____ M; $[SCN^-] =$ _____ M; $[FeSCN^{2+}] =$ _____ M

Step 5 What is the value of K_c for the reaction? (Use Eq. 2 and the results of Step 4.)

$K_c =$ _____

(continued on following page)

2. Optional   Assume that the reaction studied in Problem 1 is $Fe^{3+}(aq) + 2\ SCN^-(aq) \rightleftharpoons Fe(SCN)_2^+(aq)$. Find K_c for this reaction, given the data in Problem 1.

a. Formulate the expression for K_c for the alternate reaction just cited.

b. Find K_c as you did in Problem 1; take due account of the fact that two moles SCN^- are used up per mole $Fe(SCN)_2^+$ formed as you carry out the calculations on the remainder of this page:

Step 1 Results are as in Problem 1.

Step 2 How many moles of $Fe(SCN)_2^+$ are in the mixture at equilibrium? (This will be the same as your answer for Step 2 in Problem 1.)

_____ moles $Fe(SCN)_2^+$

How many moles of Fe^{3+} and SCN^- are used up in making the $Fe(SCN)_2^+$?

_____ moles Fe^{3+}; _____ moles SCN^-

Step 3 How many moles of Fe^{3+} and SCN^- remain in solution at equilibrium? Use the results of Steps 1 and 2, noting that moles SCN^- at equilibrium = original moles $SCN^- - (2 \times$ moles $Fe(SCN)_2^+$).

_____ moles Fe^{3+}; _____ moles SCN^-

Step 4 What are the concentrations of Fe^{3+}, SCN^-, and $Fe(SCN)_2^+$ at equilibrium? (Use Eq. 3 and the results of Step 3.)

$[Fe^{3+}] =$ _____ M; $[SCN^-] =$ _____ M; $[Fe(SCN)_2^+] =$ _____ M

Step 5 Calculate K_c based on the assumption that the alternate reaction occurs. [Use the answer to 2(a), above.]

$K_c =$ _____ ▪

Experiment 24

The Standardization of a Basic Solution and the Determination of the Molar Mass of an Acid

When a solution of a strong acid is mixed with a solution of a strong base, a chemical reaction occurs that can be represented by the following net ionic equation:

$$H^+(aq) + OH^-(aq) \rightarrow H_2O(\ell)$$

This is called a neutralization reaction, and chemists use it extensively to change the acidic or basic properties of solutions. The equilibrium constant for the reaction is on the order of 10^{14} at room temperature, so that the reaction can be considered to proceed completely to the right, using up whichever of the ions is present in the lesser amount and leaving the solution either acidic or basic, depending on whether H^+ or OH^- ion was in excess.

Since the reaction is essentially quantitative, it can be used to determine the concentrations of acidic or basic solutions. A frequently used procedure involves the titration of an acid with a base. In the titration, a basic solution is added from a buret to a measured volume of acid solution until the number of moles of OH^- ion added is just equal to the number of moles of H^+ ion present in the acid. At that point the volume of basic solution that has been added is measured.

Recalling the definition of the molarity, M_A, of species A, we have

$$M_A = \frac{\text{moles of A}}{\text{liters of solution, } V} \quad \text{or} \quad \text{moles of A} = M_A \times V \tag{1}$$

At the equivalence point of the titration,

$$\text{moles } H^+ \text{ originally present} = \text{moles } OH^- \text{ added} \tag{2}$$

So, by Equation 1,

$$M_{H^+} \times V_{acid} = M_{OH^-} \times V_{base} \tag{3}$$

Therefore, if the molarity of either the H^+ or the OH^- ion in its solution is known, the molarity of the other ion can be found from the titration.

In most acid-base titrations, a chemical called an indicator is used: it changes color at the endpoint of the titration, at a specific pH. The equivalence point is nearly identical to the titration's endpoint, provided the indicator is chosen carefully. The indicators used in acid-base titrations are weak organic acids or bases that change color when they are neutralized. One of the most common indicators is phenolphthalein, which is colorless in acid solutions but becomes red when the pH of the solution becomes 9 or higher.

When a solution of a strong acid is titrated with a solution of a strong base, the pH at the endpoint will be about 7. At the endpoint a drop of acid or base added to the solution will change its pH by several pH units, so that phenolphthalein can be used as an indicator in such titrations. If a weak acid is titrated with a strong base, the pH at the equivalence point is somewhat higher than 7, perhaps 8 or 9, and phenolphthalein is still a very satisfactory indicator. If, however, a solution of a weak base such as ammonia is titrated with a strong acid, the pH will be a unit or two less than 7 at the equivalence point, and phenolphthalein will not be as good an indicator for that titration as, for example, methyl red, whose color changes from red to yellow as the pH changes from about 5 to 6. Ordinarily, indicators will be chosen so that their color change occurs at about the pH at the equivalence point of a given acid-base titration.

In this experiment you will determine the molarity of OH^- ion in an NaOH solution by titrating that solution against a standardized solution of HCl. Since in these solutions one mole of acid in solution furnishes one mole of H^+ ion and one mole of base produces one mole of OH^- ion, $M_{HCl} = M_{H^+}$ in the acid solution, and $M_{NaOH} = M_{OH^-}$ in the basic solution. Therefore, the titration will allow you to find M_{NaOH} as well as M_{OH^-}.

In the second part of this experiment you will use your standardized NaOH solution to titrate a sample of a pure solid organic or inorganic acid. By titrating a weighed sample of the unknown acid with your standardized NaOH solution you can, by Equation 2, find the number of moles H^+ ion that your sample can furnish.

If your acid has one acidic hydrogen atom in the molecule, with formula HB, then the number of moles of acid will equal the number of moles of H^+ that react during the titration. The molar mass of the acid, MM, will equal the number of grams of acid that contain one mole of H^+ ion.

$$MM = \frac{\text{grams of acid}}{\text{moles of } H^+ \text{ ion furnished}} \tag{4}$$

Many acids release one mole of H^+ ion per mole of acid on titration with NaOH solution. Such acids are called monoprotic. Acetic acid, $HC_2H_3O_2$, is a classic example of a monoprotic acid (only the first H atom in the formula is acidic). Like all organic acids, acetic acid is weak, in that it only ionizes to a small extent in aqueous solution.

Some acids contain more than one acidic hydrogen atom per molecule, and have the general formula H_2B or H_3B. Sulfurous acid, H_2SO_3, is an example of an inorganic diprotic acid. Maleic acid, $H_2C_4H_2O_4$, is a diprotic organic acid. If we should titrate samples of these acids with a solution of a strong base like NaOH, it would take 2 moles of OH^- ion, or 2 moles of NaOH, to neutralize 1 mole of acid, since each mole of acid would release two moles of H^+ ion. If you don't know the formula of an acid, you can't be sure it is monoprotic, so you can only calculate the mass of acid that will react with one mole of OH^- ion. That is the mass capable of releasing 1 mole of H^+ ion and is called the *equivalent* mass of the acid. The molar mass and the equivalent mass are related by simple equations. Since sulfurous acid gives up 2 moles of H^+ ion in titration with NaOH, the molar mass must equal *twice* the equivalent mass, which gives up one mole of H^+ ion.

To simplify matters, in this experiment we will only use monoprotic acids, with formula HB, so 1 mole of acid will react with 1 mole of NaOH, and you can find the molar mass of your acid by Equation 4.

Experimental Procedure

WEAR YOUR SAFETY GLASSES WHILE PERFORMING THIS EXPERIMENT

Note: This experiment is relatively long unless you know precisely what to do. Study the experiment carefully before coming to class, so that you don't have to spend a lot of time finding out what the experiment is all about.

Obtain two burets and a sample of solid unknown acid.

A. Standardization of NaOH Solution

Into a small graduated cylinder draw about seven milliliters of the stock 6 M NaOH solution provided in the laboratory and dilute to about four hundred milliliters with deionized water in a 500-mL Florence flask. Stopper the flask tightly and mix the solution thoroughly at intervals over a period of at least 15 minutes before using the solution.

Draw into a clean, *dry* 125-mL Erlenmeyer flask about seventy-five milliliters of standardized HCl solution (0.10 M, which means between 0.09 M and 0.11 M; a more precise value for the concentration will be provided to you) from the stock solution on the reagent shelf. This amount should provide all the standardized acid you will need; do not waste it. Record the precise molarity of the HCl.

Prepare your buret for the titration by the following procedure, which is described in greater detail in Appendix IV. (The purpose of this procedure is to make sure that the solution in each buret has the same molarity as it has in the container from which it was poured.) Clean the two burets and rinse them with deionized water. Then rinse one buret three times with a few milliliters of the HCl solution, in each case thoroughly wetting the walls of the buret with the solution and then letting it out through the stopcock. Fill the buret with HCl; open the stopcock momentarily to fill the tip. Proceed to clean and fill the other buret with your NaOH solution in a similar manner. Put the acid buret, A, on the left side of your buret clamp, and the base buret, B, on the right side. Check to see that your burets do not leak and that there are no air bubbles in either buret tip. Read and record the levels in the two burets to ± 0.01 mL.

Draw about twenty-five milliliters of the HCl solution from the buret into a clean 250-mL Erlenmeyer flask; add to the flask about twenty-five milliliters of deionized H_2O and two or three drops of phenolphthalein indicator solution. Place a white sheet of paper under the flask to aid in the detection of any color change. Add the NaOH solution intermittently from its buret to the solution in the flask, noting the pink phenolphthalein color that appears and disappears as the drops hit the liquid and are mixed with it. Swirl the liquid in the flask gently and continuously as you add the NaOH solution. When the pink color begins to persist, slow down the rate of addition of NaOH. In the final stages of the titration add the NaOH drop by drop until the entire solution turns a pale pink color that will persist for about thirty seconds. If you go past the endpoint and obtain a red solution, add a few drops of the HCl solution to remove the color, and then add NaOH 1 drop at a time until the pink color persists. Carefully record the *final* readings on the HCl and NaOH burets.

To the 250-mL Erlenmeyer flask containing the titrated solution, add about ten milliliters more of the standard HCl solution. Titrate this as before with the NaOH to an endpoint, and carefully record both buret readings once again. To this solution add about ten milliliters more HCl and titrate a third time with NaOH.

You have now completed three titrations, with *total* HCl volumes of about twenty-five, thirty-five, and forty-five milliliters. Using Equation 3, calculate the molarity of your base, M_{OH}, for each of the three titrations. In each case, use the *total volumes* of acid and base that were added up to that point. At least two of these molarities should agree to within 1%. If they do, proceed to the next part of the experiment. If they do not, repeat these titrations until two calculated molarities do agree. Calculate the mean value of M_{OH} and the standard deviation. (See Appendix IX.)

B. Determination of the Molar Mass of an Acid

Weigh the vial containing your solid acid on the analytical balance to ± 0.0001 g. Carefully pour out about half the sample into a clean but not necessarily dry 250-mL Erlenmeyer flask. Weigh the vial again, accurately. Add about fifty milliliters of deionized water and two or three drops of phenolphthalein to the flask. The acid may be relatively insoluble, so don't worry if it doesn't all dissolve.

Refill your NaOH buret with the (now standardized) NaOH solution. Add the standard HCl to your HCl buret until it is about half full. Read both levels carefully and record them.

Titrate the solution of the solid acid with NaOH. As the acid is neutralized it will tend to dissolve in the solution. If the acid appears to be relatively insoluble, add NaOH until the pink color persists, and then swirl to dissolve the solid. If the solid still will not dissolve, and the solution remains pink, add about twenty-five milliliters of ethanol to increase the solubility. If you go past the endpoint, add HCl as necessary. The final pink endpoint should appear on addition of 1 drop of NaOH. Record the final levels in the NaOH and HCl burets.

Pour the rest of your acid sample into a clean 250-mL Erlenmeyer flask, and weigh the empty vial accurately. Titrate this sample of acid as before with the NaOH and HCl solutions.

If you use HCl in these titrations, and you probably will, the calculations needed are a bit more complicated than in the standardization of the NaOH solution. To find the number of moles of H^+ ion in the solid acid, you must subtract the number of moles of HCl used from the number of moles of NaOH. For a back-titration, which is what we have in this case:

$$\text{moles } H^+ \text{ in solid acid} = \text{moles } OH^- \text{ in NaOH soln.} - \text{moles } H^+ \text{ in HCl soln.} \tag{5}$$

For volumes in milliliters, this equation takes the form:

$$\text{moles } H^+ \text{ in solid acid} = \frac{M_{NaOH} \times V_{NaOH}}{1000} - \frac{M_{HCl} \times V_{HCl}}{1000} \tag{6}$$

Optional **C. Determination of K_a for the Unknown Acid**

Your instructor will tell you in advance if you are to do this part of the experiment. If you do this part, you will need to *save* one of the titrated solutions from Part B. Given that titrated solution, it is easy to prepare one in which $[H^+]$ ion is equal to K_a for your acid.

A weak monoprotic acid, HB, dissociates in water solution according to the equation:

$$HB(aq) \rightleftharpoons H^+(aq) + B^-(aq) \tag{7}$$

The equilibrium constant for this reaction is called the acid dissociation constant for HB and is given the symbol K_a. A solution of HB will obey the equilibrium condition given by the equation:

$$K_a = \frac{[H^+][B^-]}{[HB]} \tag{8}$$

Your acid is a weak monoprotic acid, so it will obey both of the equations above.

In the titration in Part B you converted a solution of HB into one containing NaB by adding sufficient NaOH to reach the endpoint. If, to your titrated solution, you now add some HCl, it will convert some of the B^- ions in the NaB solution back to HB. If the number of moles of HCl you add equals *one-half* the number of moles of H^+ in your original sample, then *half* of the B^- ions will be converted to HB. In the resulting solution, [HB] will equal $[B^-]$, and so, by Equation 8,

$$K_a = [H^+] \quad \text{in the half-neutralized solution} \tag{9}$$

Using the approach we have outlined, use your titrated solution to prepare one in which [HB] equals $[B^-]$. Measure the pH of that solution with a pH meter, using the procedure described in Appendix IV. From that value, calculate K_a for your unknown acid. ▪

Take it Further (Optional): Using the procedure in this experiment, find the mass percent acetic acid in household vinegar. What is the molarity of the acid?

DISPOSAL OF REACTION PRODUCTS. The reaction products in this experiment may be diluted and poured down the drain.

Experiment 24

Data and Calculations: The Standardization of a Basic Solution and the Determination of the Molar Mass of an Acid

A. Standardization of NaOH Solution

	Trial 1	Trial 2	Trial 3
Initial reading, HCl buret	_____ mL		
Final reading, HCl buret	_____ mL	_____ mL	_____ mL
Initial reading, NaOH buret	_____ mL		
Final reading, NaOH buret	_____ mL	_____ mL	_____ mL

B. Determination of the Molar Mass of an Unknown Acid

Mass of vial plus contents	_____ g
Mass of vial plus contents less Sample 1	_____ g
Mass less Sample 2 (empty vial)	_____ g

	Trial 1	Trial 2
Initial reading, NaOH buret	_____ mL	_____ mL
Final reading, NaOH buret	_____ mL	_____ mL
Initial reading, HCl buret	_____ mL	_____ mL
Final reading, HCl buret	_____ mL	_____ mL

Processing the Data

A. Standardization of NaOH Solution

	Trial 1	Trial 2	Trial 3
Total volume of HCl	_____ mL	_____ mL	_____ mL
Total volume of NaOH	_____ mL	_____ mL	_____ mL

(continued on following page)

Molarity, M_{HCl}, of standardized HCl   _____ M

Molarity, M_{H^+}, in standardized HCl   _____ M

By Equation 3,

$$M_{H^+} \times V_{acid} = M_{OH^-} \times V_{base} \quad \text{or} \quad M_{OH^-} = M_{H^+} \times \frac{V_{HCl}}{V_{NaOH}} \qquad (3)$$

Use Equation 3 to find the molarity, M_{OH^-}, of the NaOH solution. Note that the volumes do not need to be converted to liters, since we use the volume ratio.

	Trial 1	**Trial 2**	**Trial 3**

M_{OH^-} _____ M _____ M _____ M   (should agree within 1%)

The molarity of the NaOH will equal M_{OH^-}, since one mole NaOH → one mole OH⁻.

Mean molarity of NaOH solution, M_{NaOH} _____ M   Standard deviation _____ M

B. Determination of the Molar Mass of the Unknown Acid

	Trial 1	**Trial 2**
Mass of sample	_____ g	_____ g
Volume of NaOH used	_____ mL	_____ mL
Moles NaOH $= \dfrac{V_{NaOH} \times M_{NaOH}}{1000}$	_____	_____
Volume of HCl used	_____ mL	_____ mL
Moles HCL $= \dfrac{V_{HCl} \times M_{HCl}}{1000}$	_____	_____
Moles H⁺ in sample (use Eq. 5)	_____	_____
$MM = \dfrac{\text{grams acid}}{\text{moles H}^+}$	_____ g/mol	_____ g/mol
Unknown #		_____

Optional   C. Determination of K_a for the Unknown Acid

Moles H⁺ in sample (from Part B)   _____

Moles HCl to be added _____   Volume of HCl added _____ mL

pH of half-neutralized solution _____ K_a of acid _____ ▪

Experiment 24

Advance Study Assignment: Molar Mass of an Acid

1. 6.9 mL of 6.0 M NaOH are diluted with water to a volume of 400.0 mL. You are asked to find the molarity of the resulting solution.

   a. First find out how many moles of NaOH are present in 6.9 mL of 6.0 M NaOH. Use Equation 1. Note that the volume must be in liters.

   _____ moles

   b. Since the total number of moles of NaOH is not changed on dilution, the molarity after dilution can also be found by Equation 1, using the final volume of the solution. Calculate that molarity.

   _____ M

2. In an acid-base titration, 22.13 mL of an NaOH solution are needed to neutralize 22.44 mL of a 0.0997 M HCl solution. To find the molarity of the NaOH solution, we can use the following procedure:

   a. First note the value of M_{H^+} in the HCl solution.

   _____ M

   b. Find M_{OH^-} in the NaOH solution. (Use Eq. 3.)

   _____ M

   c. Obtain M_{NaOH} from M_{OH^-}.

   _____ M

(continued on following page)

3. A 0.5891-g sample of an unknown monoprotic acid requires 38.50 mL of 0.1011 M NaOH for neutralization to a phenolphthalein endpoint. There are 0.29 mL of 0.0997 M HCl used for back-titration.

    a. How many moles of OH^- are used? How many moles of H^+ from HCl?

                    _____ moles OH^- _____ moles H^+

    b. How many moles of H^+ are there in the solid acid? (Use Eq. 5.)

                    _____ moles H^+ in solid

    c. What is the molar mass of the unknown acid? (Use Eq. 4.)

                    _____ g/mol

Experiment 25

pH Measurements— Buffers and Their Properties

$\mathbf{O}$ne of the more important properties of an aqueous solution is its concentration of hydrogen ion. The H^+ ion (or, more precisely, the H_3O^+ ion) has a great effect on the solubility of many inorganic and organic species, on the nature of complex metallic cations found in solutions, and on the rates of many chemical reactions. It is important that we know how to measure the concentration of hydrogen ion and understand its effect on solution properties.

For convenience, the concentration of H^+ ion is frequently expressed as the pH of the solution rather than as molarity. The pH of a solution is defined by the following equation:

$$pH = -\log[H^+] \tag{1}$$

where the logarithm is taken to the base 10. If $[H^+]$ is 1×10^{-4} moles per liter, the pH of the solution is 4.0. If the $[H^+]$ is 5×10^{-2} M, the pH is 1.3.

Basic solutions can also be described in terms of pH. In aqueous solutions the following equilibrium relation will always be obeyed:

$$[H^+] \times [OH^-] = K_w = 1 \times 10^{-14} \text{ at } 25°C \tag{2}$$

In deionized water $[H^+]$ equals $[OH^-]$, so, by Equation 2, $[H^+]$ must be 1×10^{-7} M. Therefore, the pH of deionized water is 7. Solutions in which $[H^+] > [OH^-]$ are said to be acidic and will have a pH < 7; if $[H^+] < [OH^-]$, the solution is basic and its pH > 7. A solution with a pH of 10 will have a $[H^+]$ of 1×10^{-10} M and a $[OH^-]$ of 1×10^{-4} M.

We measure the pH of a solution experimentally in two ways. In the first of these we use a chemical called an indicator, which is sensitive to pH. These substances have colors that change over a relatively short pH range (about two pH units) and can, when properly chosen, be used to determine roughly the pH of a solution. Two very common indicators are litmus, usually used on paper, and phenolphthalein, the most common indicator in acid-base titrations. Litmus changes from red to blue as the pH of a solution goes from about six to about eight. Phenolphthalein changes from colorless to red as the pH goes from 8 to 10. A given indicator is useful for quantitatively determining pH only in the region in which it changes color. Indicators are available for measurement of pH in all the important ranges of acidity and basicity. By matching the color of a suitable indicator in a solution of known pH with that in an unknown solution, we can determine the pH of the unknown to within about 0.3 pH units.

The other method for finding pH is with a device called a pH meter. In this device two electrodes, one of which is sensitive to $[H^+]$, are immersed in a solution. The potential between the two electrodes is related to the pH. The pH meter is designed so that the scale will directly furnish the pH of the solution. A pH meter gives much more precise measurement of pH than does a typical indicator and is ordinarily used when an accurate determination of pH is needed.

Some acids and bases undergo substantial ionization in water, and are called strong because of their essentially complete ionization in reasonably dilute solutions. Other acids and bases, because of incomplete ionization (often far less than even 1% in 0.10 M solution), are called weak. Hydrochloric acid, HCl, and sodium hydroxide, NaOH, are typical examples of a strong acid and a strong base. Acetic acid, $HC_2H_3O_2$, and ammonia, NH_3, are classic examples of a weak acid and a weak base.

A weak acid will ionize according to the Law of Chemical Equilibrium:

$$HB(aq) \rightleftharpoons H^+(aq) + B^-(aq) \tag{3}$$

At equilibrium,

$$\frac{[H^+][B^-]}{[HB]} = K_a \qquad (4)$$

K_a is a constant characteristic of the acid HB; in solutions containing HB, the product of concentrations in the equation will remain constant at equilibrium independent of the manner in which the solution was made. A similar relation can be written for solutions of a weak base.

The value of the ionization constant K_a for a weak acid can be found experimentally in several ways. In Experiment 24 we used a very simple method for finding K_a. In general, however, we need to find the concentrations of each of the species in Equation 4 by one means or another. In this experiment we will determine K_a for a weak acid in connection with our study of the properties of those solutions we call buffers.

Salts that can be formed by the reaction of strong acids and bases—such as NaCl, KBr, or $NaNO_3$—ionize completely but do not react with water when in solution. They form neutral solutions with a pH of about seven. When dissolved in water, salts of *weak* acids or *weak* bases furnish ions that tend to react to some extent with water, producing molecules of the weak acid or base and liberating some OH^- or H^+ ion to the solution.

If HB is a weak acid, the B^- ion produced when NaB is dissolved in water will react with water to some extent, according to the equation

$$B^-(aq) + H_2O(\ell) \rightleftharpoons HB(aq) + OH^-(aq) \qquad (5)$$

Solutions of sodium acetate, $NaC_2H_3O_2$, the salt formed by reaction of sodium hydroxide with acetic acid, will be slightly *basic* because of the reaction of $C_2H_3O_2^-$ ion with water to produce $HC_2H_3O_2$ and OH^-. Because of the analogous reaction of the NH_4^+ ion with water to form H_3O^+ ion, solutions of ammonium chloride, NH_4Cl, will be slightly *acidic*.

Salts of most transition metal ions are acidic. A solution of $CuSO_4$ or $FeCl_3$ will typically have a pH equal to 5 or lower. The salts are completely ionized in solution. The acidity comes from the fact that the cation is hydrated [e.g., $Cu(H_2O)_4^{2+}$ or $Fe(H_2O)_6^{3+}$]. The large + charge on the metal cation attracts electrons from the O—H bonds in water, weakening them and producing some H^+ ions in solution; with $CuSO_4$ solutions the reaction would be

$$Cu(H_2O)_4^{2+}(aq) \rightleftharpoons Cu(H_2O)_3OH^+(aq) + H^+(aq) \qquad (6)$$

Buffers

Some solutions, called buffers, are remarkably resistant to pH changes. Water is not a buffer, since its pH is very sensitive to addition of any acidic or basic species. Even bubbling your breath through a straw into deionized water can lower its pH by at least one pH unit, just due to the small amount of CO_2, which is acidic, in exhaled air. With a good buffer solution, you could blow exhaled air into it for half an hour and not change the pH appreciably. All living systems contain buffer solutions, since a stable pH is essential for the occurrence of many of the biochemical reactions that go on to maintain the living organism.

There is nothing mysterious about what is needed to make a buffer. All that is required is a solution containing a weak acid and its conjugate base. An example of such a solution is one containing the weak acid HB, and its conjugate base, B^- ion, which can be added by dissolving the salt NaB in water or obtained by adding a strong base to the weak acid HB.

The pH of such a buffer is established by the relative concentrations of HB and B^- in the solution. If we manipulate Equation 4, we can solve for the concentration of H^+ ion:

$$[H^+] = K_a \times [HB]/[B^-] \qquad (4a)$$

This equation is often further modified when working with buffers, by taking the negative base 10 logarithm of both sides:

$$pH = pK_a + \log [B^-]/[HB] \qquad (4b)$$

where pK_a equals $-\log K_a$. This equation is called the Henderson-Hasselbach equation, which has many biological applications. Equation 4b tells us what quantities fix the pH of an HB/B^- buffer at a given value.

If we are working with a weak acid, HB, whose K_a equals 1×10^{-5}, its pK_a equals 5.0. If we mix equal volumes of 0.10 M HB and 0.10 M NaB, the pH of the resulting solution will equal the pK_a of the acid, in this case 5.0, since the concentrations of HB and B^- are equal in the final solution, and the base 10 logarithm of 1 is 0. $[H^+]$ in this buffer will be 1×10^{-5} M.

The H^+ ion in the solution has to come from ionization of HB, but the amount is tiny, so we can assume that *in any buffer the weak acid and its conjugate base do not react appreciably with one another when their solutions are mixed.* Their relative concentrations can be calculated from the way the buffer was put together.

Using Equation 4b, you should be able to answer several questions regarding buffers. For example, how does the pH of buffer solution change if you dilute it with water? How does it change if you add an HB solution? What happens if you add a solution of NaB, containing the conjugate base? What is the pH range over which a buffer would be useful, if you assume that the ratio of $[B^-]$ to $[HB]$ must lie between 10:1 and 1:10? How sensitive is the pH of water to the addition of a strong acid, like HCl, or addition of a strong base, like NaOH? What happens when you add a strong acid or a strong base to a buffer? Why does the pH of a buffer resist changing when small amounts of a strong acid or strong base are added?

Some of these questions you can answer by just looking at Equation 4b. For others you need to realize that, although a weak acid and its conjugate base don't react with each other, in a buffer a weak acid like HB will react quantitatively (completely) with a strong base, like NaOH:

$$HB(aq) + OH^-(aq) \rightarrow B^-(aq) + H_2O(\ell) \tag{7}$$

and in a buffer a weak base like NaB will react quantitatively (completely) with a strong acid like HCl:

$$B^-(aq) + H^+(aq) \rightarrow HB(aq) \tag{8}$$

In Reaction 7, a small amount of NaOH will not raise the pH very much, since the OH^- ion is essentially soaked up by the acid HB, producing some B^- ion and increasing the value of $[B^-]/[HB]$, but not destroying the buffer.

Reactions 7 and 8 can also be used to *prepare* a buffer from a solution of HB by addition of NaOH, or from a solution of NaB by addition of HCl.

In this experiment you will determine the approximate pH of several solutions by using acid-base indicators. Then you will find the pH of some other solutions with a pH meter. In the rest of the experiment you will carry out some reactions that will allow you to answer all the questions we have raised about buffers. Finally, you will prepare one or two buffers having specific pH values.

Experimental Procedure

WEAR YOUR SAFETY GLASSES WHILE
PERFORMING THIS EXPERIMENT

You may work in pairs on the first three parts of this experiment.

A. Determination of pH by the Use of Acid-Base Indicators

To each of five small (13 $\times$ 100 mm) test tubes add about one milliliter of 0.10 M HCl (in these tubes, a depth of about a half inch). To each tube add one or two drops of one of the indicators mentioned in Table 25.1, one indicator to a tube. Note the color of the solution you obtain in each case. By comparing the colors you observe with the information in Table 25.1, estimate the pH of the solution to within a range of one pH unit, say 1 to 2, or 4 to 5. In making your estimate, note that the color of an indicator is most indicative of pH in the region where the indicator is changing color.

Repeat the procedure with each of the following solutions:

$$\text{0.10 M NaH}_2\text{PO}_4 \quad \text{0.10 M HC}_2\text{H}_3\text{O}_2 \quad \text{0.10 M ZnSO}_4$$

Record the colors you observe and the pH range for each solution.

Table 25.1

Useful pH Ranges for Some Common Acid-Base Indicators								
	Useful pH range (approximate)							
Indicator	0	1	2	3	4	5	6	7
Methyl violet	yellow ⟶ violet							
Thymol blue		red ⟶ yellow						
Methyl yellow			red ⟶ yellow					
Congo red				violet ⟶ orange-red				
Bromocresol green				yellow ⟶ blue				

B. Measurement of the pH of Some Typical Solutions

In the rest of this experiment we will use pH meters to find pH. Your instructor will show you how to operate your meter. The electrodes may be fragile, so use due caution when handling the electrode probe. See Appendix IV for a discussion of pH meters.

Using a 25-mL sample in a 150-mL beaker, measure and record the pH of a 0.10 M solution of each of the following substances:

$$NaCl \quad Na_2CO_3 \quad NaC_2H_3O_2 \quad NaHSO_4$$

Rinse the electrode probe in deionized water between measurements. After you have completed a measurement, add 1 drop or two of bromocresol green to the solution and record the color you obtain.

Some of the solutions are nearly neutral; others are acidic or basic. For each solution having a pH less than 6 or greater than 8, write a net ionic equation that explains qualitatively why the observed pH value is reasonable.

Then write a rationale for the colors obtained with bromocresol green with these solutions.

C. Some Properties of Buffers

On the lab bench we have 0.10 M stock solutions that can be used to make three different common buffer systems. These are

$$HC_2H_3O_2/C_2H_3O_2^- \qquad NH_4^+/NH_3 \qquad HCO_3^-/CO_3^{2-}$$

acetic acid/acetate ion   ammonium ion/ammonia   hydrogen carbonate/carbonate

The sources of the ions will be sodium and ammonium salts containing those ions. Select one of these buffer systems for your experiment.

1. Using a graduated cylinder, measure out 15 mL of the acid component of your buffer into a 100-mL beaker. The acid will be one of the following solutions: 0.10 M $HC_2H_3O_2$, 0.10 M NH_4Cl, or 0.10 M $NaHCO_3$. Rinse out the graduated cylinder with deionized water and use it to add 15 mL of the conjugate base of your buffer. Measure the pH of your mixture and record it on the report page. Calculate pK_a for the acid.

2. Add 30 mL of water to your buffer mixture, mix, and pour half of the resulting solution into another 100-mL beaker. Measure the pH of the diluted buffer. Calculate pK_a once again. Add 5 drops of 0.10 M NaOH to the diluted buffer and measure the pH again. To the other half of the diluted buffer add 5 drops of 0.10 M HCl, and again measure the pH. Record your results.

3. Make a buffer mixture containing 2 mL of the acid component and 20 mL of the solution containing the conjugate base. Mix, and measure the pH. Calculate a third value for pK_a. To this solution add 3 mL 0.10 M NaOH. Measure the pH. Explain your results.

4. Put 25 mL of deionized water into a 100-mL beaker. Measure the pH. Add 5 drops of 0.10 M HCl and measure the pH again. To that solution add 10 drops of 0.10 M NaOH, mix, and measure the pH.

5. Select a pH different from any of those you observed in your experiments. Design a buffer which should have that pH by selecting appropriate volumes of your acidic and basic components. Make up the buffer and measure its pH.

D. Preparation of a Buffer from a Solution of a Weak Acid

So far in this experiment the buffers that we used were made up from a weak acid and its conjugate base. Chemists faced with making a buffer would take a simpler approach. They would start with a solution of a weak acid with a pK_a roughly equal to the pH of the buffer that was needed. To the acid they would slowly add an NaOH solution from a buret, stirring well, while at the same time measuring the pH of the solution. When they got to the desired pH, they would stop adding the NaOH. The buffer would be ready to use.

In this part of the experiment you are to work alone. We will furnish you with a 0.50 M solution of a weak acid with a known pK_a, and the pH of the buffer you will be asked to prepare. First, dilute the acid solution to 0.10 M by adding 10. mL of the acid to 40. mL of water in a 100-mL beaker and stirring well.

Using Equation 4b, calculate the ratio of $[B^-]$ to $[HB]$ in the buffer to be prepared. Once you know this ratio, use it to calculate how much 0.10 M NaOH you will have to add to 20. mL of your 0.10 M acid solution to produce your buffer. The NaOH reacts with HB, converting it to B^-, so if we add y mL of the NaOH, the value of $[B^-]/[HB]$ will become equal to $y/(20. - y)$. Can you see why? Think about it, and you will soon see that such is the case. Record the volume of NaOH that you calculate should be needed on the report page.

Now actually prepare the buffer, to check your prediction. Use the buret containing 0.10 M NaOH solution that has been set up by the pH meter. Record the volume before starting to add the base, noting the pH of the acid. Slowly add the NaOH to 20. mL of the acid, stirring well and watching the pH as it slowly goes up. When you obtain a solution of the pH to be prepared, stop adding NaOH. Record the volume reading on the buret. Report the actual volume of NaOH solution required to produce your buffer, and compare it with the amount you calculated would be necessary.

Take it Further (Optional): Vitamin C (MM = 176 g/mol) is a weak acid, called ascorbic acid. Study the buffering properties of vitamin C using a solution made from a 500-mg tablet. Find K_a for vitamin C and the pH range over which it might be useful as a buffer.

> **DISPOSAL OF REACTION PRODUCTS.** When you are finished with the experiment, you may dilute the solutions with water and pour them down the drain.

Experiment 25

Data and Calculations: pH Measurements: Buffers and Their Properties

A. Determination of pH by the Use of Acid-Base Indicators

	Color with 0.10 M solution of			
Indicator	HCl	NaH$_2$PO$_4$	HC$_2$H$_3$O$_2$	ZnSO$_4$
Methyl violet	_____	_____	_____	_____
Thymol blue	_____	_____	_____	_____
Methyl yellow	_____	_____	_____	_____
Congo red	_____	_____	_____	_____
Bromocresol green	_____	_____	_____	_____
pH range	_____	_____	_____	_____

Circle the observation(s) for each solution that was (were) most useful in estimating the pH range.

B. Measurement of the pH of Some Typical Solutions

Record the pH and the color observed with bromocresol green for each of the 0.10 M solutions that were tested.

	NaCl	Na$_2$CO$_3$	NaC$_2$H$_3$O$_2$	NaHSO$_4$
pH	_____	_____	_____	_____
Color	_____	_____	_____	_____

For any two solutions having a pH less than 6 or greater than 8, write a net ionic equation to explain qualitatively why the solution has that pH.

Solution _____ Equation _____

Solution _____ Equation _____

Explain why the color observed with bromocresol green for each of the four solutions is reasonable, given the pH.

C. Some Properties of Buffers

Buffer system selected _____ HB is _____ (name the acid)

  1. pH of buffer _____ [H$^+$] _____ M pK_a (by Eq. 4b) _____

  2. pH of diluted buffer _____ [H$^+$] _____ M   pK_a _____

     pH after addition of 5 drops of NaOH   _____

     pH after addition of 5 drops of HCl   _____

     Comment on your observations in Parts 1 and 2.

(continued on following page)

3. pH of buffer in which $[HB]/[B^-] = 0.10$ _____ pK_a _____

  pH after addition of excess NaOH _____

  Explain your observations.

4. pH of deionized water _____

  pH after addition of 5 drops of HCl _____

  pH after addition of 10 drops of NaOH _____

  Explain your observations.

5. pH of buffer solution to be prepared _____

  Mean value of pK_a (as found in Parts 1, 2, and 3) _____

  $[B^-]/[HB]$ in buffer _____

  (Volume 0.10 M NaB)/(Volume 0.10 M HB) needed in buffer _____

  Volume 0.10 M NaB used _____ mL   Volume 0.10 M HB used _____ mL

  pH of buffer you prepared _____

D. Preparation of a Buffer from a Solution of a Weak Acid

pH of buffer to be prepared _____ pK_a of acid _____

Ratio of $[B^-]/[HB]$ required in buffer _____

Volume of 0.10 M NaOH calculated _____ mL

pH of acid solution before titration _____

Initial volume reading of NaOH _____ mL

Final volume reading of NaOH _____ mL

Volume of NaOH actually required _____ mL

How does that volume compare to your calculated value?

Experiment 25

Advance Study Assignment: pH Measurements—Buffers and Their Properties

1. A solution of a weak acid was tested with the indicators used in this experiment. The colors observed were as follows:

Methyl violet	violet	Congo red	orange-red
Thymol blue	yellow	Bromocresol green	blue-green
Methyl yellow	yellow		

   What is the approximate pH of the solution?

   _____

2. The ammonia (NH_3) molecule is the conjugate base of the NH_4^+ (called "ammonium") ion, a weak acid. An aqueous solution of NH_3 has a pH of 11.6. Write the net ionic equation for the reaction that makes an aqueous solution of NH_3 basic. (The answer looks very similar to Eq. 5 in the introduction to this experiment.)

3. The pH of a 0.010 M HOBr solution is 5.3.

   a. What is $[H^+]$ in that solution?

   _____ M

   b. What is $[OBr^-]$? What is [HOBr]? (Where do the H^+ and OBr^- ions come from?)

   _____ M; _____ M

   c. What is the value of K_a for HOBr? What is the value of pK_a?

   _____ _____

(continued on following page)

4. Formic acid, HFor, has a K_a value of 1.8×10^{-4}. A student is asked to prepare a buffer having a pH of 3.85 from a solution of formic acid and a solution of sodium formate having the same molarity. How many milliliters of the NaFor solution should she add to 20.0 mL of the HFor solution to make the buffer? (See discussion of buffers.)

_____ mL

5. How many mL of 0.10 M NaOH should the student add to 20.0 mL of 0.10 M HFor if she wishes to prepare a buffer with a pH of 3.85, the same as in Problem 4?

_____ mL

Experiment 26

Determination of the Solubility Product of $Ba(IO_3)_2$

In several of the earlier experiments in this manual we used the concept of the solubility product in various applications. Every insoluble inorganic salt has an associated number, called its solubility product, that describes its behavior in mixtures where it is present as a solid, quantifying a condition on the equilibrium concentrations of the ions in a mixture in which the salt is present. The equation expressing the condition follows from the chemical formula of the solid and the balanced chemical reaction describing the dissolution of the solid. For barium iodate, $Ba(IO_3)_2$, the dissolution reaction is:

$$Ba(IO_3)_2(s) \rightleftharpoons Ba^{2+}(aq) + 2\,IO_3^{-}(aq) \tag{1}$$

Recognizing that the pure solid is left out of the equilibrium expression but must be present for the equilibrium to hold, we write the equation for the solubility product, K_{sp}, as

$$K_{sp} = [Ba^{2+}]\,[IO_3^{-}]^2 \tag{2}$$

where the concentrations of the barium and iodate ions are the values in moles per liter in the solution in which solid barium iodate is present. Equation 2 essentially states that at a given temperature, in a mixture of barium and iodate ions in equilibrium with solid barium iodate, the product of the molarity of the Ba^{2+} ion times the molarity of the IO_3^{-} ion squared cannot vary, no matter how you change the ion concentrations in the system. If one ion concentration goes up, the other must go down, keeping the ion concentration product in Equation 2 equal to K_{sp}. That's rather amazing. For barium iodate, K_{sp} has a small value, less than 10^{-7}. So, if one ion has a typical concentration, say 0.1 M, the other must have a very low concentration. If you add 0.10 M $BaCl_2$ to a small volume of a solution of 0.05 M KIO_3, you can be sure that a reaction will occur, in which $Ba(IO_3)_2$ precipitates, where the final solution will contain almost no iodate ion, and the product of the ion concentrations in Equation 2 has a fixed value.

This behavior is seen in many analyses of practical importance. In Experiment 7, in which a solution of chloride ion is titrated with a solution of $AgNO_3$, containing silver and nitrate ions, when the Ag^+ ions first meet the Cl^- ions, insoluble AgCl precipitates, since the product, K_{sp}, of their concentrations cannot be very large. As the silver nitrate is slowly added, AgCl precipitates until the Cl^- ion is essentially gone, and the Ag^+ ion concentration can finally go up. At the endpoint, silver dichlorofluoresceinate will form on the surface of the silver chloride particles.

In this experiment we will be examining the solubility behavior of $Ba(IO_3)_2$ in various mixtures of $Ba(NO_3)_2$ and KIO_3 solutions, for the purpose of ultimately measuring the solubility product (K_{sp}) of barium iodate and observing whether it indeed remains constant.

WEAR YOUR SAFETY GLASSES WHILE PERFORMING THIS EXPERIMENT

Experimental Procedure

This experiment can give excellent results if it is done carefully. It is probably the most difficult experiment in the manual to do properly. You will be working with small amounts of the chemical reactants, so small errors will be important.

A. Preparing the Reaction Mixtures

In this experiment we will precipitate solid $Ba(IO_3)_2$ under several different conditions, with varying proportions of the reagents used. We will analyze the solutions remaining after the precipitation for their iodate ion concentrations at equilibrium, and from those values and the way the mixtures were prepared, we can find the concentrations of the barium and iodate ions, and the value of K_{sp} for barium iodate.

From the stock solutions that are available, measure out about thirty-five milliliters of 0.0200 M $Ba(NO_3)_2$ into a small dry beaker. To a second small dry beaker add about twenty milliliters of 0.0350 M KIO_3. Use the labeled graduated cylinders next to the reagent bottles for measuring out these solutions.

Differentiate five medium (18 × 150 mm) dry test tubes, either by labeling them or by noting their locations in your test tube rack.

Into Test Tube 1, carefully pipet 1.00 mL of 0.0350 M KIO_3. Add 2.00, 3.00, 4.00, and 5.00 mL of that same reagent to Test Tubes 2, 3, 4, and 5, respectively.* Then, to each of the test tubes add 5.00 mL of 0.0200 M $Ba(NO_3)_2$, using your 5-mL pipet. Finally, pipet 6.00, 5.00, 4.00, 3.00, and 2.00 mL of deionized water into Test Tubes 1 through 5, respectively. The final volume in each tube should be 12.00 mL. The compositions of the mixtures are summarized in Table 26.1.

Table 26.1

Volumes of Reagents Used in Precipitating $Ba(IO_3)_2$ [mL]			
Test Tube	**0.0200 M $Ba(NO_3)_2$**	**0.0350 M KIO_3**	**H_2O**
1	5.00	1.00	6.00
2	5.00	2.00	5.00
3	5.00	3.00	4.00
4	5.00	4.00	3.00
5	5.00	5.00	2.00

With a stirring rod, stir each solution for 30 seconds, up and down as well as swirling, wiping the rod on a paper towel before proceeding to the next solution. Touch the wall of the tube occasionally as you stir.

Look at each tube. In most of them you should see a crystalline white precipitate, which slowly settles out. It will be most noticeable in the higher number tubes, but it should be present as a slight cloudiness in Test Tubes 1 and 2. There will only be a tiny amount of precipitate, only a few milligrams, so the amount is not impressive. It is most easily seen if you look through the tube toward the light, where you may observe tiny particles, very slowly settling out. The crystallization occurs slowly, so give it time.

Put a stopper in each of the tubes in which there is a precipitate, and slowly rotate the tubes so that the liquid goes from end to end twenty times or so to ensure thorough mixing and help bring the system to equilibrium.

(For any tubes in which you observe only a clear liquid, and no solid, repeat the stirring operation, scratching the walls of the tube as you stir, which may get things started. If that still doesn't do the job, touch your stirring rod to the sample of pure solid barium iodate in the lab and swirl the rod in the reluctant liquid. Then remove the rod, insert a stopper in the tube, and shake it with vigor. Sooner or later you will get the cloudiness to form.)

Repeat the rotation step, turning each tube end to end to do what you can to make the liquid homogeneous and bring it to equilibrium with the precipitate. You can't overstir, so be patient. Give yourself five minutes with this. It will be time well spent.

Finally, let the test tubes stand for 15 minutes to let the solid precipitate particles settle out completely, leaving a clear solution above the solids.

While you are waiting for the precipitates to settle out, you can begin the calculations you will need later. Fill in as many blanks as you can on the report page or in a spreadsheet. From the way you made your mixtures,

---

* You may use your 10-mL graduated pipet for these additions. (The easiest way to do this is to fill the pipet to the 0.00 level. With the pipet in Test Tube 1, let the level fall to 1.00. In Tube 2, let the level fall to 3.00, thereby adding 2.00 mL to the Tube. In Tube 3, let the level fall to 6.00, and in Tube 4 let it go to 10.00. Refill the pipet to the 5.00 level and in Tube 5 let it fall to 10.00.)

you can determine the moles of Ba^{2+} and IO$_3^-$ that were initially present in each tube. Once you have completed the analysis of the solutions at equilibrium, you can finish calculating K_{sp}.

B. Preparing the Equilibrium Mixtures for Analysis

The mixtures in your test tubes at this point should contain essentially clear solutions above the precipitates that formed. You now need to remove an aliquot from each of those solutions, which should now be at equilibrium, without disturbing the solid at the bottom of the test tubes. This is the most difficult part of the experiment, and must be done very carefully. It is done with a 5-mL pipet. First, rinse and drain that pipet with deionized water. Then blow out the water remaining in the tip.

Clamp Test Tube 1 on a ring stand, so that you can work on it without it moving. Immerse the 5-mL pipet about half-way down the clear solution, and hold it there with your hand. Compress a suction bulb slightly, and attach it gently to the top of the pipet. Slowly release the pressure on the bulb, and draw the solution into the pipet until it is near the top of the pipet. Remove the bulb and quickly put your finger on the top of the pipet to hold the liquid in place. Let liquid flow back into the test tube, very slowly, until the level is at the marking line on the pipet. Remove the pipet from the test tube and let the contents flow into a clean, dry test tube. Repeat this process for each of the five test tubes, using five labeled, clean, and dry test tubes to receive the 5-mL aliquots. Remember to rinse, drain, and blow out the 5-mL pipet between each one.

This operation is easier said than done. If something goes wrong while you are drawing the liquid into the pipet, remove the bulb and let the liquid go back into the test tube. After the precipitate re-settles to the bottom, you can try again. To increase your chances of accomplishing this transfer properly, you can practice doing it with water in a clamped test tube.

C. Analyzing the Equilibrium Solutions

Having separated 5 mL of each of the five equilibrium solutions, it is now possible to analyze for the concentration of the IO$_3^-$ ion in those solutions. We do this by reducing iodate ion to I$_2$, which is colored and can be analyzed spectroscopically. The reaction we will use is:

$$IO_3^-(aq) + 5\,I^-(aq) + 6\,H^+(aq) \rightarrow 3\,H_2O(\ell) + 3\,I_2(aq)* \tag{3}$$

To each of the test tubes add, by pipet, 1.00 mL of 1.0 M KI, 1.00 mL of 1.0 M HCl, and 3.00 mL of deionized water, bringing the volume of the five solutions to 10.00 mL. During this step the color of the solution will become orange as the reaction proceeds. Stir each mixture well with a stirring rod, drying the rod on a paper towel before you go on to the next tube.

Go to the spectrophotometer in the lab. The wavelength setting should be 500 nm; set the zero and 100% readings if they have not yet been adjusted. If you have not used the spectrophotometer before, read Appendix IV on how to use it to measure light absorption. Pick up a spectrophotometer tube, called a cuvet, and if it is not dry, shake it to remove as much liquid as you can. Then add about a milliliter of the orange liquid from Tube 1, shake to mix it with the old liquid, and then pour out the cuvet into a waste beaker. Rinse it again with another roughly one-milliliter portion of the solution from Tube 1, and drain. Finally, fill the cuvet up to the measuring level with the solution from Tube 1. (If your cuvet starts out dry, this is all you have to do.) Put the cuvet into the spectrophotometer and read the absorbance of the solution. The concentration of the IO$_3^-$ ion in the *equilibrium 12-mL mixture* will equal the value on the graph in the lab. (The graph is drawn to take into account the dilution you made.) Record the absorbance and concentration for Tube 1. Repeat this measurement process, *including the rinsing steps*, for the other four tubes you worked with, and record your values for those solutions.

---

* Given the excess of I$^-$, the actual dominant iodine species in solution is I$_3^-$, but this reaction does quantitatively describe the consumption of the reactants.

Pour all of the solutions and solids in the 10 test tubes you used in the waste container. Rinse out the tubes with deionized water.

D. Processing the Data

From the volume and concentration of the KIO$_3$ you used in making the five mixtures of reagents, you have probably already found the number of moles of iodate ion *originally* in each mixture. From the concentration of the IO$_3^-$ ion that we measured for each of the *equilibrium* solutions we calculate the number of moles of that ion in those solutions, with their original 12-mL volumes. The difference between the original number of moles and the equilibrium number of moles must be the number of moles of iodate ion in the precipitate at the bottom of the tube. The number of moles of barium in the precipitate must equal *half* of that value, given the formula of barium iodate, Ba(IO$_3$)$_2$. Why?

The number of moles of barium ion that remain in the solution will equal the difference between the number in the original mixture and the number you found were in the precipitate in each of the mixtures. Knowing the volume of the equilibrium solution, 12.00 mL, you can calculate the concentration of the Ba^{2+} ion in each mixture. Record the results of these calculations on the report page or in your spreadsheet. Finally, using the concentrations you have recorded, find the values of K_{sp} for each of the five mixtures. Calculate the mean value of K_{sp} and the standard deviation. (See Appendix IX.)

Experiment 26

Data and Calculations: The Solubility Product of $Ba(IO_3)_2$

A. Preparing the Reaction Mixtures

The calculations are outlined in the Experimental Procedure section. To use a spreadsheet, set up a table like the one below.

Test Tube Number	1	2	3	4	5
mL of 0.0350 M KIO_3	_____	_____	_____	_____	_____
mL of 0.0200 M $Ba(NO_3)_2$	_____	_____	_____	_____	_____
Total volume in mL	_____	_____	_____	_____	_____

C. Analyzing the Equilibrium Solutions

	1	2	3	4	5
Absorbance of soln.	_____	_____	_____	_____	_____
$[IO_3^-]$ in soln.	_____	_____	_____	_____	_____

D. Processing the Data

	1	2	3	4	5
Initial moles Ba^{2+}	_____	_____	_____	_____	_____
Initial moles IO_3^-	_____	_____	_____	_____	_____
Moles IO_3^- in equil. soln. (12 mL)	_____	_____	_____	_____	_____
Moles IO_3^- precipitated	_____	_____	_____	_____	_____

(continued on following page)

Test Tube Number	1	2	3	4	5
Moles Ba^{2+} precipitated	_____	_____	_____	_____	_____
Moles Ba^{2+} in equil. soln. (12 mL)	_____	_____	_____	_____	_____
$[Ba^{2+}]$ in equil. soln.	_____	_____	_____	_____	_____
K_{sp} for $Ba(IO_3)_2$	_____	_____	_____	_____	_____

Mean value of K_{sp} _____   Standard deviation of K_{sp} _____

What was the mass of solid barium iodate in the precipitate in Test Tube 1? $MM\ Ba(IO_3)_2 = 487.14$ g/mol

_____ mg

Does $[Ba^{2+}]$ vary as expected in the experiment? Explain.

What is your best value for the solubility of $Ba(IO_3)_2$ in water?

Experiment 26

Advance Study Assignment: Determination of the Solubility Product
of $Ba(IO_3)_2$

1. State in words what is meant by the solubility product equation for $Ba(IO_3)_2$:

$$K_{sp} = [Ba^{2+}] [IO_3^-]^2$$

2. If 10. mL of 0.10 M $Ba(NO_3)_2$ is mixed with 10. mL of 0.10 M KIO_3, a precipitate forms. Which ion will still be present at appreciable concentration in the equilibrium mixture if K_{sp} for barium iodate is very small? Indicate your reasoning. What would that concentration be?

_____ _____ moles/L

3. Lead bromide, $PbBr_2$, is moderately soluble, with K_{sp} equal to 1.86×10^{-5}.

   a. Predict the solubility of lead bromide in pure water, based on this K_{sp} value. (How many moles of solid $PbBr_2$ could be completely dissolved in one liter of solution?) *Ignore the effect of any other equilibria that may impact the solubility (for example, interactions between Pb^{2+} and OH^-). Your result will not match literature values for the actual solubility of $PbBr_2$.*

_____ moles/L

(continued on following page)

b. What would the predicted solubility of $PbBr_2$ be in 0.10 M NaBr? (How many moles of solid $PbBr_2$ could be completely dissolved in 1 L of solution?) *Again, ignore all other equilibria; assume that the K_{sp} expression for $PbBr_2$ completely describes the situation.*

_____ moles/L

c. The difference in the results obtained in 3(a) and 3(b) is caused by what is known as the common ion effect. State in words what the common ion effect predicts.

Experiment 27

Relative Stabilities of Complex Ions and Precipitates Prepared from Solutions of Copper(II)

In aqueous solution, typical cations, particularly those produced from atoms of the transition metals, do not exist as free ions but rather consist of the metal ion in combination with some water molecules. Such cations are called complex ions. The water molecules, usually two, four, or six in number, are bound chemically to the metallic cation, but often rather loosely, with the electrons in the chemical bonds being furnished by one of the unshared electron pairs from the oxygen atoms in the H_2O molecules. Copper ion in aqueous solution may exist as $Cu(H_2O)_4^{2+}$, with the water molecules arranged in a square around the metal ion at the center.

If a hydrated cation such as $Cu(H_2O)_4^{2+}$ is mixed with other species that can, like water, form coordinate covalent bonds with Cu^{2+}, those species, called ligands, may displace one or more H_2O molecules and form other complex ions containing the new ligands. For instance, NH_3, a reasonably good coordinating species, may displace H_2O from the hydrated copper ion, $Cu(H_2O)_4^{2+}$, to form $Cu(H_2O)_3NH_3^{2+}$, $Cu(H_2O)_2(NH_3)_2^{2+}$, $Cu(H_2O)(NH_3)_3^{2+}$, or $Cu(NH_3)_4^{2+}$. At even moderate concentrations of NH_3, essentially all the H_2O molecules around the copper ion are replaced by NH_3 molecules, forming the copper ammonia complex ion $Cu(NH_3)_4^{2+}$.

Coordinating ligands differ in their tendencies to form bonds with metallic cations, so that in a solution containing a given cation and several possible ligands, an equilibrium will develop in which most of the cations are coordinated with those ligands with which they form the most stable bonds. There are many kinds of ligands, but they all share the common property that they possess an unshared pair of electrons that they can donate to form a coordinate covalent bond with a positively charged metal ion. In addition to H_2O and NH_3, other uncharged coordinating species include CO and ethylenediamine; some common anions that can form complexes include OH^-, Cl^-, CN^-, SCN^-, and $S_2O_3^{2-}$.

As you know, when solutions containing metallic cations are mixed with other solutions containing ions, precipitates are sometimes formed. When a solution of 0.10 M copper nitrate is mixed with a little 1.0 M NH_3 solution, a precipitate forms and then dissolves in excess ammonia. The formation of the precipitate helps us understand what is occurring as NH_3 is added. The precipitate is hydrous copper hydroxide, $Cu(OH)_2(H_2O)_2(s)$, formed by reaction of the hydrated copper ion $Cu(H_2O)_4^{2+}$ with the small amount of hydroxide ion present in the NH_3 solution. The fact that this reaction occurs means that even at very low OH^- ion concentration $Cu(OH)_2(H_2O)_2(s)$ is a more stable species than $Cu(H_2O)_4^{2+}$ ion.

Addition of more NH_3 causes the solid to redissolve. The copper species then in solution cannot be the hydrated copper ion. (Why?) It must be some other complex ion and is, indeed, the $Cu(NH_3)_4^{2+}$ ion. The implication of this observation is that the $Cu(NH_3)_4^{2+}$ ion is also more stable in NH_3 solution than is the hydrated copper ion. To deduce in addition that the copper ammonia complex ion is also more stable in general than $Cu(OH)_2(H_2O)_2(s)$ is not warranted, since under the conditions in the solution $[NH_3]$ is much larger than $[OH^-]$, and given a higher concentration of hydroxide ion, $[OH^-]$, the solid hydrous copper hydroxide might possibly precipitate even in the presence of substantial concentrations of NH_3.

To resolve this, you might add a little 1.0 M NaOH solution to the solution containing the $Cu(NH_3)_4^{2+}$ ion. If you do this you will find that $Cu(OH)_2(H_2O)_2(s)$ does indeed precipitate. We can conclude from these observations that $Cu(OH)_2(H_2O)_2(s)$ is more stable than $Cu(NH_3)_4^{2+}$ in solutions in which the ligand concentrations (OH^- and NH_3) are roughly equal.

The copper-containing species and their amounts that will be present in a system depend, as we have just seen, on the conditions in the system. We cannot say in general that one species will be more stable than another; the stability of a given species depends in large measure on the kinds and concentrations of other species that are also present with it.

Another way of looking at the matter of stability is through equilibrium theory. Each of the copper species we have mentioned can be formed in a reaction between the hydrated copper ion $Cu(H_2O)_4^{2+}(aq)$ and a complexing or precipitating ligand; each reaction will have an associated equilibrium constant, which we might call a formation constant for that species. The pertinent formation reactions and their constants for the copper species we have been considering are listed here:

$$Cu(H_2O)_4^{2+}(aq) + 4\,NH_3(aq) \rightleftharpoons Cu(NH_3)_4^{2+}(aq) + 4\,H_2O(\ell) \qquad K_1 = 5 \times 10^{12} \qquad \textbf{(1)}$$

$$Cu(H_2O)_4^{2+}(aq) + 2\,OH^-(aq) \rightleftharpoons Cu(OH)_2(H_2O)_2(s) + 2\,H_2O(\ell) \qquad K_2 = 2 \times 10^{19} \qquad \textbf{(2)}$$

The formation constants for these reactions do not involve $[H_2O]$ terms, which are essentially constant in aqueous systems and are included in the magnitude of K in each case. The large size of each formation constant indicates that the tendency for the hydrated copper ion to react with the ligands listed is very high.

In terms of these data, let us compare the stability of the $Cu(NH_3)_4^{2+}$ complex ion with that of solid $Cu(OH)_2(H_2O)_2$. This is most readily done by considering the reaction:

$$Cu(NH_3)_4^{2+}(aq) + 2\,OH^-(aq) + 2\,H_2O(\ell) \rightleftharpoons Cu(OH)_2(H_2O)_2(s) + 4\,NH_3(aq) \qquad \textbf{(3)}$$

We can find the value of the equilibrium constant for this reaction by noting that it is the sum of Reaction 2 and the reverse of Reaction 1. By the Law of Multiple Equilibrium, K for Reaction 3 is given by the equation

$$K_3 = \frac{K_2}{K_1} = \frac{2 \times 10^{19}}{5 \times 10^{12}} = 4 \times 10^6 = \frac{[NH_3]^4}{[Cu(NH_3)_4^{2+}][OH^-]^2} \qquad \textbf{(4)}$$

From the expression in Equation 4, we can calculate that in a solution in which the NH_3 and OH^- ion concentrations are both about one molar,

$$[Cu(NH_3)_4^{2+}] = \frac{1}{4 \times 10^6} = 2.5 \times 10^{-7}\,M \qquad \textbf{(5)}$$

Since the concentration of the copper ammonia complex ion is very low, the vast majority of copper(II) in the system will exist as the solid hydroxide. In other words, the solid hydroxide is more stable under such conditions than is the ammonia complex ion. That is exactly what we observed when we treated the hydrated copper ion with ammonia and then with an equivalent amount of hydroxide ion.

From the experimental behavior of the copper ion, we can conclude that since the solid hydroxide is the species that exists when copper ion is exposed to equal concentrations of ammonia and hydroxide ion, the hydroxide is more stable under those conditions, *and* the equilibrium constant for the formation of the hydroxide is larger than the constant for the formation of the ammonia complex. So by determining which species is present when a cation is in the presence of equal ligand concentrations, we can assess relative stability under such conditions and can rank the formation constants for the possible complex ions and precipitates in order of their increasing magnitudes.

In this experiment you will carry out formation reactions for a group of complex ions and precipitates involving the Cu^{2+} ion. You can make these species by mixing a solution of $Cu(NO_3)_2$ with solutions containing NH_3 or anions, which may form either precipitates or complex ions by reaction with $Cu(H_2O)_4^{2+}$, the cation present in aqueous solutions of copper(II) nitrate. By examining whether the precipitates or complex ions formed by the reaction of hydrated copper(II) ion with a given species can, on addition of a second ligand, be dissolved or transformed to another species, you will be able to rank the relative stabilities of the precipitates and complex ions made from Cu^{2+} with respect to one another, and thus rank the equilibrium formation constants for each species in order of increasing magnitude. The species to be reacted with Cu^{2+} ion in aqueous solution are NH_3, Cl^-, OH^-, $C_2O_4^{2-}$, S^{2-}, NO_2^-, and PO_4^{3-}. In each case the test for relative stability will be made in the presence of essentially equal concentrations of the two ligands. Once you have completed your ranking of the known species, you will test an unknown species and incorporate it into your list.

Experimental Procedure

Obtain an unknown and eight small (13×100 mm) test tubes.

Add about one milliliter of 0.10 M $Cu(NO_3)_2$ solution to each of the test tubes (a half-inch depth).

To one of the test tubes add about one milliliter of 1.0 M NH_3, drop by drop. Note whether a precipitate forms initially, and if it dissolves in excess NH_3. Shake the tube sideways to mix the reagents. In the NH_3-NH_3 space in the Table of Observations on the report page, write a P in the upper left-hand corner if a precipitate is formed initially. In the rest of the space, write the formula and the color of the species that was present with an excess of NH_3. If the solution is clear, rather than cloudy, with no precipitate at the bottom, the copper ion will be in a complex. Cu(II) will always have a coordination number of 4, so the formula with NH_3 would be $Cu(NH_3)_4^{2+}$. If a precipitate is present, it will be neutral, and in the case of NH_3 it would be a hydroxide with the formula $Cu(OH)_2$. (There should, in principle, be two H_2O molecules in the formula, making it $Cu(OH)_2(H_2O)_2$, but these are usually omitted.) Add 1.0 mL of 1.0 M NH_3 to the rest of the test tubes.

Now you will test the stability of the species present in excess NH_3 relative to those which might be present with other precipitating or coordinating anions. Add, drop by drop, 1.0 mL of a 1.0 M solution of each of the anions in the horizontal row of the Table to the test tubes you have just prepared, one solution to a test tube. Note any changes that occur in the appropriate spaces in the Table. A change in color or the formation of a precipitate implies that a reaction has occurred between the added anion and the copper-containing species originally present. As before, put a P in the upper left-hand corner if a precipitate initially forms on addition of the anion solution. In the rest of the space, write the formula and color of the species present when the anion is in excess, that is, the copper-containing species that is stable in the presence of equal concentrations of NH_3 and the added anion. Again, in complex ions, Cu(II) will normally be attached to four ligands; copper(II) precipitates will be neutral. If a new copper-containing species forms on addition of the second reagent, its formula should be given. If no change occurs, the original copper-containing species is more stable, so put its formula in that space.

Repeat the above series of experiments, using 1.0 M Cl^- as the species originally added to the $Cu(NO_3)_2$ solution. In each case record the color of any precipitates or solutions formed on addition of the reagents in the horizontal row, and the formula of the copper-containing species that is most stable when an excess of both Cl^- ion and the added anion is present in the solution. Since these reactions are reversible, it is not necessary to retest Cl^- with NH_3, since the same results would be obtained as when the NH_3 solution was tested with Cl^- solution.

Repeat the series of experiments for each of the anions in the vertical column in the Table of Observations, omitting those tests where decisions as to relative stabilities are already clear. Where both anions produce copper-containing precipitates, it will be helpful to check the effect of the addition of the other anion to those precipitates. When complete, your table should have at least 28 entries.

Examine your table and decide on the relative stabilities of all the copper-containing species you observed to be present in all the spaces of the Table. There should be seven such species; rank them as best you can in order of increasing stability. There is only one correct ranking, and you should be prepared to defend your choices. Although we did not, in general, prepare the species by direct reaction of $Cu(H_2O)_4^{2+}$ with the added ligand or precipitating anion, the equilibrium formation constants of the direct formation reactions for those copper-containing species will have magnitudes that increase in the same order as do the relative stabilities you have established for the copper-containing species.

When you are satisfied that your stability ranking order for the seven copper-containing species is correct, carry out the necessary tests on your unknown to determine its proper position in the list. Your unknown may be one of the Cu(II) species you have already observed, or it may be a different species, present in an excess of its ligand or precipitating anion. If your unknown contains a precipitate, shake it well before using a portion of it to make a test.

DISPOSAL OF REACTION PRODUCTS.   Dispose of all reaction products in the waste container provided or as otherwise directed by your instructor.

Optional **Alternate Procedure Using Microscale Approach**

Obtain an unknown and two plastic well plates (4×6).

Align the plates into a 6×8 well configuration, with six wells across the top. To each of the six wells in the top row of the plate add 6 drops of 0.10 M $Cu(NO_3)_2$. Use a Beral pipet for this and all other additions of reagents. To the six wells add 6 drops of 1.0 M solutions containing the species in the horizontal row at the top of the Table of Observations; one species to a well, starting with 1.0 M NH_3 in the first well and ending with 1.0 M $NaNO_2$ in the sixth (we will omit S^{2-} and come back to it later). Mix the reagents by moving the well plate back and forth on the top of the lab bench. You will notice a difference in the appearance of the mixtures in each of the six wells. Each contains the species that is most stable when Cu(II) is mixed with one of the reagents. Report the color and the formula of those six species along the *diagonal* blocks in the Table, using the directions in the first two paragraphs in the ordinary procedure. The contents of these wells will serve as a reference during the rest of the experiment.

To establish the relative stabilities of the six species, we need to mix them with each of the other species in some systematic way and see what happens. We will first test them against NH_3. To do this, in the second row of wells make the same mixtures as you did in the first row. Then, add 6 drops of 1.0 M NH_3 to the second through sixth wells in that row (we do not need to test NH_3 with NH_3, since that would tell us nothing new). Mix the reagents, and note any changes that occurred on addition of NH_3. In some of the wells, changes will occur, indicating a reaction to form a more stable copper-containing species has occurred. Use the wells in the top row for comparison. Record your observations in the blocks in the first row of the Table, noting the color and formula of the more stable Cu(II) species in each case.

Now make stability tests for Cl^-, using wells in the third row of the plate. This time start with the third well, since we already know how Cl^- and NH_3 stack up from the last set of tests. (It doesn't make any difference if we change the order in which the reagents are added.) The four wells on the right end of the third row should be prepared in the same way as the corresponding wells in the first row. To each of those wells add 6 drops of 1.0 M NaCl, mix, note any changes, and report your observations as before in the blocks in the second row of the Table.

Continue along these lines, making stability tests with 1.0 M NaOH in the bottom row of the top plate. Report your observations as before. Then, in the top three rows of the lower well plate, make the same sort of tests with 1.0 M $K_2C_2O_4$, 1.0 M Na_2HPO_4, and 1.0 M KNO_2, respectively. When you are done, you should have entries in all of the blocks in the upper right half of the Table except for those involving S^{2-}.

To establish the stability of the Cu(II) species with respect to the sulfide ion, S^{2-}, add a drop or two of 0.10 M $(NH_4)_2S$ to each of the wells in the top row of the top plate. Although the sulfide solution is dilute, its effect on the contents of the wells, and on the nose, should be clear. Fill in the last column of the Table.

Given your entries in the Table, you should be able to decide on the relative stabilities of the seven Cu(II) species studied in this experiment. Repeat any tests that appear ambiguous, changing the order in which reagents are added if necessary. List the seven copper-containing species in order of increasing stability. Then carry out the necessary tests on your unknown to establish its position in the list. The unknown may be one of the Cu(II) species you have already studied, or it may be different. If the unknown contains a precipitate, stir it up before making a test.

DISPOSAL OF REACTION PRODUCTS. Pour the contents of all of the wells into the waste container unless directed otherwise by your instructor. ■

Experiment 27

Observations and Analysis: Relative Stabilities of Complex
Ions and Precipitates Prepared from
Solutions of Copper(II)

Table of Observations

	NH_3	Cl^-	OH^-	$C_2O_4^{2-}$	PO_4^{3-}	NO_2^-	S^{2-}	Unknown
S^{2-}								
NO_2^-								
PO_4^{3-}								
$C_2O_4^{2-}$								
OH^-								
Cl^-								
NH_3								

(continued on following page)

Determination of Relative Stabilities

In each row of the Table of Observations you can compare the stabilities of species involving the reagent in the horizontal row with those of the species containing the reagent initially added. In the first row of the Table, the copper(II)-NH_3 species can be seen to be more stable than some of the species obtained by addition of the other reagents, and less stable than others. Examining each row, make a list of all the complex ions and precipitates you have in the Table in order of increasing stability and formation constant.

Reasons

Lowest _____   _____

_____   _____

_____   _____

_____   _____

_____   _____

Highest _____   _____

Stability of Unknown

Indicate the position your unknown would occupy in the list above.

Reasons:

Unknown # _____

Experiment 27

Advance Study Assignment: Relative Stabilities of Complex
Ions and Precipitates Prepared from
Solutions of Copper(II)

1. In testing the relative stabilities of Cu(II) species using a well plate, a student adds 6 drops of 1.0 M NH_3 to 6 drops of 0.10 M $Cu(NO_3)_2$. He observes that a blue precipitate initially forms (after the first few drops of 1.0 M NH_3), but that in excess NH_3 (five or six drops) the precipitate dissolves and the solution turns deep blue. Addition of 6 drops of 1.0 M NaOH to the deep blue solution results in the formation of a blue precipitate.

   a. What is the formula of the Cu(II) species in the deep blue solution obtained with excess NH_3?

   b. What is the formula of the blue precipitate present after addition of 1.0 M NaOH?

   c. Which species is more stable in equal concentrations of NH_3 and OH^-, the one in Part 1(a) or the one in Part 1(b)?

   _____

2. Given the following two reactions and their equilibrium constants:

$$Cu(H_2O)_4^{2+}(aq) + 4\,NH_3(aq) \rightleftharpoons Cu(NH_3)_4^{2+}(aq) + 4\,H_2O(\ell) \quad K_1 = 5 \times 10^{12}$$

$$Cu(H_2O)_4^{2+}(aq) + CO_3^{2-} \rightleftharpoons CuCO_3(s) + 4\,H_2O(\ell) \qquad K_2 = 7 \times 10^9$$

   a. Evaluate the equilibrium constant for the reaction (see discussion):

$$Cu(NH_3)_4^{2+}(aq) + CO_3^{2-}(aq) \rightleftharpoons CuCO_3(s) + 4\,NH_3(aq)$$

   _____

   b. When 1.0 M concentrations of all dissolved species are present, which is more stable, $CuCO_3$ or $Cu(NH_3)_4^{2+}$? Explain your reasoning.

Experiment 28

Determination of the Hardness of Water

One of the factors that establishes the quality of a water supply is its degree of hardness. The hardness of water is defined in terms of its content of calcium and magnesium ions. Since the analysis does not distinguish between Ca^{2+} and Mg^{2+}, and since most hardness is caused by carbonate deposits in the earth, hardness is usually reported as total parts per million calcium carbonate by weight. A water supply with a hardness of 100 parts per million would contain the equivalent of 100 g of $CaCO_3$ in one million grams of water, or 0.1 g in 1 L of water. In the days when soap was more commonly used for washing clothes, and when people bathed in tubs instead of using showers, water hardness was more often directly observed than it is now, since Ca^{2+} and Mg^{2+} form insoluble salts with soap, precipitating as a scum that sticks to clothes or to the bathtub. Detergents have the distinct advantage of being effective in hard water, and this is really what allowed them to displace soaps for laundry purposes.

Water hardness can be readily determined by titration with the chelating agent EDTA (ethylenediaminetetraacetic acid). This reagent is a weak acid that can lose four protons on complete neutralization; its structural formula is

$$HOOC-CH_2 \diagdown CH_2-COOH$$
$$N-CH_2-CH_2-N$$
$$HOOC-CH_2 \diagup CH_2-COOH$$

The four acid sites and the two nitrogen atoms all contain unshared electron pairs, so that a single EDTA ion can form a complex with up to six sites on a given cation. The complex is typically stable, and the conditions of its formation can ordinarily be controlled so that it contains EDTA and the metal ion in a 1:1 mole ratio. In a titration to establish the concentration of a metal ion, the EDTA which is added combines quantitatively with the cation to form the complex. The endpoint occurs when essentially all of the cation has reacted.

In this experiment we will standardize a solution of EDTA by titration against a standard solution made from calcium carbonate, $CaCO_3$. We will then use the EDTA solution to determine the hardness of an unknown water sample. Since both EDTA and Ca^{2+} are colorless, it is necessary to use a rather special indicator to detect the endpoint of the titration. The indicator we will use is called Eriochrome Black T, which forms a rather stable wine-red complex, $MgIn^-$, with the magnesium ion. A tiny amount of this complex will be present in the solution during the titration. As EDTA is added, it will complex free Ca^{2+} and Mg^{2+} ions, leaving the $MgIn^-$ complex alone until essentially all of the calcium and magnesium have been converted to chelates (bound to EDTA). At this point the EDTA concentration will increase sufficiently to displace Mg^{2+} from the indicator complex; the indicator reverts to an acid form, which is sky blue, and this establishes the endpoint of the titration.

The titration is carried out at a pH of 10, in an NH_3/NH_4^+ buffer, which keeps the EDTA (H_4Y) mainly in the triply deprotonated form, HY^{3-}, where it complexes the Group 2 ions very well but does not tend to react as readily with other cations, such as Fe^{3+}, that might be present as impurities in the water. Taking HY^{3-} and HIn^{2-} as the formulas for EDTA and Eriochrome Black T, respectively, the equations for the reactions which occur during the titration are:

$$\text{(main reaction) } HY^{3-}(aq) + Ca^{2+}(aq) \rightarrow CaY^{2-}(aq) + H^+(aq)$$

(The Mg^{2+} ion undergoes an analogous reaction, though it is less
favorable than that for Ca^{2+}, such that Ca^{2+} is consumed first)

$$\text{(at endpoint) } HY^{3-}(aq) + MgIn^-(aq) \rightarrow MgY^{2-}(aq) + HIn^{2-}(aq)$$
$$\text{wine red} \text{sky blue}$$

Since the indicator requires a trace of Mg^{2+} to operate properly, we will add a little magnesium ion to each solution and titrate it as a blank.

Experimental Procedure

Obtain a 50-mL buret, a 250-mL volumetric flask, and 25- and 50-mL pipets.

Put a bit more than half a gram of calcium carbonate in a small 50-mL beaker and weigh the beaker and contents on the analytical balance. Using a spatula, transfer 0.4 g (between 0.3 g and 0.5 g) of the carbonate to a 250-mL beaker and weigh the small 50-mL beaker again, precisely determining the mass of the $CaCO_3$ sample in the 250-mL beaker by difference.

Add 25 mL of deionized water to the large beaker and then, *slowly*, about forty drops of 6 M HCl. Cover the beaker with a watch glass and allow the reaction to proceed until all of the solid carbonate has dissolved. Rinse the walls of the beaker down with deionized water from your wash bottle and heat the solution until it just begins to boil. (Note that dissolved gases, particularly CO_2, will evolve from the solution as it is heated to boiling. Make sure you actually boil the water: don't stop just because you see these escaping gas bubbles!) Add 50 mL of deionized water to the beaker and carefully transfer the solution, using a stirring rod as a pathway, to the volumetric flask. Rinse the beaker several times with small portions of deionized water and transfer each portion to the flask. All of the Ca^{2+} originally in the beaker should then be in the volumetric flask. Fill the volumetric flask to the horizontal mark with deionized water, adding the last few milliliters 1 drop at a time. Stopper the flask and mix the solution thoroughly by inverting the flask at least twenty times over a period of several minutes.

Clean your buret thoroughly and draw about two-hundred milliliters of stock EDTA solution into a dry 250-mL Erlenmeyer flask. Rinse the buret with a few milliliters of the solution at least three times. Drain through the stopcock and then fill the buret with the EDTA solution.

Determine a blank by adding 25 mL deionized water and 5 mL of the pH 10 buffer to a 250-mL Erlenmeyer flask. Add a small amount of solid Eriochrome Black T indicator mixture from the stock bottle. You need only a small portion, about twenty-five milligrams, just enough to cover the end of a small spatula. The solution should turn blue; if the color is weak, add a bit more indicator. Add 15 drops of 0.03 M $MgCl_2$, which should contain enough Mg^{2+} to turn the solution wine red. Read the buret to ± 0.01 mL and record the reading on the report page, then add EDTA from the buret to the solution until the last tinge of purple disappears. The color change is rather slow, so titrate slowly near the endpoint. Only a few milliliters will be needed to titrate the blank. Read the buret again to determine the volume of titrant required for the blank, recording it in the report page. The blank is the volume of EDTA necessary to cause the indicator to change color. When there is calcium present in a solution, the EDTA reacts first with the calcium, and then, when all the calcium is gone, it begins to react with the indicator. The total volume of EDTA required is the sum of the amount needed to react with the calcium and that required to change the color of the indicator. Therefore the blank volume must be subtracted from the total EDTA volume used in each titration that determines calcium content, and the report page will help you remember to do so. Save the blank solution as a color reference for the endpoint in all your titrations.

Pipet three 25.0-mL portions of the Ca^{2+} solution in the volumetric flask into clean 250-mL Erlenmeyer flasks. To each flask add 5 mL of the pH 10 buffer, a small amount of indicator, and 15 drops of 0.03 M $MgCl_2$. Titrate the solution in one of the flasks until its color matches that of your reference solution; the endpoint is a reasonably good one, and you should be able to hit it within a few drops if you are careful. Read the buret and record the actual volumes you read on the report page; you will correct for the blank later. Refill the buret, read it, and record the volume on the report page; then proceed to titrate the second solution, then the third.

Your instructor will furnish you with a sample of water for hardness analysis. Since the concentration of Ca^{2+} is probably lower than that in the standard calcium solution you prepared, pipet 50.0 mL of the water sample for each titration. As before, add some indicator, 5 mL of pH 10 buffer, and 15 drops of 0.03 M $MgCl_2$ before titrating. Carry out as many titrations as necessary to obtain two volumes of EDTA that agree within 3%. If the volume of EDTA required in the first titration is low due to the fact that the water is not very hard, increase the volume of the water sample so that in succeeding titrations, it takes at least 20 mL of EDTA to reach the endpoint.

Take it Further (Optional): Find the mass percent calcium carbonate in an antacid tablet or a sample of limestone.

DISPOSAL OF REACTION PRODUCTS. The chemical waste from this experiment may be diluted with water and poured down the sink.

Experiment 28

Data and Calculations: Determination of the Hardness of Water

Mass of beaker
plus $CaCO_3$ _____ g

Volume Ca^{2+}
solution prepared   _____ mL

Mass of beaker
less sample   _____ g

Molarity of Ca^{2+} _____ M

Mass of $CaCO_3$
sample   _____ g

Moles Ca^{2+} in each
aliquot titrated   _____ moles

Number of moles $CaCO_3$ in sample
(Formula mass $= 100.1$ g/mol)   _____ moles

Standardization of EDTA Solution

Determination of blank:

Initial buret
reading   _____ mL

Final buret
reading   _____ mL

Volume
of blank   _____ mL

Titration	I	II	III
Initial buret reading	_____ mL	_____ mL	_____ mL
Final buret reading	_____ mL	_____ mL	_____ mL
Volume of EDTA	_____ mL	_____ mL	_____ mL
Volume of EDTA used to titrate blank	_____ mL	_____ mL	_____ mL
Volume of EDTA used to titrate Ca^{2+}	_____ mL	_____ mL	_____ mL

Molarity of EDTA $=$

$\dfrac{\text{moles } Ca^{2+} \text{ in aliquot} \times 1000}{\text{volume of EDTA required (mL)}}$ _____ M   _____ M   _____ M

Mean molarity of EDTA: _____ M   Standard deviation in molarity of EDTA: _____ M

(continued on following page)

Determination of Water Hardness

Unknown # of water sample _____

Titration	I	II	III
Volume of water used	_____ mL	_____ mL	_____ mL
Initial buret reading	_____ mL	_____ mL	_____ mL
Final buret reading	_____ mL	_____ mL	_____ mL
Volume of EDTA	_____ mL	_____ mL	_____ mL
Volume of EDTA used to titrate blank	_____ mL	_____ mL	_____ mL
Volume of EDTA required to titrate water	_____ mL	_____ mL	_____ mL
Moles of EDTA per liter of water = moles of $CaCO_3$ per liter of water	_____ mol/L	_____ mol/L	_____ mol/L
Grams of $CaCO_3$ per liter of water	_____ g/L	_____ g/L	_____ g/L
Water hardness (1 ppm = 1 mg/L)	_____ ppm	_____ ppm	_____ ppm

Mean water hardness _____ ppm   Standard deviation in water hardness _____ ppm

Experiment 28

Advance Study Assignment: Determination of the Hardness of Water

1. A 0.4943-g sample of $CaCO_3$ is dissolved in 12 M HCl and the resulting solution is diluted to 250.0 mL in a volumetric flask.

   a. How many moles of $CaCO_3$ are used (formula mass = 100.1 g/mol)?

   _____ moles

   b. What is the molarity of the Ca^{2+} in the 250.0 mL of solution?

   _____ M

   c. How many moles of Ca^{2+} are in a 25.00-mL aliquot of the solution in 1(b)?

   _____ moles

2. 25.00-mL aliquots of the solution from Problem 1 are titrated with EDTA to the Eriochrome Black T endpoint. A blank containing a small measured amount of Mg^{2+} requires 2.77 mL of the EDTA to reach the endpoint. An aliquot to which the same amount of Mg^{2+} is added requires 39.04 mL of the EDTA to reach the endpoint.

   a. How many milliliters of EDTA are needed to titrate the Ca^{2+} ion in the aliquot?

   _____ mL

   b. How many moles of EDTA are there in the volume obtained in 2(a)?

   _____ moles

   c. What is the molarity of the EDTA solution?

   _____ M

(continued on following page)

3. A 100.-mL sample of hard water is titrated with the EDTA solution in Problem 2. The same amount of Mg^{2+} is added as previously, and the volume of EDTA required is 31.84 mL.

a. What volume of EDTA is used in titrating the Ca^{2+} in the hard water?

_____ mL

b. How many moles of EDTA are there in that volume?

_____ moles

c. How many moles of Ca^{2+} are there in the 100. mL of water?

_____ moles

d. If the Ca^{2+} comes from $CaCO_3$, how many moles of $CaCO_3$ are there in one liter of the water? How many grams of $CaCO_3$ are present per liter of the water?

_____ mol/L

_____ g/L

e. If 1 ppm = 1 mg per liter, what is the water hardness in ppm $CaCO_3$?

_____ ppm $CaCO_3$

Experiment 29

Synthesis and Analysis of a Coordination Compound

Some of the most interesting research in inorganic chemistry has involved the preparation and properties of coordination compounds. These compounds, sometimes called *complexes*, are typically salts that contain complex ions. A complex ion contains a central metal atom to which are bonded small polar molecules or anions, called *ligands*. The bonding in the complex is through coordinate covalent bonds, bonds in which the electrons are all furnished by the ligands.

Some coordination compounds can be prepared stoichiometrically pure. That is, the atoms in the compound are present in the exact ratio given by the formula. Many supposedly pure compounds do not meet this criterion, since they contain variable amounts of water, either in hydrates or adsorbed. In this experiment we will be making a compound that is indeed stoichiometrically pure. It is not a hydrate nor does it tend to adsorb water.

The compound we will be preparing contains cobalt ion, ammonia, and chloride ions. Its formula is of the form $Co_x(NH_3)_yCl_z$. Once you have made the compound, you may, depending on the time available, analyze it for its content of one or more of these species and so determine x, y, and/or z.

In many reactions involving complex ion formation the rate of reaction is very fast, so that the thermodynamically stable form of the complex is the one produced. By changing ligand concentrations, we can quickly convert from one complex to another. In Experiment 27, in which many Cu(II) complexes are prepared, all of the complex ions undergo rapid reaction, exchanging ligands very readily. Such complexes are called *labile*. With Cu(II) ion and NH_3 in solution, we can, depending on the NH_3 concentration, have one, two, three, or four NH_3 molecules in the complex, in an equilibrium mixture.

Most, but by no means all, complexes are labile. Some, including the one you will be making in this experiment, exchange ligands only very slowly. For such species, the preparation reaction takes time, but once formed, the complex ion in solution may resist change rather dramatically. Such complexes are called *inert*.

In the complex ion you will be making, NH_3 and Cl^- may be ligands. Since the overall charge of the compound must be zero, the number of moles of Cl^- ion in a mole of compound will be determined by the charge on the cobalt ion. Some of the Cl^- ions may be in the complex and some may simply be anions in the crystal. When the compound is dissolved in water, the free anions will go into solution, but the ones in the complex will remain firmly attached to cobalt. They will not react with a precipitating agent such as Ag^+ ion, which will tend to form AgCl with any free Cl^- ions. All of the NH_3 molecules will be in the complex ion, and will not react with added H^+ ion, as they would if they were free. To ensure that we find all the Cl^- and NH_3 that is present in the compound, we will destroy the complex (break all of its coordinate covalent bonds) before attempting any analyses.

The complete analysis involves a gravimetric procedure for chloride ion, a colorimetric method for cobalt ion, and a volumetric procedure for ammonia. In some of the earlier experiments we have used these techniques (Exps. 4, 7, 23, 24, and 28), so you may be somewhat familiar with them. From the analyses we can determine the number of moles of each species present in a mole of compound. Because the mole ratio must equal the atom or molecule ratio, we can determine the formula of the compound. Although we describe the procedures for analysis of all three species in the compound, the formula can be determined (albeit with less certainty) from only two analyses, since the amount of the third species can be calculated by difference.

The synthesis of the coordination compound will take one lab period. The analysis for both chloride and cobalt content can be done in one period. Determining the amount of ammonia present in the complex will take most of one period.

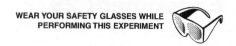
WEAR YOUR SAFETY GLASSES WHILE
PERFORMING THIS EXPERIMENT

Experimental Procedure

Carry out the procedure in either A.1 or A.2, as directed by your instructor.

A.1 Preparation of $Co_x(NH_3)_yCl_z$ (Procedure I)

Using a top-loading balance, weigh out 10 ± 0.2 g of ammonium chloride, NH_4Cl, into a 250-mL beaker and add 40 mL of deionized water. To this add 8 ± 0.2 g of cobalt(II) chloride hexahydrate, $CoCl_2 \cdot 6H_2O$, and 0.8 g activated charcoal (Norite). Heat this mixture to the boiling point, stirring to dissolve the soluble components.

Cool the beaker in water, then in an ice bath. Slowly, **in the hood**, add 40 mL of 15 M NH_3, ammonia. **CAUTION:** **This is a concentrated basic reagent with a strong, pungent odor that can knock you unconscious.** After stirring, add 50 mL of $10.\%_{mass}$ H_2O_2, hydrogen peroxide, slowly, a few milliliters at a time, while continuing to stir. **CAUTION:** **Be very careful not to spill this reagent on your skin. Use gloves.**

When the bubbling has stopped, place the 250-mL beaker in a 600-mL beaker containing about one-hundred milliliters of water at $60 \pm 10°C$ (between 50°C and 70°C). Leave the beaker in the water bath for 30 to 40 minutes, holding the temperature of the bath at $60 \pm 10°C$ and stirring occasionally.

While the mixture is reacting, prepare a fritted glass filter crucible. Fit the crucible on a filter flask; clean the crucible by pulling, under suction, about fifteen milliliters of 1.0 M HNO_3 through the disk, followed by 50 to 75 mL of deionized water. Put the crucible in an oven at 150°C for about an hour to dry. Prepare a second crucible the same way, and put it in the oven.

Once 30 to 40 minutes have elapsed, remove the 250-mL beaker from the water bath and cool it, first in a cool-water bath (use cold tap water) and then in an ice bath, until the liquid in the 250-mL beaker has a temperature below 5°C. Hold the mixture at this temperature for at least five minutes, stirring occasionally to promote crystallization of the crude product. Set up a Büchner funnel and, with suction, filter the mixture through the filter paper in the funnel. Discard the filtrate in the waste container.

Scrape the solid from the filter paper into the 250-mL beaker. Add about one-hundred milliliters of deionized water and **then, in a hood**, 5 mL of 12 M HCl. **CAUTION:** **This reagent is a concentrated acid with a choking odor**. Heat the mixture to boiling, with stirring, to dissolve the crystals. Rinse out the suction flask with deionized water, and reassemble the Büchner funnel. Holding the beaker with a pair of tongs or a folded paper towel, filter the hot mixture through the paper in the funnel, under suction. In this operation, the charcoal is removed and remains behind on the filter paper. The filtrate should be golden yellow—it contains the product we seek.

Transfer the filtrate to the (rinsed-out) 250-mL beaker and, **in the hood**, add 15 mL of 12 M HCl. Place the beaker in an ice bath and stir for five minutes or so to promote formation of the golden crystals of the product, $Co_x(NH_3)_yCl_z$. Check to ensure that the temperature is below 5°C.

Pour about fifty milliliters of ice-cold deionized water onto a piece of filter paper in the Büchner funnel. After two or three minutes, turn on the suction and pull the water into the suction flask. Empty the flask. Then filter the cold mixture containing the product through the funnel, using suction. Turn the suction off, and pour about twenty milliliters of 95% ethanol onto the crystals. Wait about ten seconds, and then turn on the suction to pull through the ethanol, which should carry with it most of the water and HCl remaining on the crystals. Draw air through the crystals for several minutes. Weigh a 100-mL beaker to ± 0.1 g. Transfer the crystals from the filter paper to the beaker. Put the beaker in your locker. Take the fritted glass crucibles out of the oven, using tongs or a folded paper towel, and put the crucibles on a paper towel in your locker.

DISPOSAL OF REACTION PRODUCTS. Discard the liquid in the suction flask in the waste container or as directed by your instructor.

A.2 Preparation of $Co_x(NH_3)_yCl_z$ (Procedure II)

There are several possible $Co_x(NH_3)_yCl_z$ compounds. This procedure will allow you to make a different one.

Using top-loading balance, weigh out 4.0 ± 0.1 g of ammonium chloride, NH_4Cl, into a 250-mL beaker, and **in the hood,** slowly add 25 mL of 15 M NH_3. **CAUTION:** **This is a concentrated basic reagent, with a strong, pungent odor that can knock you unconscious.** Cover the beaker with a watch glass. Add to the solution 8.0 ± 0.2 g of $CoCl_2 \cdot 6H_2O$, cobalt(II) chloride hexahydrate, while stirring the mixture with a stirring rod. Continue stirring until most of the solid has dissolved and a slurry of fine tan solid in a deep red solution results.

Measure out 20 mL of $10.\%_{mass}$ hydrogen peroxide, H_2O_2. **CAUTION:** **This reagent can cause severe skin burns; use gloves when handling it. In the hood,** add the H_2O_2 to the mixture in 1- to 2-mL portions while stirring. There will be some effervescence (bubbling) as oxygen gas is evolved. Stir until the bubbles of oxygen decrease significantly and all of the solid material dissolves. The initial Co^{2+} has now been oxidized to Co^{3+} and is present in the form of a complex ion. **In the hood,** measure out 25 mL of 12 M HCl. **CAUTION:** **This is a concentrated acidic reagent with a choking odor.** Add the HCl in approximately 5-mL increments while stirring the mixture. The white cloud that forms above the beaker as you do this is solid ammonium chloride, formed by a gaseous acid-base neutralization reaction, dispersed in the air.

Preferably in the hood, or at your lab bench, heat the mixture in the beaker to 75 to 85°C for about thirty minutes, stirring occasionally. Keep the beaker covered with your watch glass when not stirring. A deep purple solid should gradually form in the beaker. This is your product, which may need to be recrystallized.

While the mixture is being heated, prepare two fritted glass crucibles, according to the directions in the fourth paragraph of Part A.1 of this experiment.

Remove the beaker from the heat after 30 minutes and allow it to cool in a cool-water bath—a 400-mL beaker containing 150 mL of cold tap water—for 2 minutes. Set up a vacuum filter apparatus, using a two-part Büchner funnel. Transfer the smaller (reaction) beaker to an ice bath. Keep the reaction beaker in the ice bath until the solution is colder than 10°C.

Using #1 filter paper that has been moistened while in the Büchner funnel, filter the solution, as demonstrated by your instructor. It is not necessary to transfer all of the solid from the reaction beaker. Turn off the vacuum, then knock or scrape the contents of the filter cup, including the paper, back into the 250-mL beaker. Rinse the cup with a few milliliters of water from your wash bottle. Similarly, using tweezers, rinse any remaining solid from the filter paper into the beaker, and dispose of the paper as directed by your instructor. The filtrate solution should be transferred to a designated waste container.

It may be desirable to recrystallize your product to improve its purity. If so advised by your lab instructor, and time is available, proceed as follows:

Add water to bring the volume up to 125 mL, using the markings on the reaction beaker. Stir the solution and take the beaker to the hood. **In the hood,** add 5 mL of 15 M NH_3 to the beaker, in small portions, while stirring. Heat gently, if necessary, to dissolve all of the solid. Add 15 mL of 12 M HCl, in small portions, to the reaction beaker, while stirring. Preferably in the hood, or at your lab bench, heat the solution for 30 minutes as before but now *at approximately 60°C*, stirring occasionally and using a watch glass to cover the beaker. At the end of the heating period, place the beaker in a cool-water bath (as above) for 2 minutes. Cool further in an ice bath with occasional stirring, until the temperature falls below 10°C, and for an additional 5 minutes. You have now recrystallized the desired cobalt-containing product, which should be a violet color.

Reassemble your Büchner filter apparatus. To transfer as much of the product as possible, stir the mixture with your stirring rod just before quickly pouring out small portions of the suspension. Use a minimal amount of water from your wash bottle to transfer the remaining solid. Wash the product on the funnel with two 5-mL portions of water to aid in removing ammonium chloride. Wash with 10. mL of ethanol. Draw air through your product for 5 minutes, while weighing a watch glass to ± 0.01 g on a top-loading balance.

Transfer the crystals to the watch glass and scrape any residual product from the filter paper onto the watch glass. Weigh the watch glass again and calculate your yield. Place the watch glass in your drawer until next lab period, at which time you will be analyzing your product. Take the fritted glass crucibles out of the oven, using tongs or a folded paper towel. Put the crucibles, after cooling, in your drawer on a paper towel.

DISPOSAL OF REACTION PRODUCTS. Discard the liquid in the suction flask into the waste container or as otherwise directed by your instructor.

B. Gravimetric Determination of Chloride Content

Weigh the dried product of your synthesis in the 100-mL beaker to ± 0.0001 g on an analytical balance. (If a week has gone by, the compound should be well dried. If you are attempting to proceed on the same day you prepared the compound, you must dry it thoroughly. This could be done by putting the sample under vacuum, while keeping the sample warm. If the sample does not lose 0.0001 g in 2 minutes while on the analytical balance, you may presume that it is dry.)

Transfer 0.3 ± 0.05 g of the compound from the beaker to a labeled 250-mL beaker, and reweigh the 100-mL beaker to ± 0.0001 g. Transfer a second sample of about the same size to another labeled 250-mL beaker, and again weigh the 100-mL beaker and the remaining compound accurately. Record all masses on the report page. Cover the 100-mL beaker with a watch glass and put it in your locker.

Add 25 mL of deionized water to the sample in one of the 250-mL beakers. Add 5 mL of 6 M NaOH and stir the mixture until all of the solid has dissolved. Heat the beaker and its contents to boiling, and simmer for 3 minutes, stirring occasionally. This step will destroy the complex; the mixture will turn black because of the formation of Co_3O_4.

Once the black Co_3O_4 is formed, turn off the heat and let the beaker cool for a few minutes. Then add 6 M HNO_3 until the mixture is acidic to litmus (it will take about five milliliters). Add 1.0 mL more of the HNO_3. Add a small scoop of solid Na_2SO_3, sodium sulfite (approximately 0.2 g) and stir; this will reduce the Co_3O_4 and produce the Co^{2+} ion. The solution should clarify and turn pink. Wash down the walls of the beaker with water to bring any unreacted dark material into contact with the solution. Reheat to boiling and simmer gently for a minute or so. Then add 75 mL of deionized water.

Slowly, with stirring, add 50 mL of 0.10 M $AgNO_3$, silver nitrate. This will precipitate all of the chloride ion in the solution as silver chloride, AgCl. Heat the mixture, with occasional stirring, to the boiling point. This will help coagulate the AgCl crystals and so facilitate filtration. Remove the beaker from the heat and cover it with a watch glass. Let it cool for ten minutes or so.

While the beaker is cooling, take one of the fritted glass crucibles you prepared in the previous session from your locker and weigh it to ± 0.0001 g on the analytical balance. Then set up the filter crucible apparatus.

Carefully, with suction on, transfer *all* of the AgCl precipitate to the filter crucible. Use your wash bottle and rubber policeman to complete the transfer. Wash the AgCl, with suction on, with about ten milliliters of deionized water, followed by 5 mL of 6 M HNO_3, followed by approximately twenty more milliliters of deionized water. Keep the suction on for at least one minute after the last of the liquid has been pulled through the filter.

Put the crucible and its contents in an oven at 150°C for at least an hour. Then let it cool to room temperature in a covered beaker. When it is cool, weigh it to ± 0.0001 g on an analytical balance. Put the solid AgCl in the AgCl container.

Repeat the entire chloride analysis procedure, using the second sample of compound.

C. Colorimetric Determination of Cobalt Content

Transfer 0.5 ± 0.05 g of your compound into a clean, dry 50-mL beaker, using the same procedure as in the chloride determination. Weighings should be made to ± 0.0001 g. Make another transfer of 0.5 ± 0.05 g into another 50-mL beaker, again weighing the 100-mL beaker and its contents accurately.

Cover each beaker with a watch glass and heat one of them gently with the Bunsen burner until the solid melts, foams, and turns blue. In this process the complex is destroyed and cobalt ion is freed from the ligands. Allow the beaker and its contents to cool. Repeat this procedure with the sample in the other beaker.

To the solid in the first beaker add 5 mL of deionized water and **then** 5 mL of 6 M H_2SO_4. Heat the beaker until the liquid begins to boil and the solid is all dissolved. Wash any deposits on the watch glass into the beaker with a few milliliters of water from your wash bottle. Repeat these steps with the sample in the other beaker.

Transfer the solution from the first beaker to a clean 25-mL volumetric flask. Rinse the beaker with small portions of deionized water, adding the washings to the volumetric flask, so that all of the solution goes into the flask. Fill the flask to the mark with deionized water from your wash bottle. Stopper the flask and invert it at least twenty times to ensure that the solution inside becomes thoroughly mixed.

Pour some of the solution in the flask into a spectrophotometric test tube (three-fourths full). Measure and record the absorbance of the solution at 510 nm. From the absorbance and the calibration curve furnished by your instructor, determine the molarity of cobalt ion in the solution.

DISPOSAL OF REACTION PRODUCTS. Pour the rest of the solution in the volumetric flask in the waste container.

Rinse the flask, and use it with the solution you prepared in the other beaker. Treat that solution as you did the first one, recording its absorbance and molarity of cobalt ion on the report page.

D. Volumetric Determination of Ammonia Content

In this part of the experiment we will first decompose the complex, releasing NH_3 to the solution, as we did in the chloride analysis. We then distill off the NH_3 into another container, and titrate it against an acid. The procedure is called a Kjeldahl analysis, and is used for determining nitrogen content of many organic and inorganic substances.

Assemble the distillation apparatus shown in Figure 29.1. The details of the apparatus will vary from school to school, but you will need a 250-mL distilling flask, a distilling head, a condenser, and a receiver adapter, in addition to a 125-mL Erlenmeyer flask, which will serve as a receiving flask. The 250-mL flask should be on a wire gauze on an iron ring. The flask and condenser should be held with clamps, which are to be adjusted so that all joints are tight. On the end of the receiver adapter attach a 4-inch length of latex tubing, which should reach to the bottom of the receiving flask.

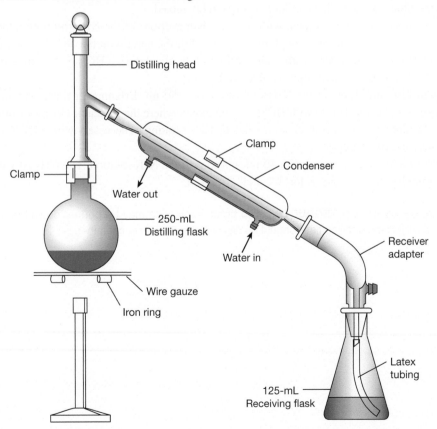

Figure 29.1 The purpose of this apparatus, used in the volumetric determination of nitrogen, is to convert the nitrogen in the analyte into ammonia gas, condense it into some water that co-distills with it, and finally trap that ammonia as ammonium ion in the acid in the receiving flask.

Put 50 mL of saturated boric acid solution in the 125-mL receiving flask; add 5 drops of bromocresol green indicator and adjust the level of the flask, if necessary, so that the latex tubing is well under the liquid surface.

Weigh out 1.0 ± 0.1 g of your compound into a beaker, as you did earlier, making the weighing to ± 0.0001 g. Disconnect the distilling flask from the head. Transfer the sample to the distilling flask, using a funnel. Rinse the beaker several times with small portions of deionized water, and add the washings to the flask. All the sample must end up in the flask. Add deionized water as necessary to give a final volume of about fifty milliliters. Swirl the flask to dissolve the sample. Add a few boiling chips and several pieces of granulated zinc. Start water flowing through the condenser, slowly. If you are working with standard taper glassware, lightly grease the lower joint of the distilling head.

Pour 50 mL of deionized water and then 40 mL of 6 M NaOH into the flask and reconnect it promptly to the distilling head. Make sure that all joints are tight, and that the top of the distilling head is stoppered. Turn on the Bunsen burner and bring the liquid in the flask to a boil. The complex will be destroyed, the liquid will turn dark as Co_3O_4 forms, and NH_3 will be driven from the solution into the receiving flask.

Adjust the burner as necessary to maintain smooth boiling and minimize bumping. The color of the indicator will change as the NH_3 is absorbed. There may be a tendency for liquid to rise in the latex tubing during the distillation; this is not serious, but if the level goes up more than a few inches, you can let in a little air by momentarily loosening one of the joints. When the volume of liquid in the receiving flask reaches 100 mL, 50 mL of the solution will have been distilled and essentially all of the NH_3 driven into the boric acid.

With the burner still heating the flask, disconnect the receiver adapter from the condenser. (If the heat is turned off first, liquid will tend to back up into the condenser.) Rinse the adapter with the tubing still in the flask, washing any liquid on the inner or outer surface into the distillate. Then turn the burner off. Pour the distillate into a 250-mL volumetric flask. Rinse the receiving flask with a few small portions of deionized water, adding the washings to the volumetric flask. Fill to the mark with deionized water. Stopper the flask and invert it at least twenty times to ensure that the liquid is thoroughly mixed.

Clean two burets. Rinse one of them with several small portions of the NH_3 solution in the volumetric flask, draining some of the solution through the stopcock. Fill the buret with the NH_3 solution and read and record the level. Rinse the other buret with small portions of the standardized HCl solution from the stock supply, and then fill that buret with that solution. Record the level.

Draw about forty milliliters of the NH_3 solution into a 250-mL Erlenmeyer flask, and add 5 drops of bromocresol green indicator. Titrate with HCl to the endpoint, where the indicator changes from blue to yellow. The actual endpoint is green and can be reached by back titrating as necessary with NH_3. Record the final levels in the NH_3 and HCl burets.

Repeat the titration once or twice with 40-mL portions of the NH_3 solution. The NH_3-HCl volume ratios from the two samples should agree within 1% if all goes well.

DISPOSAL OF REACTION PRODUCTS. The titrated solutions may be poured down the sink. The liquid remaining in the distilling flask should be discarded in the waste container.

Experiment 29

Data and Calculations: Synthesis and Analysis of a
Coordination Compound

A.1 or A.2 Preparation of $Co_x(NH_3)_yCl_z$

Procedure used _____

Mass of $CoCl_2 \cdot 6H_2O$ _____ g

Mass of 100-mL beaker   _____ g

Approximate mass of product (see Part B, line 1)   _____ g

B. Determination of Chloride Ion

Mass of 100-mL beaker plus product   _____ g

Mass after first transfer   _____ g   Mass of sample I   _____ g

Mass after second transfer   _____ g   Mass of sample II   _____ g

Mass of first filter crucible   _____ g

Mass of second filter crucible   _____ g

Mass of first crucible plus AgCl   _____ g   Mass of AgCl I   _____ g

Mass of second crucible plus AgCl   _____ g   Mass of AgCl II   _____ g

C. Determination of Cobalt Ion

Mass of 100-mL beaker plus product   _____ g

Mass after first transfer   _____ g   Mass of sample I   _____ g

Mass after second transfer   _____ g   Mass of sample II   _____ g

Absorbance of first solution   _____

Molarity of Co ion   _____ M

Absorbance of second solution   _____

Molarity of Co ion   _____ M

D. Determination of Ammonia

Mass of 100-mL beaker plus product   _____ g

Mass after transfer   _____ g   **2nd Trial**

Initial reading NH_3 buret   _____ mL   _____ mL

Final reading NH_3 buret   _____ mL   _____ mL

Initial reading HCl buret   _____ mL   _____ mL

Final reading HCl buret   _____ mL   _____ mL

Molarity of standardized HCl   _____ M

(continued on following page)

E. Calculation of Chloride Content

	Sample I	**Sample II**
Mass of AgCl	_____ g	_____ g
Moles of Cl^- in AgCl	_____ mol	_____ mol
Mass of sample	_____ g	_____ g
Moles of Cl^- per gram of sample	_____ mol/g	_____ mol/g

F. Calculation of Cobalt Content

	Sample I	**Sample II**
Molarity of cobalt ion	_____ M	_____ M
Moles cobalt in 25 mL = moles cobalt in sample	_____ mol	_____ mol
Mass of sample	_____ g	_____ g
Moles cobalt per gram of sample	_____ mol/g	_____ mol/g

G. Calculation of Ammonia Content

Mass of sample _____ g   Molarity of HCl _____ M

	Trial I	**Trial II**
Volume of NH_3 used	_____ mL	_____ mL
Volume of HCl used	_____ mL	_____ mL
Moles HCl = moles NH_3	_____ mol	_____ mol
Moles NH_3 per 250-mL NH_3 solution = moles NH_3 in sample	_____ mol	_____ mol
Moles NH_3 per gram of sample	_____ mol/g	_____ mol/g

H. Determination of the Formula of $Co_x(NH_3)_yCl_z$

For a gram of sample: Moles Cl^- _____   Moles Co ion _____   Moles NH_3 _____

Whole number ratio: $z =$ _____ $x =$ _____ $y =$ _____

Formula of compound:

Experiment 29

Advance Study Assignment: Synthesis and Analysis of a
Coordination Compound

A student prepared 7.2 g of $Co_x(NH_3)_yCl_z$. She then analyzed the compound by the procedure in this experiment.

A. In the gravimetric determination of chloride, she weighed out a 0.2988-g sample of the product compound. The following data were obtained:

Mass of crucible plus AgCl	19.0020 g
Mass of crucible	18.5228 g
Mass of AgCl	_____ g MM AgCl = 143.323 g/mol

Moles of Cl^- in AgCl   _____ mol

(Note that this is the moles of Cl present in the 0.2988 g of product compound that the student analyzed.)

Moles Cl per gram of sample   _____

B. In the colorimetric determination of cobalt, she used a sample weighing 0.4781 g. The molarity of cobalt ion in the solution from the volumetric flask was 0.071 M.

Moles of cobalt ion in 25 mL solution = moles of cobalt in sample   _____

Moles of cobalt per gram of sample   _____

C. In the volumetric determination of ammonia, the sample weighed 0.9985 g. In the titration, 0.1000 M HCl was used. She found that 40.92 mL of the NH_3 solution required 35.77 mL of the HCl to reach the endpoint.

Moles of HCl used   _____ = moles of NH_3 in 40.92 mL of NH_3 solution

Moles of NH_3 in 250.0 mL NH_3 solution   _____ = moles of NH_3 in 0.9985 g of product compound

Moles of NH_3 per gram of sample   _____

D. Calculation of the formula of $Co_x(NH_3)_yCl_z$:

In 1 g of the product compound
(1 g of sample) there are:

moles Co ion = _____

moles NH_3 = _____

moles Cl^- = _____

Dividing by smallest number,

moles Co ion _____

moles NH_3 _____

moles Cl^- _____

Formula of compound _____

(This may or may not be the formula of the compound you will be making!)

Experiment 30

Determination of Iron by Reaction with Permanganate—A Redox Titration

Potassium permanganate, $KMnO_4$, is widely used as an oxidizing agent in volumetric analysis. In acidic solution, MnO_4^- ion undergoes reduction to Mn^{2+} according to the following half-reaction:

$$8\,H^+(aq) + MnO_4^-(aq) + 5\,e^- \rightarrow Mn^{2+}(aq) + 4\,H_2O(\ell)$$

Since the MnO_4^- ion has an intense violet color while the Mn^{2+} ion is nearly colorless, the endpoint in titrations using $KMnO_4$ as the titrant can be taken to be indicated by the first permanent pink color that appears in the solution.

  $KMnO_4$ will be employed in this experiment to determine the mass percent of iron in an unknown solid containing iron(II) ammonium sulfate, $Fe(NH_4)_2(SO_4)_2 \cdot 6\,H_2O$. The titration, which involves the oxidation of Fe^{2+} ion to Fe^{3+} by permanganate ion, is carried out in sulfuric acid solution to prevent the oxidation of Fe^{2+} by oxygen in the air. The endpoint of the titration is sharpened markedly if phosphoric acid is present, because the Fe^{3+} ion produced in the titration forms an essentially colorless complex with phosphate ions.

  The moles of potassium permanganate used in the titration equals the product of the molarity of the $KMnO_4$ in the titrant and the volume of titrant used to reach the endpoint. The number of moles of iron present in the sample is obtained from the balanced chemical reaction and the amount of MnO_4^- ion reacted. The percentage by weight of iron in the solid sample follows directly. You will calculate the mean percentage of iron for your values, along with the standard deviation. (See Appendix IX.)

WEAR YOUR SAFETY GLASSES WHILE PERFORMING THIS EXPERIMENT

Experimental Procedure

Obtain a buret and an unknown iron(II) sample.

  Weigh out accurately on an analytical balance three samples of about one gram of your unknown into clean 250-mL Erlenmeyer flasks.

  Clean your buret thoroughly. Obtain about one-hundred milliliters of the standard $KMnO_4$ stock solution. Rinse the buret with a few milliliters of the $KMnO_4$ and drain it out. Repeat this process three times. Then fill the buret with the $KMnO_4$ solution.

  Prepare 150 mL of 1.0 M H_2SO_4 by pouring 25 mL of 6 M H_2SO_4 into 125 mL of H_2O, while stirring. Add 50 mL of this 1.0 M H_2SO_4 to *one* of the iron samples. The sample should dissolve completely. Without delay, titrate this iron solution with the $KMnO_4$ solution, recording the initial volume in the buret as precisely as you can. (Note: Because the color of the MnO_4^- ion is so intense, it will be difficult to see the meniscus and thus read the buret as precisely as you might with a colorless titrant. With this titrant, we usually read volumes to the nearest ± 0.05 mL by noting the position of the very top of the violet color. Whatever you choose to do, however, be consistent in how you read the initial and final readings for a given trial.) When a light-yellow color develops in the iron solution during the titration, add 3 mL of 85% H_3PO_4. **CAUTION:** **Corrosive reagent.**

  Continue the titration until you obtain the first pink color that persists for 15 to 30 seconds. Repeat the titration with the other two samples. You may add the phosphoric acid before you start titrating, if you prefer, now that you have seen what it does.

  Optional The $KMnO_4$ solution can be standardized by the method you used in this experiment. $Fe(NH_4)_2(SO_4)_2 \cdot 6\,H_2O$ is a primary standard with a molar mass equal to 392.2 g/mol. For standardization, use 0.7 ± 0.05 g samples of the primary standard, weighed to ± 0.0001 g. Draw about one-hundred milliliters of the stock $KMnO_4$ solution and titrate the samples as you did your unknown. ▪

Take it Further (Optional): Determine the mass percent of iron in iron supplement tablets that contain iron(II), ferrous ion.

DISPOSAL OF REACTION PRODUCTS. Dispose of your titrated solutions as directed by your instructor.

Name _____ Section _____

Experiment 30

Data and Calculations: Determination of Iron by Reaction with
Permanganate—A Redox Titration

Mass of unknown sample container plus contents _____ g

Mass after removing Sample I _____ g

Mass after removing Sample II _____ g

Mass after removing Sample III _____ g

Molarity of standard $KMnO_4$ solution _____ M

Sample	I	II	III
Initial buret reading	_____ mL	_____ mL	_____ mL
Final buret reading	_____ mL	_____ mL	_____ mL
Volume of $KMnO_4$ required	_____ mL	_____ mL	_____ mL
Moles $KMnO_4$ required	_____	_____	_____
Moles Fe^{2+} in sample	_____	_____	_____
Mass of Fe in sample	_____ g	_____ g	_____ g
Mass of sample	_____ g	_____ g	_____ g
Mass percent of Fe in sample	_____ $\%_{mass}$	_____ $\%_{mass}$	_____ $\%_{mass}$

Unknown # _____

Mean $\%_{mass}$ Fe _____ $\%_{mass}$

Standard deviation _____ $\%_{mass}$

(continued on following page)

Optional **Standardization of KMnO$_4$**

Sample	IV	V	VI
Mass of primary standard	_____ g	_____ g	_____ g
Moles Fe in standard	_____	_____	_____
Moles KMnO$_4$ required	_____	_____	_____
Initial buret reading	_____ mL	_____ mL	_____ mL
Final buret reading	_____ mL	_____ mL	_____ mL
Volume of KMnO$_4$ used	_____ mL	_____ mL	_____ mL
Molarity of KMnO$_4$	_____ M	_____ M	_____ M
Mean molarity	_____ M		
Standard deviation	_____ M		

Experiment 30

Advance Study Assignment: Determination of Iron by Reaction with
Permanganate—A Redox Titration

1. Write the balanced net ionic equation for the reaction between MnO_4^- ion and Fe^{2+} ion in acidic solution.

2. How many moles of Fe^{2+} ion can be oxidized by 1.85×10^{-3} moles MnO_4^- ion in the reaction in Question 1?

_____ moles

3. A solid sample containing some Fe^{2+} ion had a total mass of 0.9791 g. It required 18.2 mL of 0.02304 M $KMnO_4$ to titrate the Fe^{2+} in the dissolved sample to a pink endpoint.

   a. How many moles of MnO_4^- ion were required?

_____ moles

   b. How many moles of Fe^{2+} were present in the sample?

_____ moles

(continued on following page)

c. How many grams of iron were present in the sample?

_____ g

d. What was the mass percent of iron in the sample?

_____ %$_{mass}$

4. What is the mass percent of iron in iron(II) ammonium sulfate hexahydrate, $Fe(NH_4)_2(SO_4)_2 \cdot 6H_2O$?

_____ %$_{mass}$

Experiment 31

Determination of an Equivalent Mass by Electrolysis

In Experiment 24 we determined the molar mass of an acid by titration of a known mass of the acid with a standardized solution of NaOH. We noted that the acid could contain only one acidic hydrogen atom per molecule if we were to be able to find the molar mass. If there are two or three such acidic hydrogen atoms, we can only determine the mass of acid that will furnish one mole of H^+ ion. That is called the equivalent mass of the acid.

Experimentally, we find that the equivalent mass of an element can be related in a fundamental way to the chemical changes observed in the process known as electrolysis. As you know, some liquids, because they contain ions, are capable of conducting an electric current. If the two terminals on a storage battery, or any other source of sufficient DC voltage, are connected through metal electrodes to a conducting liquid, an electric current will seek to pass through the liquid and will actually do so if chemical reactions occur at the two metal electrodes; in this process electrolysis is said to occur, and the liquid is said to be electrolyzed.

At the electrode connected to the *negative* pole of the battery or power supply, reduction reactions have the potential to take place. In such reactions electrons are accepted by one of the species present in the liquid, which, in the experiment we shall be doing, will be an aqueous solution. The species thereby reduced will ordinarily be a metallic cation, the H^+ ion, or possibly water itself; the dominant reaction that is actually observed will be that with the most favorable combination of thermodynamics and kinetics, and will depend on the composition of the solution. In the electrolysis cell we shall study, the reduction reaction of interest will occur in a slightly acidic medium; hydrogen gas will be produced by the reduction of hydrogen ion:

$$2\,H^+(aq) + 2\,e^- \rightarrow H_2(g) \tag{1}$$

In this reduction reaction, which will occur at the negative pole, or cathode, of the cell, for every H^+ ion reduced *one* electron will be required, and for every molecule of H_2 formed, *two* electrons will be needed.

Ordinarily in chemistry we deal not with individual ions or molecules but rather with moles of substances. In terms of moles, we can say that, by Equation 1,

The reduction of one mole of H^+ ion requires one mole of electrons

The production of one mole of $H_2(g)$ requires two moles of electrons

A mole of electrons is a fundamental amount of electricity in the same way that a mole of pure substance is a fundamental amount of matter, at least from a chemical point of view. A mole of electrons is called a faraday, after Michael Faraday, who discovered the basic laws of electrolysis. The amount of a species that will react with a mole of electrons, or one faraday, is defined as the equivalent mass of that species. Since one faraday will reduce one mole of H^+ ion, we say that the equivalent mass of hydrogen is 1.008 g/mol, equal to the mass of one mole of H^+ ion [or 1/2 mole of $H_2(g)$]. To form one mole of $H_2(g)$, we would have to pass two faradays through the electrolysis cell.

In the electrolysis experiment we will perform, we will measure the volume of hydrogen gas produced under known conditions of temperature and pressure. By using the Ideal Gas Law we will be able to calculate how many moles of H_2 were formed, and hence how many faradays of electricity passed through the cell.

At the *positive* pole of an electrolysis cell (the metal electrode that is connected to the + terminal of the battery or power supply and is called the anode), oxidation reactions have the potential to take place. These are reactions in which some species will give up electrons, and they may involve again an ionic or neutral species in the solution or the metal electrode itself. In the cell that you will be studying, the pertinent oxidation reaction will be that in which the metal under study is oxidized:

$$M(s) \rightarrow M^{n+}(aq) + n\,e^- \tag{2}$$

During the course of the electrolysis the atoms in the metal electrode will be converted to metallic cations and will go into the solution. The mass of the metal electrode will decrease, depending on the amount of electricity passing through the cell and the nature of the metal. To oxidize one mole, or one molar mass, of the metal, it would take n faradays, where n is the charge on the cation that is formed. By definition, one faraday would cause one equivalent mass, EM, of metal to go into solution. The molar mass, MM, and the equivalent mass of the metal are related by the equation

$$MM = EM \times n \tag{3}$$

In an electrolysis experiment, since n is not determined independently, it is not possible to find the molar mass of a metal directly. It is possible, however, to find equivalent masses of many metals, and that will be our main purpose here. We will oxidize a sample of an unknown metal at the positive pole of an electrolysis cell, weighing the metal before and after the electrolysis and thereby determining its loss in mass. We will use the same amount of electricity (the same number of electrons) to reduce hydrogen ion at the negative pole of the electrolysis cell. From the volume of H_2 gas that is produced under known conditions we can calculate the number of moles of H_2 formed, and hence the number of faradays that passed through the cell. The equivalent mass of the metal is then calculated as the amount of metal that would be oxidized if one faraday were used. In an optional part of the experiment, your instructor may tell you the identity of the metal you used. Using Equation 3, it will be possible to determine the charge on the metallic cations that were produced during electrolysis.

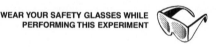

WEAR YOUR SAFETY GLASSES WHILE PERFORMING THIS EXPERIMENT

Experimental Procedure

Obtain a buret and a sample of metal unknown. Lightly sand the metal to clean it. Rinse the metal with water and then with acetone. Let the acetone evaporate. When the sample is thoroughly dry, weigh it on the analytical balance to ± 0.0001 g.

Set up the electrolysis apparatus as shown in Figure 31.1. There should be about one-hundred milliliters of 0.5 M $HC_2H_3O_2$ in 0.5 M Na_2SO_4 in the beaker with the gas buret. This will serve as the conducting solution. Insert the bare, tightly coiled end of the heavy copper wire into the open end of the buret, then lower this combination into the beaker containing the conducting solution until the buret opening is about one centimeter above the bottom of the beaker. Attach a length of rubber tubing to the upper end of the buret. Open the stopcock of the buret and, with suction, carefully draw the conducting solution up to the top of the gradations. Close the stopcock. Ensure that all of the exposed heavy copper wire is above the bottom lip of the buret opening; any wire contacting the conducting solution but not inside the buret must be covered with watertight insulation. Check the solution level after a few minutes to make sure the stopcock does not leak. Record the level.

The metal unknown will serve as the anode in the electrolysis cell. Connect the metal to the positive (+) pole of the power source with an alligator clip and immerse the metal (but not the clip!) in the conducting solution. The heavy copper wire electrode will be the cathode in the cell. Connect the dry end of that electrode to the negative (−) pole of the power source. Hydrogen gas should immediately begin to bubble from the copper cathode. Collect the gas until almost 50 mL have been produced but *before* the conducting solution level drops below the final gradation in the buret. At that point, stop the electrolysis by disconnecting the copper electrode from the power source. Record the level of the liquid in the buret. Measure and record the temperature and the atmospheric pressure in the laboratory. (In some cases a cloudiness may develop in the solution during the electrolysis. This is caused by the formation of a metal hydroxide and will have no adverse effect on the experiment.)

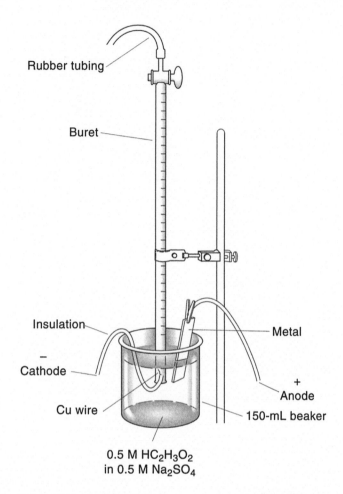

Rubber tubing

Buret

Insulation

−
Cathode

Cu wire

Metal

+
Anode

150-mL beaker

0.5 M HC$_2$H$_3$O$_2$
in 0.5 M Na$_2$SO$_4$

Figure 31.1 Obtaining accurate data with the apparatus used in this experiment requires that all of the hydrogen that bubbles off of the bare portion of the copper wire cathode be collected in the inverted buret.

Raise the buret and discard the conducting solution in the beaker in the waste container. Rinse the beaker with water, and pour in one-hundred milliliters of fresh conducting solution. Repeat the electrolysis, again generating almost 50 mL of H$_2$ and recording the initial and final liquid levels in the buret. Take the alligator clip off the metal anode and wash the anode with 0.10 M HC$_2$H$_3$O$_2$, acetic acid. Rinse the electrode in water and then in acetone. Let the acetone evaporate. Weigh the dry metal electrode to the nearest ± 0.0001 g.

DISPOSAL OF REACTION PRODUCTS.   When you are finished with the experiment, return the metal electrode and discard the conducting solution in the waste container unless directed otherwise by your instructor.

Experiment 31

Data and Calculations: Determination of an Equivalent Mass
by Electrolysis

Mass of metal anode   _____ g

Mass of anode after electrolysis   _____ g

Initial buret reading   _____ mL

Buret reading after first electrolysis   _____ mL

Buret reading after refilling   _____ mL

Buret reading after second electrolysis   _____ mL

Ambient (atmospheric) pressure, $P_{ambient}$ _____ mm Hg

Room temperature, t _____ °C

Vapor pressure of H_2O at t (consult Appendix I), VP_{H_2O} _____ mm Hg

Total volume of H_2 produced, V

                                                                                           _____ mL

Absolute temperature, T

                                                                                           _____ K

Pressure exerted by dry H_2: $P = P_{ambient} - VP_{H_2O}$
(ignore any pressure effect due to liquid levels in the buret)

                                                                                           _____ mm Hg

(continued on following page)

Moles of H_2 produced, n (use the Ideal Gas Law, $PV = nRT$)

_____ moles

Faradays of charge passed (moles of electrons)

_____

Mass lost by anode   _____ g

Equivalent mass of metal $\left(EM = \dfrac{\text{grams of metal lost}}{\text{faradays passed}}\right)$

_____ g/mol e^-

Unknown metal #   _____

Optional   Identity of metal _____ MM _____ g/mol

Charge n on cation _____ (Eq. 3) ▪

Experiment 31

Advance Study Assignment: Determination of an Equivalent Mass
by Electrolysis

1. In an electrolysis cell similar to the one employed in this experiment, a student observed that his unknown metal anode lost 0.446 g while a total volume of 98.24 mL of water-saturated hydrogen gas was being produced. The temperature in the laboratory was 23°C and the ambient (atmospheric) pressure was 729 mm Hg. You can find the vapor pressure of water in Appendix I. To calculate the equivalent mass of his metal, the student filled in the blanks below. Fill in the blanks as he did.

   $P_{H_2} = P_{ambient} - VP_{H_2O} =$ _____ mm Hg = _____ atm

   $V_{gas} =$ _____ mL = _____ L

   $T =$ _____ K

   $n_{H_2} =$ _____ moles $n_{H_2} = \dfrac{PV}{RT}$ (where $P = P_{H_2}$ and $V = V_{gas}$; show your work below)

   1 mole H_2 requires passage of _____ faradays

   Faradays passed = _____ mol of e^-

   Mass lost by metal anode = _____ g

   Grams of metal lost per faraday passed $= \dfrac{\text{grams lost}}{\text{faradays passed}} =$ _____ g/mol $e^- = EM$

(continued on following page)

The student was told that his metal anode was made of tin.

MM Sn = _____ g/mol. The charge n on the Sn ion is therefore _____. (See Eq. 3; show work below.)

2. In standard IUPAC units, the faraday is equal to 96,480 coulombs. A coulomb is the amount of electric charge passed when a current of one ampere flows for one second. Given the charge on the electron is 1.6022×10^{-19} coulombs, calculate a value for Avogadro's number.

_____

Experiment 32

Voltaic Cell Measurements

Many chemical reactions can be classified as oxidation-reduction reactions, because they involve the oxidation of one species and the reduction of another. Such reactions can conveniently be considered as the result of two half-reactions, one of oxidation and the other reduction. In the case of the oxidation-reduction reaction that would occur if a piece of metallic zinc were put into a solution of lead nitrate,

$$Zn(s) + Pb^{2+}(aq) \rightarrow Zn^{2+}(aq) + Pb(s)$$

the two half-reactions would be

$$Zn(s) \rightarrow Zn^{2+}(aq) + 2\,e^- \quad \text{oxidation half-reaction}$$

$$2\,e^- + Pb^{2+}(aq) \rightarrow Pb(s) \quad \text{reduction half-reaction}$$

The tendency for an oxidation-reduction reaction to occur can be measured if the two half-reactions are made to occur in separate regions connected by a barrier that is porous to ion movement. An apparatus, called a voltaic cell, in which such a reaction might be carried out under this condition is shown in Figure 32.1.

If we connect a voltmeter between the two electrodes, we will find that there is a voltage, or potential, between them. The magnitude of the potential is a direct measure of the chemical tendency, or more properly the thermodynamic driving force, for the spontaneous oxidation-reduction reaction to occur.

If we study several oxidation-reduction reactions, we find that the voltage of each associated voltaic cell can be considered to be the sum of a potential for the oxidation half-reaction and a potential for the reduction half-reaction. In the $Zn(s)\,|\,Zn^{2+}(aq)\,\|\,Pb^{2+}(aq)\,|\,Pb(s)$ cell we have been discussing, for example,

$$E_{cell} = E_{Zn,Zn^{2+} \text{ oxidation half-reaction}} + E_{Pb^{2+},Pb \text{ reduction half-reaction}} \qquad \textbf{(1)}$$

By convention, the negative electrode in a voltaic cell is taken to be the one from which electrons are emitted (i.e., where oxidation occurs). The negative electrode is the one that is connected to the minus pole of the voltmeter when the voltage is measured.

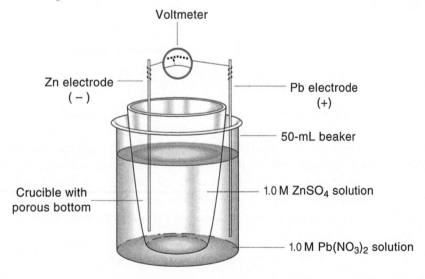

Figure 32.1 The configuration of the voltaic cell used in this experiment is simple, but it works well provided that the porous bottom of the crucible has very small pores, preventing macroscopic liquid flow between the chambers, and that this porous membrane is rinsed well between varying cell chemistries.

Since any cell potential is the sum of two half-reaction potentials, it is not possible, by measuring overall cell potentials, to determine individual absolute half-reaction potentials. However, if a half-reaction potential is arbitrarily assigned to one half-reaction, other electrode potentials can be given definite values relative to that, based on the assigned value. The usual procedure is to assign a value of 0.0000 volts to the standard potential for the half-reaction

$$2\,H^+(aq) + 2\,e^- \rightarrow H_2(g); \qquad E^0_{H^+,H_2\ red} = 0.0000\ V$$

For the $Zn(s)\,|\,Zn^{2+}(aq)\,\|\,H^+(aq)\,|\,H_2(g)$ cell, the measured potential is 0.76 V, and the zinc electrode is negative. Zinc metal is therefore oxidized, and the overall cell reaction must be

$$Zn(s) + 2\,H^+(aq) \rightarrow Zn^{2+}(aq) + H_2(g); \qquad E^0_{cell} = +0.76\ V$$

Given this information, we can readily find the standard potential for half-reaction in which Zn is oxidized to Zn^{2+}:

$$E^0_{cell} = E^0_{Zn,Zn^{2+}\ oxid} + E^0_{H^+,H_2\ red}$$

$$+0.76\ V = E^0_{Zn,Zn^{2+}\ oxid} + 0.00\ V; \qquad E^0_{Zn,Zn^{2+}\ oxid} = +0.76\ V$$

If the potential for a half-reaction is known, the potential for the reverse half-reaction can be obtained by simply reversing the sign. For example:

$$\text{if} \quad E^0_{Zn,Zn^{2+}\ oxid} = +0.76\ V, \quad \text{then} \quad E^0_{Zn^{2+},Zn\ red} = -0.76\ V$$

$$\text{if} \quad E^0_{Pb^{2+},Pb\ red} = +y\ V, \quad \text{then} \quad E^0_{Pb,Pb^{2+}\ oxid} = -y\ V$$

In the first part of this experiment you will measure the voltages of several different cells. By arbitrarily assigning the potential of a particular half-reaction to be 0.00 V, you will be able to calculate the (relative) potentials corresponding to all of the various half-reactions that occurred in your cells.

In our discussion so far we have not considered the possible effects of such system variables as temperature, potential at the liquid–liquid junction, size of the metal electrodes, and the concentrations of solute species. Although temperature and liquid junction potentials do have a definite effect on cell potentials, taking account of their influence involves complex thermodynamic concepts and is usually not of concern in an elementary course. The size of a metal electrode has no appreciable effect on electrode potential, although it does relate directly to the capacity of the cell to produce useful electrical energy. In this experiment, we will operate the cells so that they deliver essentially no energy but exert their maximum possible potentials; under these conditions the size of the electrodes will not matter.

The effect of solute ion concentrations is important and can be described relatively easily. Consider the general overall cell reaction

$$a\,A(s) + b\,B^+(aq) \rightarrow c\,C(s) + d\,D^{2+}(aq) \tag{2}$$

At 25°C, the overall cell potential depends on the solute concentrations according to the Nernst Equation, which is

$$E_{cell} = E^0_{cell} - \frac{0.0592\ V}{n} \log \frac{[D^{2+}]^d}{[B^+]^b} \tag{3}$$

where E^0_{cell} is a constant for a given reaction and is called the standard cell potential, while n is the number of electrons being transferred in the electrode half-reactions whose sum yields the overall cell reaction.

By Equation 3 you can see that the measured cell potential, E_{cell}, will equal the standard cell potential if the molarities of D^{2+} and B^+ are both unity, or, if d equals b, if the molarities are simply equal to each other. We will carry out experiments under such conditions that the cell potentials you observe will be very close to those predicted by the standard potentials given in the tables in your chemistry text.

Considering the $Cu(s)\,|\,Cu^{2+}(aq)\,\|\,Ag^+(aq)\,|\,Ag(s)$ cell as a specific example, the observed cell reaction would be

$$Cu(s) + 2\,Ag^+(aq) \rightarrow Cu^{2+}(aq) + 2\,Ag(s) \tag{4}$$

For this cell, Equation 2 takes the form

$$E_{cell} = E^0_{cell} - \frac{0.0592\ V}{2} \log \frac{[Cu^{2+}]}{[Ag^+]^2} \tag{5}$$

In this equation n is 2 because in the cell reaction two electrons being transferred in each of the two half-reactions that sum to provide the overall cell reaction. E^0 would be the cell potential when the copper and silver salt solutions are both 1.0 M, since the logarithmic term is then equal to zero.

If we decrease the Cu^{2+} concentration, keeping that of Ag^+ at 1.0 M, the potential of the cell will go up by 0.03 volts for every factor of ten by which we decrease $[Cu^{2+}]$. Ordinarily it is not convenient to change ion concentrations by several orders of magnitude, so concentration effects in voltaic cells are generally relatively small. However, if we added a complexing or precipitating species to the copper salt solution, the value of $[Cu^{2+}]$ would drop drastically, and the voltage change would be appreciable. In this experiment, we will illustrate this effect by using NH_3 to complex the Cu^{2+}. Using Equation 3, we can calculate $[Cu^{2+}]$ in the solution of its complex ion. (It is very low!)

In an analogous experiment we will determine the solubility product of AgCl. In this case we will surround the Ag electrode in a $Cu(s)|Cu^{2+}(aq)\|Ag^+(aq)|Ag(s)$ cell with a solution of known Cl^- ion concentration that is saturated with AgCl. From the measured cell potential, we can use Equation 5 to calculate the (very small) value of $[Ag^+]$ that remains unprecipitated in the chloride-containing solution.

In the two experiments in which we change the cation concentrations, first Cu^{2+} and then Ag^+, we end up with systems in equilibrium. In the first case the Cu^{2+} ion is in equilibrium with $Cu(NH_3)_4^{2+}$ and NH_3. Using the cell potential, we can calculate $[Cu^{2+}]$. From the way we made up the system we can calculate $[Cu(NH_3)_4^{2+}]$ and $[NH_3]$. If we wish to do so, we can use these data to find the dissociation constant for the $Cu(NH_3)_4^{2+}$ ion. Many equilibrium constants are found by experiments of this sort.

In the last experiment we have an equilibrium system containing Ag^+, Cl^-, and AgCl(s). Here we can use the cell potential to find $[Ag^+]$. There isn't much free Ag^+ present in solution, but there is a tiny bit, and the cell potential allows us to determine how much. We know $[Cl^-]$ from the way we made up the mixture in the crucible. From these data we can easily calculate K_{sp} for AgCl:

$$AgCl(s) \rightleftharpoons Ag^+(aq) + Cl^-(aq) \qquad K_{sp}= [Ag^+][Cl^-] \qquad \textbf{(6)}$$

WEAR YOUR SAFETY GLASSES WHILE
PERFORMING THIS EXPERIMENT

Experimental Procedure

You may work in pairs in this experiment.

A. Cell Potentials

In this experiment you will be working with these seven electrode systems:

$Ag^+(aq)|Ag(s)$ $Br_2(\ell)|Br^-(aq)|Pt(s)$

$Cu^{2+}(aq)|Cu(s)$ $Cl_2(g, 1\ atm)|Cl^-|Pt(s)$

$Fe^{2+}(aq), Fe^{3+}(aq)|Pt(s)$ $I_2(s)|I^-|Pt(s)$

$Zn^{2+}(aq)|Zn(s)$

Your purpose will be to measure enough voltaic cell potentials to allow you to determine each electrode's half-reaction potential relative to a single arbitrarily chosen electrode potential. You may assume the potentials of the cells you set up will be essentially the standard potentials, even where the reagents are not in their standard states.

Using the apparatus shown in Figure 32.1, set up a voltaic cell involving any two of the electrodes in the list above. About ten milliliters of each solution should suffice for the cell. Measure the cell potential and record it along with which electrode has negative polarity. Pour both solutions into a beaker and rinse the crucible and 50-mL beaker with deionized water.

In a similar manner set up and measure the cell voltage and polarities of other cells, sufficient in number to include all of the electrode systems on the list at least once. Do not combine the silver electrode system with any of the halogen electrode systems because a precipitate will form; for the same reason, do not use the copper and iodine electrodes in the same cell. Any other combinations may be used. The raw data from this part of the experiment should be recorded in the first three columns of Table A.1 on the report page.

B. Effect of Concentration on Cell Potentials

1. Complex Ion Formation Set up the $Cu(s)\,|\,Cu^{2+}(aq)\,\|\,Ag^+(aq)\,|\,Ag(s)$ cell, using about ten milliliters of $CuSO_4$ solution in the crucible and about ten milliliters of $AgNO_3$ solution in the beaker. Measure the potential of this cell. While the potential is being measured, add about ten milliliters of 6 M NH_3 to the $CuSO_4$ solution, stirring carefully with your stirring rod. Measure and record the new cell potential when it becomes steady.

2. Determination of the Solubility Product of AgCl Remove the crucible from the cell you have just studied and discard the $Cu(NH_3)_4^{2+}$. Clean the crucible by drawing a little 6 M NH_3 through it, using the adapter and suction flask. Then draw some deionized water through it. Empty the beaker and rinse it well. Reassemble the $Cu(s)\,|\,Cu^{2+}(aq)\,\|\,Ag^+(aq)\,|\,Ag(s)$ cell, this time using the beaker for the $Cu(s)\,|\,CuSO_4(aq)$ electrode system. Put about ten milliliters of 1.0 M KCl in the crucible (in place of $AgNO_3$); immerse the Ag electrode in that solution. Add 1 drop of $AgNO_3$ solution to form a little AgCl, so that an equilibrium between Ag^+ and Cl^- is established. Measure the potential of this cell, noting which electrode is negative. In this case $[Ag^+]$ will be very low, which will decrease the potential of the cell to such an extent that its polarity may change from that observed previously.

DISPOSAL OF REACTION PRODUCTS. As you finish each voltage measurement, pour the two solutions into a beaker. At the end of the experiment, pour the contents of the beaker into the waste container, unless directed otherwise by your instructor. Once again clean the crucible by drawing a little 6 M NH_3 through it, then drawing some deionized water through to rinse it. (The ammonia forms a strong complex with Ag^+, dissolving the AgCl precipitate formed in the last experimental step.)

Experiment 32

Data and Calculations: Voltaic Cell Measurements

Table A.1. Cell Potentials

Electrode systems used in cell	Cell potential, E^0_{cell} [volts]	Negative electrode	Oxidation reaction	$E^0_{oxidation}$ in volts	Reduction reaction	$E^0_{reduction}$ in volts
1.						
2.						
3.						
4.						
5.						
6.						
7.						

Calculations

A. Noting that oxidation occurs at the *negative* pole in a cell, write the oxidation reaction taking place in each of the cells. The other electrode system must undergo reduction; write the reduction reaction that occurs in each cell.

B. Assume that $E^0_{Ag^+,Ag} = 0.00$ volts (whether in reduction or oxidation). Enter that value in the table for all of the silver electrode systems you used in your cells. Since $E^0_{cell} = E^0_{oxidation} + E^0_{reduction}$, you can calculate E^0 values for all the electrode systems in which the Ag,Ag$^+$ system was involved. Enter those values in the table.

C. Using the values and relations in Part B and taking advantage of the fact that for any given electrode system, $E^0_{oxidation} = -E^0_{reduction}$, complete the table of E^0 values. The best way to do this is to use one of the E^0 values you found in Part B in another cell with that electrode system. That potential, along with E^0_{cell}, will allow you to find the potential of the other electrode. Continue this process with other cells until all the electrode potentials have been determined.

(continued on following page)

Table A.2. Table of Electrode Potentials

In Table A.1, you should have a value for E^0_{red} or E^0_{oxid} for each of the electrode systems you have studied. Remembering that for any electrode system, $E^0_{red} = -E^0_{oxid}$, you can find the value for E^0_{red} for each system. List those potentials in the left column of the table below, in order of decreasing (more negative) value.

$E^0_{reduction}$ ($E^0_{Ag^+,Ag} = 0.00$ V)	Electrode reaction in reduction	$E^0_{reduction}$ ($E^0_{H^+,H_2} = 0.00$ V)
_____	_____	_____
_____	_____	_____
_____	_____	_____
_____	_____	_____
_____	_____	_____
_____	_____	_____
_____	_____	_____

The electrode potentials you have determined are based on setting $E^0_{Ag^+,Ag} = 0.00$ V. The usual convention in electrochemistry is to set $E^0_{H^+,H_2} = 0.00$ V under which conditions $E^0_{Ag^+,Ag\ red} = +0.80$ V. Convert from one base to the other by adding 0.80 V to each of the electrode potentials in the first column, and enter these values in the third column of the table.

Why are the values of E^0_{red} on the two bases related to each other in such a simple way?

(continued on following page)

B. Effect of Concentration on Cell Potentials

1. Complex ion formation:

Potential, E^0_{cell}, before addition of 6 M NH_3 _____ V

Potential, E_{cell}, after $Cu(NH_3)_4{}^{2+}$ formed   _____ V

Given Equation 5,

$$E_{cell} = E^0_{cell} - \frac{0.0592 \text{ V}}{2} \log \frac{[Cu^{2+}]}{[Ag^+]^2} \tag{5}$$

calculate the residual concentration of free Cu^{2+} ion in equilibrium with $Cu(NH_3)_4{}^{2+}$ in the solution in the crucible. Simplify the calculation by taking $[Ag^+]$ to be exactly $\underline{1}$ M (though in truth it is not, so your result will be an approximation).

$[Cu^{2+}] =$ _____ M

2. Solubility product of AgCl:

Potential, E^0_{cell}, of the $Cu(s)\,|\,Cu^{2+}(aq)\,\|\,Ag^+(aq)\,|\,Ag(s)$ cell (from B.1)

_____ V   Negative electrode _____

Potential, E_{cell}, with 1.0 M KCl present   _____ V   Negative electrode _____

Using Equation 5, calculate $[Ag^+]$ in the cell, where it is in equilibrium with 1.0 M Cl^- ion. (E_{cell} in Eq. 5 is the *negative* of the measured value if the polarity is not the same as in the standard cell.) Take $[Cu^{2+}]$ to be exactly $\underline{1}$ M (again, this simplifies the calculation, but makes the result an approximation).

$[Ag^+] =$ _____ M

Since Ag^+ and Cl^- in the crucible are in equilibrium with AgCl, we can find K_{sp} for AgCl from the concentration of Ag^+ and Cl^-, which we now know. Formulate the expression for K_{sp} for AgCl, and determine its value.

$K_{sp} =$ _____

Experiment 32

Advance Study Assignment: Voltaic Cell Measurements

1. A student measures the potential of a cell made up with 1.0 M $CuSO_4$ in one solution reservoir and 1.0 M Ag_2SO_4 in the other. There is a metallic copper (Cu) electrode in the $CuSO_4$ and a metallic silver (Ag) electrode in the Ag_2SO_4, and the cell is set up as shown in Figure 32.1. She finds that the potential, or voltage, of the cell, E^0_{cell}, is 0.45 V, and that the copper electrode is negative.

   a. At which electrode is oxidation occurring? _____

   b. Write the equation for the oxidation half-reaction in this cell.

   c. Write the equation for the reduction half-reaction in this cell.

   d. Write the net ionic equation for the spontaneous oxidation-reduction reaction that occurs in this cell.

2. In another cell, the potential of the copper metal | copper(II) ion electrode was found to be +0.442 V relative to a silver metal | silver ion electrode, with the copper electrode being negative.

   a. If the potential of the silver | silver ion electrode, $E^0_{Ag^+,Ag\ red}$ is taken to be 0.000 V in oxidation or reduction, what is the value of the potential for the oxidation reaction, $E^0_{Cu,Cu^{2+}\ oxid}$? Keep in mind that $E^0_{cell} = E^0_{oxid} + E^0_{red}$.

   _____ volts

   b. If $E^0_{Ag^+,Ag\ red}$ equals +0.800 V, as in standard tables of electrode potentials, what is the value of the potential of the oxidation reaction of copper, $E^0_{Cu,Cu^{2+}\ oxid}$?

   _____ volts

(continued on following page)

c. The student adds 6 M NH_3 to the $CuSO_4$ solution in this second cell until the Cu^{2+} ion is essentially all converted to $Cu(NH_3)_4^{2+}$ ion. The voltage of the cell, E_{cell}, goes up to 0.920 V and the Cu electrode is still negative. Find the residual concentration of Cu^{2+} ion in the cell. (Use Eq. 5.)

_____M

d. In Part 2(c), $[Cu(NH_3)_4^{2+}]$ is 0.05 M and $[NH_3]$ is 3 M. Given those values and the result in Part 2(c) for $[Cu^{2+}]$, calculate K_{eq} for the complex formation reaction

$$Cu(NH_3)_4^{2+}(aq) \rightleftharpoons Cu^{2+}(aq) + 4\ NH_3(aq)$$

$K_{eq} = $ _____

Experiment 33

Preparation of Copper(I) Chloride

Oxidation-reduction reactions are, like precipitation reactions, often used in the preparation of inorganic substances. In this experiment we will employ a series of such reactions to prepare one of the less commonly encountered salts of copper, copper(I) chloride. Most copper compounds contain copper(II), but copper(I) is present in a few slightly soluble or complex copper salts.

The process of synthesis of CuCl we will use begins by dissolving copper metal in nitric acid:

$$Cu(s) + 4\,H^+(aq) + 2\,NO_3^-(aq) \rightarrow Cu^{2+}(aq) + 2\,NO_2(g) + 2\,H_2O(\ell) \qquad (1)$$

The solution obtained is treated with sodium carbonate in excess, which neutralizes the remaining acid with evolution of CO_2 and precipitates Cu(II) as the carbonate:

$$2\,H^+(aq) + CO_3^{2-}(aq) \rightleftharpoons (H_2CO_3)(aq) \rightleftharpoons CO_2(g) + H_2O(\ell) \qquad (2)$$

$$Cu^{2+}(aq) + CO_3^{2-}(aq) \rightleftharpoons CuCO_3(s) \qquad (3)$$

The $CuCO_3$ will be purified by filtration and washing and then dissolved in hydrochloric acid. Copper metal added to the highly acidic solution reduces the Cu(II) to Cu(I) and is itself oxidized to Cu(I) in a disproportionation reaction. In the presence of excess chloride, the copper will be present as a $CuCl_4^{3-}$ complex ion. Addition of this solution to water destroys the complex, and white CuCl precipitates.

$$CuCO_3(s) + 2\,H^+(aq) + 4\,Cl^-(aq) \rightarrow CuCl_4^{2-}(aq) + CO_2(g) + H_2O(\ell) \qquad (4)$$

$$CuCl_4^{2-}(aq) + Cu(s) + 4\,Cl^-(aq) \rightarrow 2\,CuCl_4^{3-}(aq) \qquad (5)$$

$$CuCl_4^{3-}(aq) \xrightarrow{\;H_2O\;} CuCl(s) + 3\,Cl^-(aq) \qquad (6)$$

Because CuCl is readily oxidized, due care must be taken to minimize its exposure to air during its preparation and while it is being dried.

In an optional part of the experiment, we prove that the formula of the compound synthesized is CuCl.

Experimental Procedure

WEAR YOUR SAFETY GLASSES WHILE
PERFORMING THIS EXPERIMENT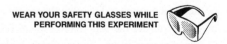

Obtain a 1.0-g sample of copper metal turnings, a Büchner funnel, and a filter flask. Weigh the copper metal on a top-loading balance to ± 0.01 g.

Put the metal in a 150-mL beaker and, **under a hood,** add 5 mL of 15 M HNO_3. **CAUTION:** **Corrosive reagent.** Brown NO_2 gas will be evolved and an acidic blue solution of $Cu(NO_3)_2$ produced. If it is necessary, you may warm the beaker gently to dissolve all of the copper. When all of the copper has gone into solution, add 50 mL of water to the solution and allow it to cool.

Weigh out about five grams of sodium carbonate in a small beaker on the top-loading balance. With your spatula, add small amounts of the Na_2CO_3 to the solution, adding more solid only as the evolution of CO_2 subsides. Stir the solution to expose it to the solid. When the acid is neutralized, a blue-green precipitate of $CuCO_3$ will begin to form. At that point, add the rest of the Na_2CO_3, stirring the mixture well to ensure complete precipitation of the copper carbonate.

Transfer the precipitate to the Büchner funnel and use suction to remove the excess liquid. Use your rubber policeman and a spray from your wash bottle to ensure complete transfer of the solid. Wash the precipitate well with deionized water with suction on; then let it remain on the filter paper with suction on for a minute or two.

Remove the filter paper from the funnel and transfer the solid $CuCO_3$ back into the 150-mL beaker. Add about ten milliliters of water and then, slowly, 30 mL of 6 M HCl to the solid, stirring continuously. When the $CuCO_3$ has all dissolved, add 1.5 g Cu turnings to the beaker and cover it with a watch glass.

Heat the mixture in the beaker to the boiling point and keep it at that temperature, just simmering, for about ten minutes. It may be that the dark-colored solution that forms will clear to a yellow color before that time is up, and if it does, you may stop heating and proceed to the next step.

While the mixture is heating, put 150 mL of deionized water in a 400-mL beaker and put the beaker in an ice bath. Cover the beaker with a watch glass. After you have heated the acidic Cu-$CuCl_2$ mixture for ten minutes, or as soon as it clarifies and its color lightens, carefully decant the hot liquid into the 400-mL beaker of cold water, taking care not to transfer any of the excess Cu metal to the beaker. White crystals of CuCl should form. Continue to cool the beaker in the ice bath to promote crystallization and to increase the yield of solid.

Cool 25 mL of deionized water, to which you have added 5 drops of 6 M HCl, in an ice bath. Put 20 mL of acetone into a small beaker. Filter the crystals of CuCl in the Büchner funnel using suction. Swirl the beaker to aid in transferring the solid to the funnel. Just as the last of the liquid is being pulled through, wash the CuCl with one-third of the acidified cold water. Rinse the last of the CuCl into the funnel with another portion of the water and use the final third to rewash the solid. Turn off the suction and add half of the acetone to the funnel; wait about ten seconds and turn the suction back on. Repeat this operation with the other half of the acetone. Draw air through the sample for a few minutes to dry it. If you have properly washed the solid, it will be pure white: if the moist compound is allowed to come into contact with air, it will tend to turn pale green, due to oxidation of Cu(I) to Cu(II). Weigh the CuCl in a previously weighed beaker to ± 0.01 g. Show your sample to your instructor for evaluation.

DISPOSAL OF REACTION PRODUCTS. The contents of the suction flask after the first filtration (of $CuCO_3$) can be discarded down the sink. The contents after the second filtration include an appreciable amount of copper, so they should be put in the waste container.

Optional **Determination of the Formula of Copper(I) Chloride**

There are many ways to prove that the formula of the copper compound you have prepared is indeed CuCl. The following procedure is easy to carry out and gives good results.

Dry the copper compound by putting it under a heat lamp or in a drying oven at 110°C for 10. minutes. Weigh out a 0.1 ± 0.04-g sample of the compound accurately on an analytical balance, using a weighed 50-mL beaker as a container. Dissolve the sample in 5 mL of 6 M HNO_3. Add about ten milliliters of deionized water, mix, and transfer the solution to a clean 100-mL volumetric flask. Use several 5-mL portions of deionized water to rinse the remaining solution from the beaker into the flask. Fill the flask to the mark with deionized water. Stopper the flask and invert it at least twenty times to ensure that the solution in the flask is thoroughly mixed.

Use a clean 10-mL pipet to transfer a 10.0-mL aliquot of the solution to a clean, dry 50-mL beaker. Add 10.0 mL of 6 M NH_3 to the beaker, using the 10-mL pipet, after you have rinsed it with 6 M NH_3. This will convert any Cu^{2+} in the solution to deep blue $Cu(NH_3)_4^{2+}$. After mixing, measure the absorbance of the solution of $Cu(NH_3)_4^{2+}$ at a wavelength of 575 nm. Determine $[Cu(NH_3)_4^{2+}]$ from a graph that is provided to you, or by comparison with a standard solution. (You can prepare the standard by accurately weighing 0.06 ± 0.02 g of copper turnings and putting them into a 50-mL beaker. Dissolve the copper in 5 mL of 6 M HNO_3, and proceed as you did with the solution of the copper compound in the previous paragraph. Measure the absorbance of the standard at 575 nm. Calculate $[Cu(NH_3)_4^{2+}]$ in the solution of the compound by assuming that Beer's Law holds.) Knowing that concentration, and the fact that the original sample was effectively diluted to a volume of 200. mL, calculate the number of moles of Cu in the sample. Then calculate the number of grams of Cu in the sample, and the number of grams of Cl by difference. Use that value to find the number of moles of Cl, and from the mole ratio Cu:Cl obtain the (empirical) formula of the compound. ■

Experiment 33

Data and Calculations: Preparation of Copper(I) Chloride

Mass of Cu sample _____ g

Mass of beaker _____ g

Mass of beaker plus CuCl _____ g

Mass of CuCl actually prepared _____ g

Theoretical yield of CuCl _____ g

Percentage yield _____ %

Optional   **Determination of the Formula of Copper(I) Chloride**

Mass of Cu(I) chloride sample _____ g

Absorbance of solution of $Cu(NH_3)_4{}^{2+}$ _____

Mass of Cu turnings _____ g

Absorbance of standard solution _____

Moles of Cu in turnings

_____ moles

All of the copper in the turnings is converted to $Cu(NH_3)_4{}^{2+}$
in a solution whose total volume would be 200 mL.

$[Cu(NH_3)_4{}^{2+}]$ in standard solution

_____ M

(continued on following page)

We will assume that Beer's Law holds, such that

$$[Cu(NH_3)_4{}^{2+}] \text{ in solution of sample} = [Cu(NH_3)_4{}^{2+}] \text{ in standard} \times \frac{\text{absorbance of sample}}{\text{absorbance of standard}}$$

$[Cu(NH_3)_4{}^{2+}]$ in solution of sample

_____ M

Moles of Cu in sample (volume = 0.200 L)   _____ moles

Grams of Cu in sample

_____ g

Grams of Cl in sample (sample mass = mass Cu + mass Cl)

_____ g

Moles of Cl in sample

_____ moles

Mole ratio Cl:Cu   _____

Empirical formula of prepared compound (rounded to nearest integers)   _____

Experiment 33

Advance Study Assignment: Preparation of Copper(I) Chloride

1. Suppose the Cu^{2+} ions in this experiment are produced by the reaction of 1.06 g of copper turnings with excess nitric acid. How many moles of Cu^{2+} are produced?

_____ mol

2. Why isn't hydrochloric acid used in a direct reaction with copper to prepare the $CuCl_2$ solution?

3. How many grams of metallic copper are required to react with the number of moles of Cu^{2+} calculated in Problem 1 to form the CuCl? The overall reaction can be taken to be:

$$Cu^{2+}(aq) + 2\ Cl^-(aq) + Cu(s) \rightarrow 2\ CuCl(s)$$

_____ g

4. What is the theoretical maximum mass of CuCl that can be prepared from the reaction sequence of this experiment, using 1.06 g of Cu turnings to prepare the Cu^{2+} solution?

_____ g

(continued on following page)

5. A sample of the compound prepared in this experiment, weighing 0.0989 g, is dissolved in HNO_3, and diluted to a volume of 100. mL. A 10.0-mL aliquot of that solution is mixed with 10.0 mL of 6 M NH_3. The $[Cu(NH_3)_4{}^{2+}]$ in the resulting solution is found to be 5.12×10^{-3} M.

    a. How many moles of Cu were in the original sample, which had been effectively diluted to a volume of 200. mL?

                                               _____ moles

    b. How many grams of Cu were in the sample?

                                               _____ g

    c. How many grams of Cl were in the sample? How many moles?

                                _____ g _____ moles

    d. What is the empirical formula of the copper chloride compound?

Experiment 34

Development of a Scheme for Qualitative Analysis

In many of the previous experiments in this manual you were asked to find out how much of a given species is present in a sample. You may have determined the concentration of chloride in an unknown solution, the molarity of a sodium hydroxide solution, and the amount of calcium ion in a sample of hard water. All of these experiments fall into that part of chemistry called quantitative analysis.

Sometimes chemists are interested more in the nature of the species in a sample than the amount of those species. In that sort of problem we find out what the sample contains but not how much. For example, in Experiment 12 students are asked to determine which alkaline earth halide is present in a solution. That experiment involves qualitative analysis. This and the next four experiments deal with the qualitative analysis of solutions containing various anions and metallic cations.

The procedures in qualitative analyses of this sort involve using precipitation reactions to remove the cations sequentially from a mixture. If the precipitate can contain only one cation under the conditions that prevail, that precipitate serves to prove the presence of that cation. If the precipitate may contain several cations, it can be dissolved and further resolved in a series of steps that may include acid-base, complex ion formation, redox, or other precipitation reactions. The ultimate result is a resolution of the sample into fractions that can contain only one cation, whose presence is established by formation of a characteristic precipitate or a colored complex ion.

In this experiment you will be asked to develop a scheme for the qualitative analysis of four cations, using this systematic approach. The behavior of the cations toward a set of common test reagents differs from one cation to another and furnishes the basis for their separation.

The cations we will study are Ba^{2+}, Mg^{2+}, Cd^{2+}, and Al^{3+}. The test reagents we will use are 1.0 M Na_2SO_4, 1.0 M Na_2CO_3, 6 M NaOH, 6 M NH_3, and 6 M HNO_3. These reagents furnish anions or molecules that will precipitate or form complex ions with the cations, so you may observe the formation of insoluble sulfates, carbonates, and hydroxides (the last with either NaOH or NH_3). In addition, you may form complexes with OH^- or NH_3 when 6 M NaOH or 6 M NH_3 are added to the cation-containing solutions. The complexes may be quite stable: stable enough to prevent the precipitation of an otherwise insoluble salt on addition of a particular anion. The complexes, however, are all unstable in excess acid and can be broken down by the addition of 6 M HNO_3, releasing the cation for further reactions.

In the first part of this experiment you will observe the behavior of the four cations in the presence of one or more of the reagents we have listed. On the basis of your observations you can set up the scheme for identifying the cations in a mixture. After testing the scheme with a known containing all of the cations, you will be given an unknown to analyze.

WEAR YOUR SAFETY GLASSES WHILE PERFORMING THIS EXPERIMENT

Experimental Procedure

To four small (13 × 100 mm) test tubes add 1.0 mL of 0.10 M solutions of the nitrates or chlorides of Ba^{2+}, Mg^{2+}, Cd^{2+}, and Al^{3+}, one solution to a test tube (in the specified test tubes, a depth of one centimeter corresponds to a volume of about one milliliter). To each solution add 1 drop of 1.0 M Na_2SO_4, and stir with your stirring rod. Keep the rod in a 400-mL beaker of deionized water between tests. If a precipitate forms, write the formula of the precipitate in the box on the report page corresponding to the cation-SO_4^{2-} pair, to indicate that the sulfate of that cation is insoluble in water. Then add 1.0 mL more of the Na_2SO_4 to each test tube. If a precipitate were to dissolve upon stirring in the face of additional Na_2SO_4, it would indicate that the cation forms a complex ion with sulfate ion. Under such conditions, you should put the formula of the complex ion in the box.

In this experiment you can assume that any cations that form complex ions have a coordination number of four. To any precipitates or complex ions, add 6 M HNO_3, drop by drop, until the solution is acidic to litmus (blue to red). If the precipitate dissolves, note that with an A, meaning that the precipitate dissolves in acid. In the case of complexes, the precipitate that originally formed may reprecipitate if the ligands react with acid, and then dissolve again when the solution becomes acidic. If it reprecipitates, indicate that with a P, and as before use an A if the precipitate dissolves when the solution becomes acidic.

Pour the contents of the four tubes into a beaker and rinse the tubes with deionized water. Repeat the tests, first with 1.0 M Na_2CO_3, then with 6 M NaOH, and finally with 6 M NH_3. Although in most cases you will not observe formation of complex ions, you will see a few, and it is important not to miss them. One drop of reagent typically will produce a precipitate if the cation-anion compound is insoluble in water. The complex may be very stable and form readily in excess reagent. You may not get a precipitate reforming when you add HNO_3. If the solution becomes acidic, and you saw no precipitate or a faint one, slowly add 1.0 M Na_2CO_3. If the carbonate comes down, write its formula in the box.

When your table is complete you should be able to use it to state whether the sulfate, carbonate, and hydroxide of each of the cations is insoluble in water, and whether it dissolves in acid. You should also know whether the cation forms a complex ion with SO_4^{2-}, CO_3^{2-}, OH^-, or NH_3, and whether the complex is destroyed by acid.

Now, given the solubility and reaction data you have obtained, your challenge is to devise a step-by-step procedure for establishing whether each cation is present in a given mixture. If you think about it for a while, several possible approaches should occur to you. As you construct your scheme, there are a few things to keep in mind besides the solubility and reaction data.

1. To separate a precipitate from a solution, you can use a centrifuge. Decant the solution into a test tube for use in further steps. Consult the "Separation of Precipitates from Solution" section of Appendix IV for further information.

2. If a precipitate can contain only one cation, its presence serves to prove the presence of that cation. If it may involve more than one cation, it must be further resolved. In that case, the precipitate must be washed free of any cations that did not precipitate in that step, because those cations would possibly interfere with later steps. To clean a precipitate, wash it twice with a 1:1 mixture of water and the precipitating reagent, stirring before centrifuging out the wash liquid.

3. pH is important. Your original tests were made with the cations in a neutral solution. If you want to obtain a precipitate, the solution must have a pH where that precipitate can form. 6 M HNO_3 or 6 M NaOH can be used to bring a mixture to a pH of about seven.

When you have your separation scheme in mind, describe it by constructing a flow diagram. The design of a flow diagram is discussed in the Advance Study Assignment. In the flow diagram the formulas of all species involving the cations should be given at the beginning and end of each step. Reagents are shown alongside the line connecting reactants and products.

When you have completed your flow diagram, test your scheme with a known solution containing all four cations. If your scheme works, show your flow diagram to your instructor, who will give you an unknown to analyze.

DISPOSAL OF REACTION PRODUCTS. On completing the experiment, pour the contents of the beaker used for reaction products into the waste container unless directed otherwise by your instructor.

Experiment 34

Observations and Analysis: Development of a Scheme
for Qualitative Analysis

Table of Solubility Properties

	Ba^{2+}	Mg^{2+}	Cd^{2+}	Al^{3+}
1.0 M Na_2SO_4				
1.0 M Na_2CO_3				
6 M NaOH				
6 M NH_3				

Key: No entry = soluble on mixing reagents
Top formula = precipitate insoluble in water
Second formula = complex ion that forms in excess reagent
Bottom formula = carbonate that precipitates from acidified complex
A = precipitate dissolves in acid
P = precipitate reforms when complex is slowly acidified

Flow diagram for separation scheme:

(continued on following page)

Observations on known:

Observations on unknown:

Cations present in unknown: _____ _____ _____ _____

Unknown # _____

Experiment 34

Advance Study Assignment: Development of a Scheme
for Qualitative Analysis

Qualitative analysis schemes can be summarized by flow diagrams. The flow diagram for a scheme that might be used to analyze a mixture that could contain Cu^{2+}, Pb^{2+}, and Sn^{2+} is shown below:

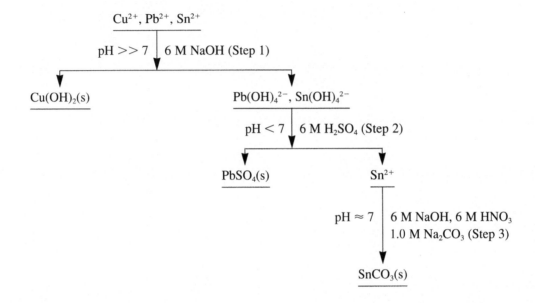

The meaning of the flow diagram is almost obvious. In the first step, 6 M NaOH is added in excess. $Cu(OH)_2$ precipitates (if Cu is present in the sample), and Pb^{2+} and Sn^{2+} remain in solution because of the formation of the hydroxo-complex ions indicated. Any $Cu(OH)_2$ precipitate formed is removed by centrifuging, leaving complex ions still in the solution. In Step 2, the solution is made acidic with H_2SO_4, and $PbSO_4$ precipitates after the complex is destroyed by the H^+ ion in the acid (if Pb is present in the sample). Any $PbSO_4$ that precipitates is removed by centrifuging. In Step 3, pH control is used to bring the pH to about seven. Then addition of Na_2CO_3 precipitates $SnCO_3$ (if Sn is present in the sample).

1. Construct the flow diagram for the following separation scheme for a solution containing Ag^+, Ni^{2+}, and Zn^{2+}. The steps in the procedure are as follows:

   Step 1. Add 6 M HCl to precipitate Ag^+ as AgCl. Ni^{2+} and Zn^{2+} are not affected. Centrifuge out the AgCl.

   Step 2. Add 6 M NaOH in excess, precipitating $Ni(OH)_2(s)$ and converting Zn^{2+} to the $Zn(OH)_4^{2-}$ complex ion. Centrifuge out the $Ni(OH)_2$.

   Step 3. Neutralize the solution with 6 M HCl (add HCl until the pH is about seven). Add 1.0 M Na_2CO_3, precipitating $ZnCO_3$.

   Use the following page for your flow diagram.

Experiment 35

Spot Tests for Some Common Anions

There are two broad categories of problems in analytical chemistry. Quantitative analysis deals with the determination of the amounts of certain species present in a sample; there are several experiments in this manual involving quantitative analysis, and you probably have performed some of them. The other area of analysis, called qualitative analysis, has a more limited purpose, establishing whether given species are or are not present in detectable amounts in a sample. Relatively few of the experiments in this manual deal with problems in qualitative analysis.

We can carry out the qualitative analysis of a sample in various ways. Probably the simplest approach, which we will use in this experiment, is to test for the presence of each possible component by adding a reagent that will cause the component, if it is in the sample, to react in a characteristic way. This method involves a series of "spot" tests, one for each component, carried out on separate samples of the unknown. The difficulty with this way of doing qualitative analysis is that frequently, particularly in complex mixtures, one species may interfere with the analytical test for another. Although interferences are common, there are many ions that can, under optimum conditions at least, be identified in mixtures by simple spot tests.

In this experiment we will use spot tests for the analysis of a mixture that may contain the following commonly encountered anions in solution:

$$CO_3^{2-} \qquad PO_4^{3-} \qquad Cl^- \qquad SCN^-$$

$$SO_4^{2-} \qquad SO_3^{2-} \qquad C_2H_3O_2^- \qquad NO_3^-$$

The procedures we will use involve simple acid-base, precipitation, complex ion formation, and oxidation-reduction reactions. In each case, you should try to recognize the kind of reaction that occurs so that you can write the net ionic equation that describes it.

Experimental Procedure

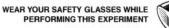
WEAR YOUR SAFETY GLASSES WHILE
PERFORMING THIS EXPERIMENT

Carry out the test for each of the anions as directed. Repeat each test using a solution made by diluting the anion solution 9:1 with deionized water; use your 10-mL graduated cylinder to make the dilution and make sure you mix well before taking the sample for analysis. In some of the tests, a boiling-water bath containing about one-hundred milliliters of water in a 150-mL beaker will be needed, so set that up before proceeding. When performing a test, if no reaction is immediately apparent, stir the mixture with your stirring rod to mix the reagents. These tests can easily be used to detect the anions at concentrations of 0.02 M or greater, but in dilute solutions approaching this limit careful observation may be required.

Test for the Presence of Carbonate Ion, CO_3^{2-}

Cautiously add 1.0 mL of 6 M HCl to 1.0 mL of 1.0 M Na_2CO_3 in a small (13 × 100 mm) test tube. With concentrated solutions, bubbles of carbon dioxide gas are immediately evolved. With dilute solutions, the effervescence will be much less obvious. Warming in the boiling-water bath, with stirring, will increase the amount of bubble formation. Carbon dioxide is colorless and odorless.

Test for the Presence of Sulfate Ion, SO_4^{2-}

Add 1.0 mL of 6 M HCl to 1.0 mL of 0.5 M Na_2SO_4. Add a few drops of 1.0 M $BaCl_2$. A white, finely divided precipitate of $BaSO_4$ indicates the presence of SO_4^{2-} ion.

Test for the Presence of Phosphate Ion, PO_4^{3-}

Add 1.0 mL of 6 M HNO_3 to 1.0 mL of 0.5 M Na_2HPO_4. Then add 1.0 mL of 0.5 M $(NH_4)_2MoO_4$ and stir thoroughly. A yellow precipitate of ammonium phosphomolybdate, $(NH_4)_3PO_4 \cdot 12\,MoO_3$, establishes the presence of phosphate. The precipitate may form slowly, particularly in more dilute solutions; if it does not appear promptly, put the test tube in a boiling-water bath for a few minutes.

Test for the Presence of Sulfite Ion, SO_3^{2-}

Sulfite ion in acid solution tends to evolve SO_2, which can be detected by its odor even at low concentrations. Sulfite ion is slowly oxidized to sulfate in moist air, so sulfite-containing solutions will usually test positive for sulfate ion.

To 1.0 mL of 0.5 M Na_2SO_3 add 1.0 mL of 6 M HCl, and mix with your stirring rod. Cautiously, sniff the rod to try to detect the acrid odor of SO_2, which is a good test for sulfite. If the odor is too faint to detect, put the test tube in the boiling-water bath for 10. seconds and sniff again.

To proceed with a chemical test, add 1.0 mL of 1.0 M $BaCl_2$ to the solution in the tube. Stir, and centrifuge out any precipitate of $BaSO_4$. Decant the clear solution into a test tube and add 1.0 mL of $3\%_{mass}$ H_2O_2, hydrogen peroxide. Stir the solution and let it stand for a few seconds. If sulfite is present, its oxidation to sulfate will cause a new precipitate of $BaSO_4$ to form.

If thiocyanate is present, it may interfere with this chemical test. In that case, add 1.0 mL of 1.0 M $BaCl_2$ to 1.0 mL of the sample. Centrifuge out any precipitate, which will contain $BaSO_3$ if sulfite is present, but will not contain thiocyanate. Wash the solid with 2 mL of water, stir, centrifuge, and discard the wash. To the precipitate add 1.0 mL of 6 M HCl and 2 mL of water, and stir. Centrifuge out any $BaSO_4$, and decant the clear liquid into a test tube. To the liquid add 1.0 mL of $3\%_{mass}$ H_2O_2. If you have sulfite present, you will observe a new precipitate of $BaSO_4$ within a few seconds.

Test for the Presence of Thiocyanate Ion, SCN^-

Add 1.0 mL of 6 M acetic acid, $HC_2H_3O_2$, to 1.0 mL of 0.5 M KSCN and stir. Add one or two drops of 0.10 M $Fe(NO_3)_3$. A deep red coloration as a result of formation of $FeSCN^{2+}$ ion is proof of the presence of SCN^- ion.

Test for the Presence of Chloride Ion, Cl^-

Add 1.0 mL of 6 M HNO_3 to 1.0 mL of 0.5 M NaCl. Add two or three drops of 0.10 M $AgNO_3$. A white, curdy precipitate of AgCl will form if chloride ion is present.

Several anions interfere with this test, because they too form white precipitates with $AgNO_3$ under these conditions. In this experiment only SCN^- ion will interfere. If the sample contains SCN^- ion, put 1.0 mL of the solution into a 30- or 50-mL beaker and add 1.0 mL of 6 M HNO_3. Boil the solution gently until its volume is decreased by half. This will destroy most of the thiocyanate. Transfer the solution to a small test tube and add 1.0 mL of 6 M HNO_3 and a few drops of 0.10 M $AgNO_3$ solution. If Cl^- is present you will get a curdy precipitate. If Cl^- is absent, you may see some cloudiness due to residual amounts of SCN^-.

Test for the Presence of Acetate Ion, $C_2H_3O_2^-$

In the general case, we would begin this test by ensuring the solution is acidic as indicated by litmus, adding 6 M HNO_3 if needed to reach that point. To demonstrate the test, we will use 1.0 mL of 0.5 M $NaC_2H_3O_2$, which is guaranteed to start out basic. Thus, 6 M HNO_3 is added until the solution turns blue litmus red, indicating that it has been made acidic. Add 6 M NH_3 until the solution is just basic to litmus, thereby assuring a nearly neutral solution. Add 1 drop of 1.0 M $BaCl_2$. If a precipitate forms, add 1.0 mL of the $BaCl_2$ solution to precipitate interfering anions. Stir, centrifuge, and decant the clear liquid into a test tube. Add 1 drop of $BaCl_2$ to make sure that precipitation is complete. To 1.0 mL of the liquid add 0.10 M KI_3, drop by drop, until the solution takes on a fairly strong rust color. Add 0.5 mL of 0.10 M $La(NO_3)_3$ and 6 drops of 6 M NH_3. Stir, and put the test tube in the boiling-water bath. If acetate is present, the mixture will darken to nearly black in a few minutes. The color is due to iodine adsorbed on the basic lanthanum acetate precipitate.

Test for the Presence of Nitrate Ion, NO_3^-

To 1.0 mL of 0.5 M $NaNO_3$ add 1.0 mL of 6 M NaOH. Then add a few granules of Al metal, using your spatula, and put the test tube in the boiling-water bath. In a few seconds, the Al-NaOH reaction will produce H_2 gas, which will reduce the NO_3^- ion to NH_3, which will come off as a gas. To detect the NH_3, hold a piece of moistened red litmus paper just above the end of the test tube. If the sample contains nitrate ion, the litmus paper will gradually turn blue, within a minute or two. Blue spots caused by effervescence are not to be confused with the blue color over all of the litmus exposed to NH_3 vapors. Cautiously sniff the vapors at the top of the tube; you may be able to detect the odor of ammonia.

If SCN^- is present, it will interfere with this test. In that case, first add 1.0 mL of 1.0 M $CuSO_4$ to 1.0 mL of the sample and put the test tube in the boiling-water bath for a minute or two. Centrifuge out the precipitate, and decant the solution into a test tube. Add 1.0 mL of 1.0 M Na_2CO_3 to remove excess Cu^{2+} ion. Centrifuge out the precipitate, and decant the solution into a test tube. To 1.0 mL of the solution add 1.0 mL of 6 M NaOH, and proceed, starting with the second sentence of this procedure.

Analysis of an Unknown

Once you have familiarized yourself with all of the tests, obtain an unknown from your instructor and analyze it by applying the tests to separate 1.0-mL portions. The unknown will contain three or four ions on the list, so your test for a given ion may be affected by the presence of others. When you think you have properly analyzed your unknown, you may, if you wish, make a "known" with the composition you found and test it to see if it behaves as your unknown did.

> **DISPOSAL OF REACTION PRODUCTS.** As you complete each test, pour the products into a beaker. When you have finished the experiment, pour the contents of the beaker into the waste container, unless directed otherwise by your instructor.

Experiment 35

Observations and Analysis: Spot Tests for Some Common Anions

Table of Spot Test Results

Ion	Result with stock solution	Result with 9:1 dilution	Result with unknown
CO_3^{2-}			
SO_4^{2-}			
PO_4^{3-}			
SO_3^{2-}			
Cl^-			
$C_2H_3O_2^-$			
SCN^-			
NO_3^-			

Unknown # _____ contains _____

Experiment 35

Advance Study Assignment: Spot Tests for Some Common Anions

1. Each of the observations in the following list was made on a different solution. Given the observations, state which ion studied in this experiment is present. If the test is not definitive, indicate that with a question mark.

   A. Addition of 6 M NaOH and Al to the solution produces a vapor that turns red litmus blue.
   Ion present:

   B. Addition of 6 M HCl produces a vapor with an acrid odor.
   Ion present:

   C. Addition of 6 M HCl produces an effervescence.
   Ion present:

   D. Addition of 6 M HNO_3 plus 1.0 M $BaCl_2$ produces a precipitate.
   Ion present:

   E. Addition of 6 M HNO_3 plus 0.10 M $AgNO_3$ produces a precipitate.
   Ion present:

   F. Addition of 6 M HNO_3 plus 0.5 M $(NH_4)_2MoO_4$ produces a precipitate.
   Ion present:

2. An unknown containing one or more of the ions studied in this experiment has the following properties:

   A. No effect is observed on addition of 6 M HNO_3.

   B A white precipitate is observed on addition of 1.0 M $BaCl_2$ to the solution from Part A.

   C. No effect is observed on addition of 0.10 M $AgNO_3$ to the solution from Part A.

   D. A yellow precipitate is observed on addition of $(NH_4)_2MoO_4$, to the solution from Part A.

On the basis of this information, which ions are present, which are absent, and which are in doubt?

Present	Absent	In doubt

3. The chemical reactions that are used in the anion spot tests in this experiment are for the most part simple precipitation or acid-base reactions. Given the information in each test procedure, try to write the net ionic equation for the key reaction in each test.

A. CO_3^{2-}

B. SO_4^{2-}

C. PO_4^{3-} (Reactants are HPO_4^{2-}, NH_4^+, MoO_4^{2-}, and H^+; products are $(NH_4)_3PO_4 \cdot 12\,MoO_3$ and H_2O; no oxidation or reduction occurs.)

D. SO_3^{2-}

E. SCN^-

F. Cl^-

G. NO_3^- [Take Al and NO_3^- to be the reactants, and NH_3 and $Al(OH)_4^-$ to be the products; the final reaction also contains OH^- and H_2O as reactants. Note that this is a redox reaction, and is most easily balanced as such. The half-reactions involved are $NO_3^-(aq) + 6\,H_2O(\ell) + 8\,e^- \rightarrow NH_3(g) + 9\,OH^-(aq)$ and $Al(s) + 4\,OH^-(aq) \rightarrow Al(OH)_4^-(aq) + 3\,e^-$.]

Experiment 36

Qualitative Analysis of Group I Cations

The objective of this experiment is the precipitation and separation of three Group I cations, Pb^{2+}, Hg_2^{2+}, and Ag^+, from a solution. Their chlorides are all insoluble in cold water, so they can be removed as a group from solution by the addition of HCl. The reactions that occur are simple precipitations:

$$Ag^+(aq) + Cl^-(aq) \rightarrow AgCl(s) \tag{1}$$

$$Pb^{2+}(aq) + 2\,Cl^-(aq) \rightarrow PbCl_2(s) \tag{2}$$

$$Hg_2^{2+}(aq) + 2\,Cl^-(aq) \rightarrow Hg_2Cl_2(s) \tag{3}$$

It is important to add enough HCl to ensure complete precipitation, but not too large an excess. In concentrated HCl solution these chlorides tend to dissolve, producing chloro-complexes such as $AgCl_2^-$.

Lead chloride is separated from the other two chlorides by heating with water. The $PbCl_2$ dissolves in hot water by the reverse of Reaction 2:

$$PbCl_2(s) \rightarrow Pb^{2+}(aq) + 2\,Cl^-(aq) \tag{4}$$

Once Pb^{2+} has been separated into solution, we can check for its presence by adding a solution of K_2CrO_4. The chromate ion, CrO_4^{2-}, gives a yellow precipitate with Pb^{2+}:

$$Pb^{2+}(aq) + CrO_4^{2-}(aq) \rightarrow PbCrO_4(s) \tag{5}$$
$$\text{yellow}$$

The other two insoluble chlorides, AgCl and Hg_2Cl_2, can be separated by adding aqueous ammonia. Silver chloride dissolves, forming the complex ion $Ag(NH_3)_2^+$:

$$AgCl(s) + 2\,NH_3(aq) \rightarrow Ag(NH_3)_2^+(aq) + Cl^-(aq) \tag{6}$$

Ammonia also reacts with Hg_2Cl_2, via a rather unusual oxidation-reduction reaction. The products include finely divided metallic mercury, which is black, and a compound of formula $HgNH_2Cl$, which is white:

$$Hg_2Cl_2(s) + 2\,NH_3(aq) \rightarrow Hg(\ell) + HgNH_2Cl(s) + NH_4^+(aq) + Cl^-(aq) \tag{7}$$
$$\text{white} \qquad\qquad\quad \text{black} \qquad \text{white}$$

As this reaction occurs, the solid appears to change color, from white to black or gray.

The solution containing $Ag(NH_3)_2^+$ needs to be further tested to establish the presence of silver. The addition of a strong acid (HNO_3) to the solution destroys the complex ion and reprecipitates silver chloride. We may consider this reaction to occur in two steps:

$$Ag(NH_3)_2^+(aq) + 2\,H^+(aq) \rightarrow Ag^+(aq) + 2\,NH_4^+(aq)$$
$$\underline{Ag^+(aq) + Cl^-(aq) \rightarrow AgCl(s)}$$
$$Ag(NH_3)_2^+(aq) + 2\,H^+(aq) + Cl^-(aq) \rightarrow AgCl(s) + 2\,NH_4^+(aq) \qquad \text{(8)}$$
$$\text{white}$$

WEAR YOUR SAFETY GLASSES WHILE PERFORMING THIS EXPERIMENT

Experimental Procedure

See Appendix IV for some suggestions regarding procedures in qualitative analysis.

Step 1. Precipitation of the Group I Cations To gain familiarity with the procedures used in qualitative analysis, we will first analyze a known Group I solution made by mixing equal volumes of 0.10 M $AgNO_3$, 0.2 M $Pb(NO_3)_2$, and 0.10 M $Hg_2(NO_3)_2$.

Add 2 drops of 6 M HCl to 1.0 mL of the known solution in a small (13×100 mm) test tube (1 mL $\approx$ 1 cm depth in the tube). Mix with your stirring rod. Centrifuge the mixture, making sure there is a blank test tube containing about the same amount of water in the opposite opening in the centrifuge. Add 1 more drop of the 6 M HCl to test for completeness of precipitation. If more precipitate forms, centrifuge again and repeat until no new precipitate forms. Decant the supernatant liquid into another test tube and save it for further tests if cations from other groups may be present. The precipitate will be white and will contain the chlorides of the Group I cations.

Step 2. Separation of Pb^{2+} Wash the precipitate with 1 or 2 mL of deionized water. Stir, centrifuge, and decant the liquid, which may be discarded.

Add 2 mL of deionized water from your wash bottle to the precipitate in the test tube, then place the test tube in a 250-mL beaker that is about half full of boiling water. Leave the test tube in the boiling-water bath for a minute or two, stirring occasionally with a glass rod. This will dissolve most of the $PbCl_2$ but not the other two chlorides. Centrifuge the hot mixture, and decant the hot liquid into a test tube. Set aside the remaining precipitate for further tests later on (right now, in Step 3, you will work with the liquid).

Step 3. Identification of Pb^{2+} Add 1 drop of 6 M acetic acid and a few drops of 0.10 M K_2CrO_4 to the *liquid* solution (NOT THE SOLID) decanted off in Step 2. If Pb^{2+} is present, a bright yellow precipitate of $PbCrO_4$ will form.

Step 4. Separation and Identification of Hg_2^{2+} To the *precipitate* from Step 2, add 1.0 mL of 6 M NH_3 and stir thoroughly. Centrifuge the mixture and decant the liquid into a test tube. A gray or black precipitate left behind, produced by the reaction of Hg_2Cl_2 with ammonia, proves the presence of Hg_2^{2+}.

Step 5. Identification of Ag^+ Add 6 M HNO_3 to the solution from Step 4 until it is acidic to litmus paper. (This will require about one milliliter of 6 M HNO_3.) Test for acidity by dipping the end of your stirring rod in the solution, then touching it to a piece of blue litmus paper (which will turn red in acidic solution). If Ag^+ is present, it will precipitate as white AgCl as the solution is acidified.

Step 6. When you have completed the tests on the known solution, obtain an unknown and analyze it for the possible presence of Ag^+, Pb^{2+}, and Hg_2^{2+}.

DISPOSAL OF REACTION PRODUCTS. All reaction products in this experiment should be dealt with as directed by your instructor.

Flow Diagrams

It is possible to summarize the directions for analysis of the Group I cations in a flow diagram. In the diagram, successive steps in the procedure are linked with arrows. Reactant cations or reactant substances containing the ions are at one end of each arrow and products formed are at the other end. Reagents and conditions used to carry out each step are placed alongside the arrows. A partially completed flow diagram for the Group I ions follows:

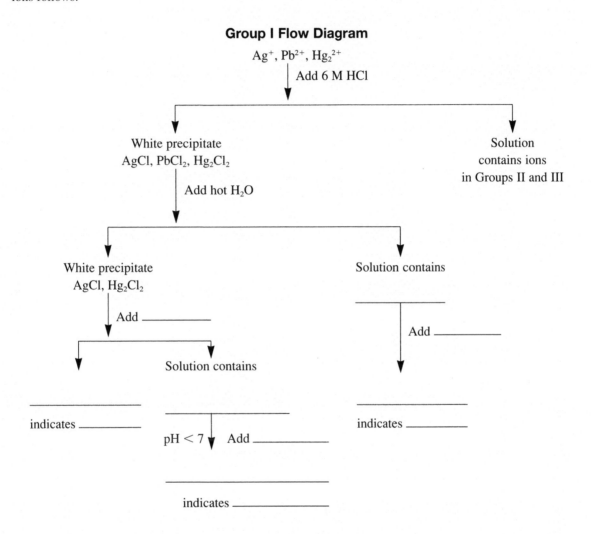

Group I Flow Diagram

Ag^+, Pb^{2+}, Hg_2^{2+}

Add 6 M HCl

White precipitate
$AgCl$, $PbCl_2$, Hg_2Cl_2

Solution
contains ions
in Groups II and III

Add hot H_2O

White precipitate
$AgCl$, Hg_2Cl_2

Solution contains

Add ―――――

Add ―――――

Solution contains

―――――― indicates ―――――

―――――― indicates ―――――

$pH < 7$ ↓ Add ――――――

―――――― indicates ――――――

You will find it useful to construct flow diagrams for each of the cation groups. You can use such diagrams in the laboratory as brief guides to procedure, and you can use them to record your observations on your known and unknown solutions.

Name _____ **Section** _____

Experiment 36

Observations and Analysis: Qualitative Analysis of Group I Cations

Flow Diagram for Group I

On the flow diagram above, indicate the actual observations made upon the known and the unknown. For the known, simply note any deviations from the expected results per the flow diagram, labeling them with "KNOWN." For the unknown, clearly indicate the path through the flow diagram that you followed to reach your conclusions.

Unknown # _____   Ions reported present _____ _____ _____

Experiment 36

Advance Study Assignment: Qualitative Analysis of Group I Cations

1. At the start of the report page for this experiment, draw a complete flow diagram for the separation and identification of the ions in Group I. (Copy and complete the partial flow diagram provided at the end of the instructions.)

2. Why would it be unwise to simply add 10 drops of 6 M HCl immediately in Step 1? What could go wrong?

3. A solution may contain Ag^+, Pb^{2+}, and/or Hg_2^{2+}. A white precipitate forms on addition of 6 M HCl. None of the precipitate dissolves in hot water. The precipitate turns gray upon addition of ammonia. Which of the Group I ions are present, which are absent, and which remain undetermined? State your reasoning. *Note:* For "paper unknowns" such as this one, confirmatory tests are usually not included, and you do not need to provide any. The information provided here, without confirmatory tests, is sufficient to clearly indicate the presence or absence of some of the Group I ions studied in this experiment, while leaving others in doubt.

Present _____

Absent _____

In doubt _____

(continued on following page)

4. You are given an unknown solution that contains only one of the Group I cations and no other metallic cations. Develop the simplest procedure you can think of to determine which cation is present. Draw a flow diagram showing the procedure and the observations to be expected at each step with each of the possible cations. The information in Appendix IIA should be helpful.

Experiment 37

Qualitative Analysis of Group II Cations

T‌he sulfides of the Group II ions are insoluble at a pH of 0.5. They include Bi^{3+}, Sn^{4+}, Sb^{3+}, and Cu^{2+}. If a solution containing these ions is adjusted to this pH and then saturated with H_2S, then Bi_2S_3, SnS_2, Sb_2S_3, and CuS will precipitate. The reaction with Bi^{3+} is typical:

$$2\ Bi^{3+}(aq) + 3\ H_2S(aq) \rightarrow Bi_2S_3(s) + 6\ H^+(aq) \qquad (1)$$
$$\text{black}$$

Saturation with H_2S could be achieved by simply bubbling the gas from a generator through the solution. A more convenient method, however, is to heat the acid solution after adding a small amount of thioacetamide. This compound, CH_3CSNH_2, hydrolyzes when heated in aqueous solution to liberate H_2S:

$$CH_3CSNH_2(aq) + 2\ H_2O(\ell) \rightarrow H_2S(aq) + CH_3COO^-(aq) + NH_4^+(aq) \qquad (2)$$

Using thioacetamide as the precipitating reagent has the advantage of minimizing odor problems and giving denser precipitates.

The four insoluble sulfides can be separated into two subgroups by extracting with a solution of sodium hydroxide. The sulfides of tin and antimony dissolve, forming hydroxo-complexes:

$$SnS_2(s) + 6\ OH^-(aq) \rightarrow Sn(OH)_6^{2-}(aq) + 2\ S^{2-}(aq) \qquad (3)$$

$$Sb_2S_3(s) + 8\ OH^-(aq) \rightarrow 2\ Sb(OH)_4^-(aq) + 3\ S^{2-}(aq) \qquad (4)$$

Since Cu^{2+} and Bi^{3+} do not readily form hydroxo-complexes, CuS and Bi_2S_3 do not dissolve in solutions of NaOH.

The solution containing the $Sb(OH)_4^-$ and $Sn(OH)_6^{2-}$ complex ions is treated with HCl and thioacetamide. The H^+ ions of the strong acid HCl destroy the hydroxo-complexes; the free cations then reprecipitate as the sulfides. The reaction with $Sn(OH)_6^{2-}$ may be written as follows:

$$Sn(OH)_6^{2-}(aq) + 6\ H^+(aq) \rightarrow Sn^{4+}(aq) + 6\ H_2O(\ell)$$
$$\underline{Sn^{4+}(aq) + 2\ H_2S(aq) \rightarrow SnS_2(s) + 4\ H^+(aq)}$$
$$Sn(OH)_6^{2-}(aq) + 2\ H^+(aq) + 2\ H_2S(aq) \rightarrow SnS_2(s) + 6\ H_2O(\ell) \qquad (5)$$
$$\text{tan}$$

The $Sb(OH)_4^-$ ion behaves in a similar manner, being converted first to Sb^{3+} and then to Sb_2S_3. The Sb_2S_3 and SnS_2 are then dissolved as chloro-complexes in hydrochloric acid and their presence is confirmed by appropriate tests.

The confirmatory test for tin takes advantage of the two oxidation states, +2 and +4, of the metal. Aluminum is added to reduce Sn^{4+} to Sn^{2+}:

$$2\ Al(s) + 3\ Sn^{4+}(aq) \rightarrow 2\ Al^{3+}(aq) + 3\ Sn^{2+}(aq) \qquad (6)$$

The classic test for the presence of Sn^{2+} in this solution is to add a solution containing mercury(II) chloride, $HgCl_2$, which would bring about another oxidation-reduction process:

$$Sn^{2+}(aq) + 2\ Hg^{2+}(aq) + 2\ Cl^-(aq) \rightarrow Sn^{4+}(aq) + Hg_2Cl_2(s) \qquad (7)$$
$$\text{white}$$

The appearance of a white precipitate of insoluble Hg_2Cl_2 would confirm the presence of tin. We no longer use this test due to the hazards and disposal challenges associated with mercury compounds. Instead, we use an organic compound called Janus Green (JG). This compound is easily reduced by a strong reducing agent, such as Sn^{2+}, and a color change from blue to either violet-red or colorless (depending upon the relative concentrations) indicates the presence of Sn^{2+}:

$$Sn^{2+}(aq) + 2\,JG(aq) \rightarrow Sn^{4+}(aq) + 2\,JG^{-}(aq) \qquad (8)$$
$$\text{blue} \qquad\qquad\qquad\qquad \text{violet-red}$$
$$\text{or}$$
$$Sn^{2+}(aq) + JG(aq) \rightarrow Sn^{4+}(aq) + JG^{2-}(aq) \qquad (9)$$
$$\text{blue} \qquad\qquad\qquad\qquad \text{colorless}$$

The Sb^{3+} ion is difficult to confirm in the presence of Sn^{4+}; the colors of the sulfides of these two ions are similar. To prevent interference by Sn^{4+}, the solution to be tested for Sb^{3+} is first treated with oxalic acid. This forms a very stable oxalato complex with Sn^{4+}, $Sn(C_2O_4)_3^{2-}$. Treatment with H_2S then gives a bright-orange precipitate of Sb_2S_3 if antimony is present:

$$2\,Sb^{3+}(aq) + 3\,H_2S(aq) \rightarrow Sb_2S_3(s) + 6\,H^{+}(aq) \qquad (10)$$
$$\text{bright}$$
$$\text{orange}$$

As pointed out earlier, CuS and Bi_2S_3 are insoluble in NaOH solutions, and they do not dissolve in hydrochloric acid. However, these two sulfides can be brought into solution by treatment with the oxidizing acid HNO_3. The reactions that occur are of the oxidation-reduction type. The NO_3^- ion is reduced, usually to NO_2; S^{2-} ions are oxidized to elementary sulfur, and the cation, Cu^{2+} or Bi^{3+}, is brought into solution. The reactions are:

$$CuS(s) + 4\,H^{+}(aq) + 2\,NO_3^-(aq) \rightarrow Cu^{2+}(aq) + S(s) + 2\,NO_2(g) + 2\,H_2O(\ell) \qquad (11)$$

$$Bi_2S_3(s) + 12\,H^{+}(aq) + 6\,NO_3^-(aq) \rightarrow 2\,Bi^{3+}(aq) + 3\,S(s) + 6\,NO_2(g) + 6\,H_2O(\ell) \qquad (12)$$

The two ions, Cu^{2+} and Bi^{3+}, are easily separated by the addition of aqueous ammonia. The Cu^{2+} ion is converted to the deep-blue complex $Cu(NH_3)_4^{2+}$:

$$Cu^{2+}(aq) + 4\,NH_3(aq) \rightarrow Cu(NH_3)_4^{2+}(aq) \qquad (13)$$
$$\text{aqua} \qquad\qquad\qquad \text{deep blue}$$

The reaction of ammonia with Bi^{3+} is quite different. The OH^- ions produced by the reaction of NH_3 with water precipitate Bi^{3+} as $Bi(OH)_3$. We can consider the reaction as occurring in two steps:

$$3\,NH_3(aq) + 3\,H_2O(\ell) \rightleftharpoons 3\,NH_4^+(aq) + 3\,OH^-(aq)$$
$$\underline{Bi^{3+}(aq) + 3\,OH^-(aq) \rightarrow Bi(OH)_3(s)}$$
$$Bi^{3+}(aq) + 3\,NH_3(aq) + 3\,H_2O(\ell) \rightarrow Bi(OH)_3(s) + 3\,NH_4^+(aq) \qquad (14)$$
$$\text{white}$$

To confirm the presence of Bi^{3+}, the precipitate of $Bi(OH)_3$ is dissolved by treating with hydrochloric acid:

$$Bi(OH)_3(s) + 3\,H^{+}(aq) \rightarrow Bi^{3+}(aq) + 3\,H_2O(\ell) \qquad (15)$$
$$\text{white}$$

The solution formed is poured into deionized water. If Bi^{3+} is present, a white precipitate of bismuth oxychloride, BiOCl, will form:

$$Bi^{3+}(aq) + H_2O(\ell) + Cl^-(aq) \rightarrow BiOCl(s) + 2\,H^{+}(aq) \qquad (16)$$
$$\text{white}$$

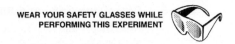

Experimental Procedure

Step 1. Adjustment of pH Prior to Precipitation Pour a 1.0 mL sample of the "known" solution for Group II, containing equal volumes of 0.10 M solutions of the nitrates or chlorides of Sn^{4+}, Sb^{3+}, Cu^{2+}, and Bi^{3+}, into a small (13 × 100 mm) test tube. Prepare a boiling-water bath, using a 250-mL beaker about two-thirds full of water. Use another beaker full of water as the storage and rinsing place for your stirring rods. Prepare a pH test paper by making ten spots of methyl violet indicator on a piece of filter paper, about 1 drop to a spot. Let the paper dry in the air.

The Group II known will be very acidic because of the presence of HCl, which is necessary to keep the salts of Bi^{3+}, Sn^{4+}, and Sb^{3+} in solution. Add 6 M NH_3, drop by drop, until the solution, after stirring, produces a violet spot on the pH test paper; test for pH by dipping your stirring rod in the solution and touching it to the paper. (A precipitate will probably form during this step because of the formation of insoluble salts of the Group II cations.) Then add 1 drop of 6 M HCl for each milliliter of solution. This should bring the pH of the solution close to 0.5. Test the pH, using the test paper. At a pH of 0.5, methyl violet will have a blue-green color. Compare your spot with that made by putting 1 drop of 0.3 M HCl (pH = 0.5) on the test paper. Adjust the pH of your solution as necessary by adding HCl or NH_3, until the pH test gives about the right color on the paper. If you have trouble deciding on the color, centrifuging out the precipitate may help. When you have established the proper pH, add 1.0 mL of 1.0 M thioacetamide to the solution and stir.

Step 2. Precipitation of the Group II Sulfides Heat the test tube in the boiling-water bath for at least five minutes. **CAUTION:** **Small amounts of H_2S will be liberated; this gas is toxic, so avoid inhaling it unnecessarily.**

In the presence of Group II ions, a precipitate will form: typically, its color will be initially light, gradually darkening, and finally approaching black. Continue to heat the tube for at least two minutes after the color has stopped changing. Cool the test tube under the water tap and let it stand for a minute or so. Centrifuge out the precipitate and decant the solution, which will contain any Group III ions that may be present, into a test tube. Test the solution for completeness of precipitation by adding 2 drops of thioacetamide and letting it stand for a minute. If a precipitate forms, add a few more drops of thioacetamide and heat again in the boiling-water bath; then combine the two batches of precipitate. Wash the precipitate with 2 mL of 1.0 M NH_4Cl solution and stir thoroughly. Centrifuge, then discard the wash into a waste beaker.

Step 3. Separation of the Group II Sulfides into Two Subgroups To the precipitate from Step 2, add 2 mL of 1.0 M NaOH. Heat in the boiling-water bath, with stirring, for two minutes. Any SnS_2 or Sb_2S_3 should dissolve. The residue will typically be dark and may contain CuS and/or Bi_2S_3. Centrifuge and decant the yellow liquid into a test tube. Wash the precipitate twice, each time using 2 mL of deionized water and a few drops of 1.0 M NaOH. Stir, centrifuge, and decant, discarding the liquid each time into a waste beaker. To the precipitate add 2 mL of 6 M HNO_3 and put the test tube aside (for Step 8). At this point the tin and antimony are present in the solution as complex ions, and the copper and bismuth are in the sulfide precipitate.

Step 4. Reprecipitation of SnS_2 and Sb_2S_3 To the yellow liquid from Step 3 add 6 M HCl drop by drop until the mixture is just acidic to litmus. Upon acidification, the tin and antimony will again precipitate as orange sulfides. Add 5 drops of 1.0 M thioacetamide and heat in the boiling-water bath for two minutes to complete the precipitation. Centrifuge the tube and decant off the liquid, which may be discarded.

Step 5. **Dissolving SnS₂ and Sb₂S₃ in Acid Solution** Add 2 mL of 6 M HCl to the precipitate from Step 4 and heat in the boiling-water bath for a minute or two to dissolve the precipitate. Transfer the solution to a 30-mL beaker and bring it to a gentle boil for about a minute to drive out the H_2S. Add 1.0 mL of 6 M HCl and 1.0 mL of water, then pour the liquid back into a small test tube. If there is an insoluble residue, centrifuge it out and decant the liquid into a test tube.

Step 6. **Confirmation of the Presence of Tin** Pour half of the solution from Step 5 into a test tube and add 2 mL of 6 M HCl and a 1.0-cm length of 24-gauge aluminum wire. Heat the test tube in the boiling-water bath to promote reaction of the Al and production of H_2. In this reducing medium, any tin present will be converted to Sn^{2+} and any antimony to the metal, which will appear as black specks. Heat the tube for two minutes after *all* of the wire has reacted. Centrifuge out any solid and decant the liquid into a test tube. To the liquid, add 1 drop of Janus Green indicator solution. The presence of tin will be established by the blue solution added either turning to a violet/red color or yielding a colorless solution, depending upon the amount of tin present. The reaction is evidence of the reducing power of Sn^{2+}.

Step 7. **Confirmation of the Presence of Antimony** To the other half of the solution from Step 5, add 1.0 M NaOH to bring the pH to 0.5. Add 2 mL of water and about half a gram of oxalic acid, then stir until no more crystals dissolve. Oxalic acid forms a very stable complex with the Sn^{4+} ion. Add 1.0 mL of 1.0 M thioacetamide and put the test tube in the boiling-water bath. The formation of a red-orange precipitate of Sb_2S_3 confirms the presence of antimony.

Step 8. **Dissolving the CuS and Bi₂S₃** Heat the test tube containing the precipitate from Step 3 in the boiling-water bath. Any sulfides that have not already dissolved should go into solution in a minute or two, possibly leaving some insoluble sulfur residue. Continue heating until no further reaction appears to occur, at least two minutes after the initial changes. Centrifuge and decant the solution, which may contain Cu^{2+} and/or Bi^{3+}, into a test tube. Discard any solid residue.

Step 9. **Confirmation of the Presence of Copper** To the solution from Step 8 add 6 M NH_3 dropwise until the mixture is just basic to litmus. Add 0.5 mL more. Centrifuge out any white precipitate that forms, and decant the liquid into a test tube. If the liquid is deep blue, the color is due to the $Cu(NH_3)_4^{2+}$ ion, and copper is present.

Step 10. **Confirmation of the Presence of Bismuth** Wash the precipitate from Step 9, which probably contains bismuth, with 1.0 mL of water and 0.5 mL of 6 M NH_3. Stir, centrifuge, and discard the wash. To the precipitate add 0.5 mL of 6 M HCl and 0.5 mL of water. Stir to dissolve any $Bi(OH)_3$ that is present. Add the solution, drop by drop, to 400 mL of water in a 600-mL beaker. A white cloudiness, caused by the slow precipitation of BiOCl, confirms the presence of bismuth.

Step 11. When you have completed the analysis of your known, obtain a Group II unknown and test it for the presence of Sn^{4+}, Sb^{3+}, Cu^{2+}, and Bi^{3+}.

Take it Further (Optional): Experimentally establish the presence of copper in a mineral supplement tablet. Estimate the amount of copper present by comparison with a suitable standard.

DISPOSAL OF REACTION PRODUCTS. As you complete each part of this experiment, put the waste products in a beaker. At the end of the experiment, pour the contents of the beaker into a waste container, unless directed otherwise by your instructor.

Experiment 37

Observations and Analysis: Qualitative Analysis of Group II Cations

Flow Diagram for Group II

On the flow diagram above, indicate the actual observations made upon the known and the unknown. For the known, simply note any deviations from the expected results, per the flow diagram, labeling them with "KNOWN." For the unknown, clearly indicate the path through the flow diagram that you followed to reach your conclusions.

Unknown # _____   Ions reported present _____ _____ _____ _____

Name _____ Section _____

Experiment 37

Observations and Analysis: Qualitative Analysis of Group II Cations

Flow Diagram for Group II

Experiment 37

Advance Study Assignment: Qualitative Analysis of Group II Cations

1. Prepare a complete flow diagram for the separation and identification of the Group II cations and put it on the report page for this experiment.

2. Write balanced net ionic equations for the following reactions:

   a. Precipitation of tin(IV) sulfide with H_2S

   b. The confirmatory test for antimony

   c. The dissolution of Bi_2S_3 in hot nitric acid

   d. The confirmatory test for copper

3. A solution that may contain Cu^{2+}, Bi^{3+}, Sn^{4+}, and/or Sb^{3+} ions is treated with thioacetamide in an acid medium. The black precipitate that forms is partly soluble in strongly alkaline solution. The precipitate that remains is soluble in 6 M HNO_3 and gives only a blue solution on treatment with excess NH_3. The alkaline solution, when acidified, produces an orange precipitate. On the basis of this information, which ions are present, which are absent, and which are still in doubt? State your reasoning. *Note:* For "paper unknowns" such as this one, confirmatory tests are usually not included, and you do not need to provide any. The information provided here, without confirmatory tests, is sufficient to clearly indicate the presence or absence of some of the Group II ions studied in this experiment, while leaving others in doubt.

   Present _____

   Absent _____

   In doubt _____

Experiment 38

Qualitative Analysis of Group III Cations

The four ions that we will consider from Group III are Cr^{3+}, Al^{3+}, Fe^{3+}, and Ni^{2+}. The first step in their analysis involves treating a solution containing them with sodium hydroxide, NaOH, and sodium hypochlorite, NaOCl. The OCl^- ion oxidizes Cr^{3+} to CrO_4^{2-}:

$$2\ Cr^{3+}(aq) + 3\ OCl^-(aq) + 10\ OH^-(aq) \rightarrow 2\ CrO_4^{2-}(aq) + 3\ Cl^-(aq) + 5\ H_2O(\ell) \qquad (1)$$
$$\text{blue-gray} \qquad\qquad\qquad\qquad\qquad\qquad \text{yellow}$$

The chromate ion, CrO_4^{2-}, stays in solution. The same is true of the hydroxo-complex ion $Al(OH)_4^-$, formed by the reaction of Al^{3+} with excess OH^-:

$$Al^{3+}(aq) + 4\ OH^-(aq) \rightarrow Al(OH)_4^-(aq) \qquad (2)$$

In contrast, the other two ions in the group form insoluble hydroxides under these conditions:

$$Ni^{2+}(aq) + 2\ OH^-(aq) \rightarrow Ni(OH)_2(s) \qquad (3)$$
$$\text{green} \qquad\qquad\qquad\qquad \text{green}$$

$$Fe^{3+}(aq) + 3\ OH^-(aq) \rightarrow Fe(OH)_3(s) \qquad (4)$$
$$\text{yellow} \qquad\qquad\qquad\qquad \text{red}$$

The Ni^{2+} and Fe^{3+} ions, unlike Al^{3+}, do not readily form hydroxo-complexes. Unlike Cr^{3+}, they do not have a stable higher oxidation state, and so are not oxidized by OCl^-.

To separate aluminum from chromium, the solution containing CrO_4^{2-} and $Al(OH)_4^-$ is first acidified. This destroys the hydroxo-complex of aluminum:

$$Al(OH)_4^-(aq) + 4\ H^+(aq) \rightarrow Al^{3+}(aq) + 4\ H_2O(\ell) \qquad (5)$$

Treatment with aqueous ammonia then gives a white gelatinous precipitate of aluminum hydroxide. We can think of this reaction as occurring in two steps:

$$3\ NH_3(aq) + 3\ H_2O(\ell) \rightleftharpoons NH_4^+(aq) + 3\ OH^-(aq)$$
$$\underline{Al^{3+}(aq) + 3\ OH^-(aq) \rightarrow Al(OH)_3(s)}$$
$$Al^{3+}(aq) + 3\ NH_3(aq) + 3\ H_2O(\ell) \rightarrow Al(OH)_3(s) + 3\ NH_4^+(aq) \qquad (6)$$
$$\text{white}$$

The concentration of OH^- in dilute NH_3 is too low to form the $Al(OH)_4^-$ complex ion by Reaction 2.

Equation 6 represents the classic confirmatory test for aluminum. Because the precipitate is occasionally difficult to see, a few drops of the organic reagent aluminon were often added. The aluminon would absorb onto the $Al(OH)_3$ and give it a reddish color. This test was not always reliable, occasionally giving false positives. An improved confirmatory test involves dissolving the $Al(OH)_3$ precipitate in dilute acetic acid, followed by the addition of a few drops of a solution of the organic reagent catechol violet, a weak acid which can be represented as HCV. With the improved test, the Al^{3+} ions, if present, interact with the deprotonated anionic form of catechol violet, CV^-, to form a blue complex, $AlCV^{2+}$:

$$Al(OH)_3(s) + 3\ HAc(aq) \rightarrow Al^{3+}(aq) + 3\ Ac^-(aq) + 3\ H_2O(\ell) \qquad (7)$$

$$Al^{3+}(aq) + HCV(aq) \rightarrow AlCV^{2+}(aq) + H^+(aq) \qquad (8)$$
$$\text{blue}$$

The CrO_4^{2-} ion remains in solution after Al^{3+} has been precipitated. It can be tested for by precipitation as yellow $BaCrO_4$ by the addition of $BaCl_2$ solution:

$$Ba^{2+}(aq) + CrO_4^{2-}(aq) \rightarrow BaCrO_4(s) \qquad (9)$$
$$\text{light yellow}$$

The precipitate of $BaCrO_4$ is dissolved in acid; the solution formed is then treated with hydrogen peroxide, H_2O_2. A deep blue color is produced, because of the presence of a peroxo-compound, probably CrO_5. The reaction may be represented by the overall equation:

$$2\,BaCrO_4(s) + 4\,H^+(aq) + 4\,H_2O_2(aq) \rightarrow 2\,Ba^{2+}(aq) + 2\,CrO_5(aq) + 6\,H_2O(\ell) \qquad (10)$$
$$\text{light yellow} \qquad\qquad\qquad\qquad\qquad \text{deep blue}$$

The mixed precipitate of $Ni(OH)_2$ and $Fe(OH)_3$ formed by Reactions 3 and 4 is dissolved by adding a strong acid, HNO_3. An acid-base reaction occurs:

$$Ni(OH)_2(s) + 2\,H^+(aq) \rightarrow Ni^{2+}(aq) + 2\,H_2O(\ell) \qquad (11)$$
$$\text{green} \qquad\qquad\qquad \text{green}$$

$$Fe(OH)_3(s) + 3\,H^+(aq) \rightarrow Fe^{3+}(aq) + 3\,H_2O(\ell) \qquad (12)$$
$$\text{red} \qquad\qquad\qquad \text{yellow}$$

At this point, the Ni^{2+} and Fe^{3+} ions are separated by adding ammonia. The Ni^{2+} ion is converted to the deep blue complex $Ni(NH_3)_6^{2+}$, which stays in solution:

$$Ni^{2+}(aq) + 6\,NH_3(aq) \rightarrow Ni(NH_3)_6^{2+}(aq) \qquad (13)$$
$$\text{green} \qquad\qquad\qquad \text{deep blue}$$

while the Fe^{3+} ion, which does not readily form a complex with NH_3, is reprecipitated as red-brown $Fe(OH)_3$:

$$3\,NH_3(aq) + 3\,H_2O(\ell) \rightleftharpoons 3\,NH_4^+(aq) + 3\,OH^-(aq)$$
$$\underline{Fe^{3+}(aq) + 3\,OH^-(aq) \rightarrow Fe(OH)_3(s)}$$
$$Fe^{3+}(aq) + 3\,NH_3(aq) + 3\,H_2O(\ell) \rightarrow Fe(OH)_3(s) + 3\,NH_4^+(aq) \qquad (14)$$
$$\text{yellow} \qquad\qquad\qquad\qquad\qquad \text{red-brown}$$

The confirmatory test for nickel in the solution is made by adding an organic reagent, dimethylglyoxime, $C_4H_8N_2O_2$. This gives a deep rose-colored precipitate with nickel:

$$Ni^{2+}(aq) + 2\,C_4H_8N_2O_2(aq) \rightarrow Ni(C_4H_7N_2O_2)_2(s) + 2\,H^+(aq) \qquad (15)$$
$$\text{green} \qquad\qquad\qquad\qquad \text{deep rose}$$

We can confirm Fe^{3+} by dissolving the precipitate of $Fe(OH)_3$ in HCl (Reaction 10) and adding KSCN solution. If iron(III) is present, the blood-red $FeSCN^{2+}$ complex ion will form:

$$Fe^{3+}(aq) + SCN^-(aq) \rightarrow FeSCN^{2+}(aq) \qquad (16)$$
$$\text{yellow} \qquad\qquad\qquad \text{deep red}$$

WEAR YOUR SAFETY GLASSES WHILE PERFORMING THIS EXPERIMENT

Experimental Procedure

Step 1. If you are testing a solution from which the Group II ions have been precipitated (Experiment 37), remove the excess H_2S and excess acid by boiling the solution until the volume is reduced to about one milliliter. Remove any sulfur residue by centrifuging the solution.

If you are working on the analysis of Group III cations only, prepare a known solution containing Fe^{3+}, Al^{3+}, Cr^{3+}, and Ni^{2+} by mixing together 0.5-mL portions of each of the appropriate 0.10 M solutions containing those cations.

Step 2. Oxidation of Cr(III) to Cr(VI) and Separation of Insoluble Hydroxides Add 1.0 mL of 6 M NaOH to 1.0 mL of the known solution in a 30-mL beaker. Boil very gently for a minute, stirring to minimize bumping. Remove heat, and slowly add 1.0 mL of 1.0 M NaOCl, sodium hypochlorite. Swirl the beaker for 30 seconds, using your tongs if necessary. Then boil the mixture gently for a minute. Add 0.5 mL of 6 M NH$_3$ and let stand for 30 seconds. Then boil for another minute. Transfer the mixture to a test tube and centrifuge out any remaning solid, which contains iron and nickel hydroxides. Decant the solution, which contains chromium and aluminum [CrO$_4^{2-}$ and Al(OH)$_4^-$] ions, into a test tube (for use in Step 3). Wash the solid twice with 2 mL of water and 0.5 mL of 6 M NaOH; after mixing, centrifuge each time, discarding the wash. Add 1.0 mL of water and 1.0 mL of 6 M H$_2$SO$_4$ to the solid and put the test tube aside (for Step 6).

Step 3. Separation of Al from Cr Acidify the solution from Step 2 by adding 6 M acetic acid slowly until, after stirring, the mixture is definitely acidic to litmus. If necessary, transfer the solution to a 50-mL beaker and boil it to reduce its volume to about three milliliters. Pour the solution into a test tube. Add 6 M NH$_3$, drop by drop, until the solution is basic to litmus, and then add 0.5 mL in excess. Stir the mixture for a minute or so to bring the system to equilibrium. If aluminum is present, a light, translucent, gelatinous white precipitate of Al(OH)$_3$ should be floating in the clear (possibly yellow) solution. Centrifuge out the solid, and transfer the liquid, which may contain CrO$_4^{2-}$, into a test tube (for use in Step 5).

Step 4. Confirmation of the Presence of Aluminum Wash the precipitate from Step 3 with 3 mL of water once or twice, while warming the test tube in a boiling-water bath and stirring well. Centrifuge and discard the wash each time. Dissolve the precipitate in 2 drops of 6 M HC$_2$H$_3$O$_2$, acetic acid: no more, no less. Add 3 mL of water and 2 drops of catechol violet reagent, and stir. If Al^{3+} is present, the solution will turn blue.

Step 5. Confirmation of the Presence of Chromium If the solution from Step 3 is yellow, chromium is probably present; if it is colorless, chromium is absent. To the solution add 0.5 mL of 1.0 M BaCl$_2$. In the presence of chromium you obtain a finely divided yellow precipitate of BaCrO$_4$, which may be mixed with a white precipitate of BaSO$_4$. Put the test tube in a boiling-water bath for a few minutes; then centrifuge out the solid and discard the liquid. Wash the solid with 2 mL of water; centrifuge and discard the wash. To the solid add 0.5 mL of 6 M HNO$_3$, and stir to dissolve the BaCrO$_4$. Add 1.0 mL of water, stir the orange solution, and add 2 drops of 3%$_{mass}$ H$_2$O$_2$, hydrogen peroxide. A deep blue color, which may fade rapidly, is confirmatory evidence for the presence of chromium.

Step 6. Separation of Iron and Nickel Returning to the precipitate from Step 2, stir to dissolve the solid in the H$_2$SO$_4$. If necessary, warm the test tube in a boiling-water bath to complete the dissolution process. Then add 6 M NH$_3$ until the solution is basic to litmus. At that point iron will precipitate as red-brown Fe(OH)$_3$. Add 1.0 mL more of the NH$_3$, and stir to bring the nickel into solution as the Ni(NH$_3$)$_6^{2+}$ ion. Centrifuge and decant the liquid into a test tube. Save the precipitate (for Step 8).

Step 7. Confirmation of the Presence of Nickel If the solution from Step 6 is blue, nickel is probably present. To that solution add 0.5 mL of dimethylglyoxime reagent. Formation of a rose-red precipitate confirms the presence of nickel.

Step 8. Confirmation of the Presence of Iron Dissolve the precipitate from Step 6 in 0.5 mL of 6 M HCl. Add 2 mL of water and stir. Then add 2 drops of 0.5 M KSCN. Formation of a deep red solution of FeSCN^{2+} definitively confirms the presence of iron.

Step 9. When you have completed your analysis of the known solution, obtain a Group III unknown and test it for the possible presence of Fe^{3+}, Al^{3+}, Cr^{3+}, and/or Ni^{2+}.

DISPOSAL OF REACTION PRODUCTS. As you complete each step in the procedure, put the waste products into a beaker. When you are finished with the experiment, pour the contents of the beaker into a waste container, unless otherwise directed by your instructor.

Experiment 38

Observations and Analysis: Qualitative Analysis of Group III Cations

Flow Diagram for Group III

On the flow diagram above, indicate the actual observations made upon the known and the unknown. For the known, simply note any deviations from the expected results, per the flow diagram, labeling them with "KNOWN." For the unknown, clearly indicate the path through the flow diagram that you followed to reach your conclusions.

Unknown # _____   Ions reported present _____ _____ _____ _____

Experiment 38

Advance Study Assignment: Qualitative Analysis of Group III Cations

1. Prepare a complete flow diagram for the separation and identification of the ions in Group III and put it on the report page for this experiment.

2. Write balanced net ionic equations for the following reactions:

   a. The oxidation of Cr^{3+} to CrO_4^{2-} by ClO^- in alkaline solution (ClO^- is converted to Cl^-)

   b. The dissolution of $Ni(OH)_2$ in nitric acid

   c. The confirmatory test for Ni^{2+}

   d. The confirmatory test for Fe^{3+}

3. A solution may contain any of the four Group III cations considered in this experiment. Treatment of the solution with ClO^- in alkaline medium yields a yellow solution and a colored precipitate. The acidified solution is unaffected by treatment with NH_3. The colored precipitate dissolves in nitric acid; addition of excess NH_3 to this acidic solution produces only a blue solution. On the basis of this information, which Group III cations are present, absent, or still in doubt? State your reasoning. *Note:* For "paper unknowns" such as this one, confirmatory tests are usually not included, and you do not need to provide any. The information provided here, without confirmatory tests, is sufficient to clearly indicate the presence or absence of some of the Group III ions studied in this experiment, while leaving others in doubt.

   Present _____

   Absent _____

   In doubt _____

Experiment 39

Identification of a Pure Ionic Solid

In the previous three experiments we have seen how we can identify which anions, or cations, are actually present in an unknown solution. In this experiment you will be asked to determine the cation and anion that are present in a solid sample of a pure ionic salt. The possible cations will include those we studied in Experiments 36, 37, and 38:

$$Ag^+ \quad Pb^{2+} \quad Hg_2^{2+} \quad Sn^{2+} \quad Sb^{3+}$$
$$Cu^{2+} \quad Bi^{3+} \quad Cr^{3+} \quad Al^{3+} \quad Ni^{2+} \quad Fe^{3+}$$

The anions we will consider are those in Experiment 35:

$$CO_3^{2-} \quad PO_4^{3-} \quad Cl^- \quad SCN^- \quad SO_4^{2-} \quad SO_3^{2-} \quad C_2H_3O_2^- \quad NO_3^-$$

There are 88 possible compounds that you might be given to identify. There are, of course, many ionic compounds that are not included, but the set of unknowns is reasonably representative. At first sight it might seem that the best approach would be to carry out the procedures for analysis of Groups I, II, and III, in succession, and stop when you get to the cation in your unknown. That approach should work, but it would take longer than necessary, and would require that you have a solution of the compound to work with. Given a knowledge of the properties of the possible compounds, an experienced chemist would carry out some preliminary tests to determine the solubility properties of the sample and try to use those to significantly narrow down the number of compounds she needed to consider as possibilities. The solubility of your unknown salt in various solvents, plus its color, and its odor (should it have one), may afford some very useful information.

Most of the possible compounds are not soluble in water but may dissolve in strong acids, like 6 M HCl, or strong bases, like 6 M NaOH, or solutions containing ligands that form stable complex ions with the cation in the compound. Some important complexing ligands that may dissolve otherwise insoluble solids containing the cations we are considering include:

$$NH_3 \quad OH^- \quad Cl^- \quad C_2H_3O_2^-$$

Nearly every compound in our set of 88 will dissolve in, or react with, at least one of the above species in solution.

Before beginning the analysis, it will be helpful to note some of the properties of the cations that distinguish them from one another. We have given those properties in Appendix IIA. You should read that appendix and complete the Advance Study Assignment before coming to lab. You will need the information in the appendix to complete your analysis.

WEAR YOUR SAFETY GLASSES WHILE PERFORMING THIS EXPERIMENT

DISCARD ALL REACTION PRODUCTS IN A 250-mL BEAKER

Experimental Procedure

In this experiment you will be given two unknowns to identify. The cation in your first sample will be colored or will produce colored solutions. The second unknown will be colorless and somewhat more difficult to identify. In both cases the analysis will involve determining the solubility of the solid in water and some acidic or basic solutions. You should have about a half gram of finely divided solid available. Read the section on qualitative analysis at the end of Appendix IV if you haven't already. Prepare a boiling-water bath and a bath for rinsing your stirring rods.

A. Identifying the Cation in a Colored Ionic Solid

The solvents we will consider are the following:

1) Water	2) 6 M HNO_3	3) 6 M HCl
4) 6 M NaOH	5) 6 M NH_3	6) 6 M H_2SO_4

Determine the solubility of your colored solid in as many of the solvents as you need to find one in which the solid will dissolve. Use a small sample of the solid, about fifty milligrams: enough to cover a 2-mm-diameter circle on the end of a small spatula. Put the solid in a small (13 × 100 mm) test tube and add about one milliliter of water. Stir the solid with a stirring rod for a minute or so, noting any color changes in the solid or the solution. If the solid does not dissolve, put the tube in your boiling-water bath for a few minutes. If at this point you have a solution, add 5 mL more water and stir. From the color of the solution you should be able to make a good guess as to which cation is in your unknown. (The information in Appendices II and IIA should prove helpful.) Use about one milliliter of your solution to carry out the confirmatory test for that cation, as directed in Experiment 37 or 38. Make sure that you set the pH and any other conditions for the test properly before making a decision. If your guess is confirmed, you may proceed to analysis for the anion. If you were wrong, perform the confirmatory test for another cation that seems likely. It should not be difficult to identify the cation that is present.

If your sample does not dissolve in water, try 6 M HNO_3 as a solvent, using the same procedure as with water (about one milliliter of reagent and fifty milligrams of solid). If the sample goes into solution, either cold or after being heated in the boiling-water bath, add 5 mL of water, stir, and proceed to identify the cation present. With HNO_3 you may get some effervescence. This indicates that the anion is either carbonate or sulfite. If it is carbonate, odorless CO_2 gas will be given off, while if it is sulfite then gaseous SO_2, with its characteristic sharp odor, will be released. Nearly all the possible compounds in our set that are colored are also soluble in either hot water or hot HNO_3. If you by chance have one that will not dissolve in those reagents, carry out the solubility test using 6 M HCl as the solvent. That should work if the first two solvents have not. In this part of the experiment, you should not need to use the three solvents on the second row of the list above.

B. Anion Analysis of a Colored Ionic Solid

Before beginning analysis for the anion in your unknown, you should find it helpful to know that some anions are much more common than others in the chemicals that are readily available from chemical supply houses. It is easiest for chemists to work with chemicals that are soluble in water, so that is what chemical supply houses tend to sell. Nitrates, chlorides, sulfates, and acetates are often soluble and are the most common salts we find on the shelves. Thiocyanates and sulfites are sometimes soluble but are usually found as sodium or potassium salts, and so are not likely to be in the compounds in the set we are using. Carbonates and phosphates are insoluble in water but are occasionally available. Hydroxides and oxides are relatively easy to synthesize, but they are insoluble and will not be in your unknown unless your instructor tells you otherwise.

Given the information above and the spot tests in Experiment 35, determine which anion is in your solid. For each test, use about one or two milliliters of the solution you prepared. You don't need to do all the spot tests, only those that seem reasonable in light of the previous discussion. If you used water as your solvent, any positive test for an anion should be conclusive. If you used HNO_3, a test for nitrate ion in that solution will always come out positive because of the nitrate put into the solution by the nitric acid. Similarly, you cannot carry out a meaningful test for chloride ion in an HCl solution.

Some anions may be difficult to determine by the usual spot test, since the cation present may interfere. That is the case with acetate and nitrate, where the test solution is basic and will cause the cation to precipitate. With acetates, a simple test is to add a milliliter of 6 M H_2SO_4 to about fifty milligrams of the solid, as if you were testing for solubility. Heat the solution, or slurry, in the boiling-water bath. If the sample contains acetate, acetic acid will be volatilized and can be detected by its characteristic odor. Put your stirring rod in the warm liquid, and, cautiously, sniff the rod. If there are no interfering anions, the odor of vinegar is an excellent test for acetate. With nitrates, begin by adding a milliliter of 6 M NaOH to a small sample of solid unknown. If you get a hydroxide precipitate, you can try simply proceeding with the spot test. If the cation forms a hydroxide complex, you can try the spot test for nitrate, and it may work. In many cases, the aluminum metal will reduce the cation to the metal, and you will get a black solid. With excess Al, you may still be able to detect evolved NH_3 and establish the presence of nitrate. The test for nitrate is perhaps the most difficult, and if worse comes to worst, you can treat your acidic solution of unknown with excess 1.0 M Na_2CO_3 until effervescence ceases and the solution is definitely basic; then add 0.5 mL more. Centrifuge out the solid cation-containing carbonate, which can be discarded. Acidify the decanted liquid with 6 M H_2SO_4 drop by drop until CO_2 evolution ceases, and the solution is acidic. Carefully boil this solution down to about two milliliters and use it for the nitrate test and any other anion tests that seem indicated.

C. Analysis of a Colorless Ionic Solid

The second unknown you will be given to analyze will be a white solid, which may have very limited solubility in water. It may also be insoluble in some of the acids and bases on our original list.

The procedure—at least initially—is the same as that you used with the colored sample. Check the solubility of the solid in water, both at room temperature and in the boiling-water bath. Some of the solids do dissolve. With these, and the rest of the unknowns, the solubility in the other solvents is likely to furnish clues about the cation that is present. So, carry out solubility tests with all six of the solvents we list above (again, use about one milliliter of reagent added to about fifty milligrams of solid). Before deciding on the solubility in a given solvent, make sure you give the solid time to dissolve if it is going to, with stirring, particularly in the boiling-water bath. If the solid does not dissolve, add a milliliter of water and stir; that may help. Solids containing cations that form complexes with Cl^- or OH^- are likely to dissolve in 6 M HCl or 6 M NaOH, respectively. Those with cations that form ammonia complexes will tend to dissolve in 6 M NH_3. The solubility depends on the K_{sp} of the solid and the stability of the complex, so a complexing ligand may dissolve one solid and not another. Nitrate ion and sulfate ion are not very good complexing ions, so 6 M HNO_3 and 6 M H_2SO_4 will not dissolve solids by complex ion formation. If they dissolve a solid, it is by reacting with an anion in the solid to form a weak acid or by reacting with a basic salt like BiOCl. There are a few compounds in our possible set of unknowns that will not dissolve in any of the six reagents, or may dissolve in only one reagent. Those compounds are all discussed in Appendix IIA.

When you have completed the solubility tests on your solid, compare your results with those you obtained in your Advance Study Assignment using information from Appendix IIA. You should be able to narrow down the list of possible cations to only a few, and in some cases you will be able to decide not only which cation is present but also which anion. If you can dissolve the solid in water or in acid, you can proceed to carry out a confirmatory test for any likely cations, as described in Experiments 36, 37, and 38. With the colorless unknowns the number of likely anions is even less than with the first unknown (in which nitrate and chloride were most likely). If the sample will dissolve in water, or in HNO_3 or HCl, you can carry out spot tests for any anions that are not present in the solvent. For some solids, you will need to identify the anion purely on the basis of the solubility (or insolubility!) of the salt.

On the report page, note all your observations and the name and formula of the compound you believe is in your unknown.

DISPOSAL OF REACTION PRODUCTS. When you are finished with the experiment, discard all the reaction products in your 250-mL beaker in the waste container.

Experiment 39

Observations and Analysis: Identification of a Pure Ionic Solid

A. Identifying the Cation in a Colored Ionic Solid

Color of first unknown _____

Solubility properties of unknown

_____

_____

_____

Cations that are likely components of unknown _____ _____

Confirmatory tests performed to identify cation

    Procedures used

    _____

    _____

    _____

    Results observed

    _____

    _____

    _____

_____ is present

B. Anion Analysis of a Colored Ionic Solid

Spot tests performed to identify anion

    Procedures used

    _____

    _____

    _____

    Results observed

    _____

    _____

    _____

_____ is present

    Formula of colored compound   _____

(continued on following page)

C. Analysis of a Colorless Ionic Solid

Solubility of white unknown in:

Water _____   6 M HNO_3 _____

6 M HCl _____   6 M NaOH _____

6 M NH_3 _____   6 M H_2SO_4 _____

Reasoning and results of confirmatory tests

_____

_____

_____

_____

_____

_____

_____

_____

_____

_____

_____

_____

_____

Formula of white compound _____

Experiment 39

Advance Study Assignment: Identification of a Pure Ionic Solid

Use the information in Appendix II and Appendix IIA to find answers to these questions.

1. What are the formulas and colors of the colored cations used in this experiment?

   _____

2. Which of the colorless cations

   a. will form complex ions with NH_3?   _____

   b. will form complex ions with OH^-?   _____

   c. will form complex ions with Cl^-?   _____

   d. have colored, rather than black, sulfides?   _____

      Note: Sulfides are generally black; if a sulfide color is not otherwise specified,
      you may assume it is black.

   e. have one or more water-soluble salts (salts that readily dissolve in water alone)?   _____

3. Give the formula of a white or colorless compound formed from the ions involved in this experiment that

   a. dissolves in 6 M NH_3 _____

   b. turns black when NH_3 is added   _____

   c. is soluble in 6 M NaOH but not in 6 M H_2SO_4 _____

   d. is much more soluble in hot water than in cold   _____

   e. is not soluble in any of the solvents we use in this experiment   _____

4. A colorless ionic solid is insoluble in water, and in 6 M NH_3, but will dissolve in 6 M HCl and 6 M NaOH.
   Identify two compounds that have these properties.

   _____ and _____

Experiment 40

The Ten Test Tube Mystery

Having carried out at least some of the previous experiments on qualitative analysis, you should have acquired some familiarity with the properties of the cations and anions you studied. Along with this you should at this point have had some experience with interpretation of the data in Appendix II on the solubilities of ionic substances in different media.

In this experiment we are going to ask you to apply what you have learned to a related but somewhat different kind of problem. You will be given ten numbered test tubes, each containing a solution of a single solute. You will be provided, one week in advance, a list of formulas and concentrations of each of the ten solutes that will be used. Your challenge in the laboratory will be to find out which solution is in which test tube—that is, to assign a test tube number to each of the solution compositions. You are to do this by intermixing small volumes of the solutions in the test tubes. No external reagents or acid-base indicators such as litmus are allowed. You are permitted, however, to use the odor and color of the different solute species, as well as the consequences of their reactions with each other, in attempting to identify them. Do not neglect to note heat evolved in any reactions, as well as to make use of heat to drive reactions that prove otherwise reluctant to proceed. Each student will have a different set of (test tube number)-(solution composition) correlations, and there will be several different sets of compositions as well.

Of the ten solutions, four will be common laboratory reagents. They will be 6 M HCl, 3 M H_2SO_4, 6 M NH_3, and 6 M NaOH. The other solutions will usually be 0.10 M nitrate, chloride, or sulfate solutions of the cations which have been studied in previous experiments in this manual (in Experiments 12, 36, 37, and/or 38).

To determine which solution is in each test tube, you will need to know in advance what happens when the various solutions are mixed with one another. In some cases, nothing happens that you can observe. This will often be the case when a solution containing one of the cations is mixed with a solution of another cation. When one of the reagents is mixed with a cation solution, you may get a precipitate, white or colored, and that precipitate may dissolve in excess reagent by complex ion formation. In a few cases a gas may be evolved. When one laboratory reagent is mixed with another, you may find that the resulting solution gets very hot and/or that a visible vapor is produced.

There is no way that you will be able to solve your particular test tube mystery without doing some preliminary work. You will need to know what to expect when any two of your ten solutions are mixed. You can find this out, for any pair, by scrutiny of Appendix II, by consulting your chemistry text, and by referring to various reference works on qualitative analysis.* A convenient way to tabulate the information you obtain is to set up a matrix with ten columns and ten rows, one for each solution. The key information about a mixture of two solutions is put in the space where the row for one solution and the column for the other intersect (as is done in Appendix II for various cations and anions). In that space you might put "NR", to indicate no apparent reaction on mixing of the two solutions. If a precipitate forms, write in "P", followed by "D" if the precipitate dissolves in excess reagent. If the precipitate or final solution is colored, state the color. If heat is evolved, write "H"; if a gas or smoke is formed, write in "G" or "S". Since mixing solution A with solution B is the same as mixing B with A, not all 100 spaces in the 10×10 matrix need to be filled. Actually, there are only 45 relevant pairs, since combining A with A is not very informative either.

---

* For example, the following references may be useful:

[1]*Qualitative Analysis and the Properties of Ions in Aqueous Solutions.* 2nd ed., by E. J. Slowinski and W. L. Masterton, Saunders College Publishing, 1990.

[2]*Handbook of Chemistry and Physics.* Chemical Rubber Publishing, 2017.

Because you are allowed to use the odor or color of a solution to identify it, the problem is somewhat simpler than it might first appear. In each set of ten solutions you will probably be able to identify at least two solutions simply by odor and/or color observations. Knowing the identities of those solutions, you can make mixtures with the other solutions in which one of the components is known. From the results obtained with those mixtures, and the information in the matrix, you can identify other solutions. These can be used to identify still others, until finally the entire set of ten is identified with certainty.

Let us now go through the various steps that would be involved in the solution to a somewhat simpler problem than the one you will solve. There are several steps, including constructing the reaction matrix, identifying solutions by simple observations, and devising efficient mixing tests. Let us assume we have to identify the following six solutions, in numbered test tubes:

0.10 M $NiSO_4$ 0.10 M $BiCl_3$(in 3 M HCl)

0.10 M $BaCl_2$ 6 M NaOH

0.10 M $Al(NO_3)_3$ 3 M H_2SO_4

A. Construction of a Reaction Matrix

In the matrix there will be six rows and six columns, one for each solution. The matrix should be set up as shown in Table 40.1.

Each solution contains a cation and an anion, both of which need to be considered. We need to include every possible pair of solutions: for six solutions there are 15 possible pairs. So we need 15 pieces of information as to what happens when a pair of solutions is mixed. The first pair we might consider is $Al(NO_3)_3$ and $BaCl_2$, containing Al^{3+}, NO_3^-, Ba^{2+}, and Cl^- ions. Consulting Appendix II, we see that in the Al^{3+}, Cl^- space there is an "S", meaning that $AlCl_3$ is soluble and will not precipitate when aluminum and chloride ions are mixed. Similarly, $Ba(NO_3)_2$ is soluble because all nitrates are soluble. This means that there will be no reaction when the solutions of $Al(NO_3)_3$ and $BaCl_2$ are mixed. So, in the space for $Al(NO_3)_3$-$BaCl_2$, we have written "NR". The same is true if we mix solutions of $Al(NO_3)_3$ and $NiSO_4$, so we insert "NR" in that space as well. Since $BiCl_3$ won't react with $Al(NO_3)_3$ solutions, there is also an "NR" in that space. However, when $Al(NO_3)_3$ is mixed with NaOH, one of the possible products is $Al(OH)_3$, which we see in Appendix II is insoluble in water but dissolves in acid and excess OH^- ion ("A, C", with OH^- listed as a complexing agent in the very last, rightmost column). This means that on adding NaOH to $Al(NO_3)_3$, we would initially get a precipitate of $Al(OH)_3$, "P", but that it would dissolve in excess NaOH, "D". The precipitate is white and the solution is colorless, so we have just "P" leading to "D", written as "P → D", in the space for that mixture.

Proceeding now to the $BaCl_2$ row, we don't need to consider the first two columns, because they would give no new information. If $BaCl_2$ and $NiSO_4$ were mixed, $BaSO_4$ is a possible product. In Appendix II we see that $BaSO_4$ is insoluble in all common solvents ("I"). Hence, on mixing those solutions, we would get a precipitate of $BaSO_4$, "P". Using the *Handbook of Chemistry and Physics,* or some other source, we find that it is white, so no color indication goes into that space. With $BiCl_3$ there would be no reaction; with NaOH we might get a slight precipitate ["S^-" for $Ba(OH)_2$ in Appendix II], so we have written "P?" in the appropriate space. With H_2SO_4, we would again expect to get a white precipitate of $BaSO_4$, "P".

In the row for $NiSO_4$ we note that the color of the solution is green, since in Appendix II we see that the Ni^{2+}(aq) ion is green. The first column is that for $BiCl_3$; no precipitate forms on mixing solutions of $NiSO_4$ and $BiCl_3$, so "NR" is in that space. With NaOH, $Ni(OH)_2$ would precipitate, since the entry in Appendix II at the intersection of Ni^{2+} with OH^- is not "S". From the *Handbook of Chemistry and Physics* or other sources we find that the precipitate is green, so in the space we have written "P (green)". In Appendix II the $Ni(OH)_2$ entry is "A, C", which, in conjunction with the very last (rightmost) column on complexes, tells us that the precipitate will dissolve in 6 M NH_3 but not in 6 M NaOH, as Al^{3+} does. There would be no reaction between solutions of $NiSO_4$ and H_2SO_4, so that spot in the grid gets "NR".

Table 40.1

Reaction Matrix for Six Solutions						
	$Al(NO_3)_3$	$BaCl_2$	$NiSO_4$	$BiCl_3$	NaOH	H_2SO_4
$Al(NO_3)_3$		NR	NR	NR	P → D	NR
$BaCl_2$			P	NR	P?	P
$NiSO_4$ (green)				NR	P (green)	NR
$BiCl_3$					H → P	NR
NaOH						H
H_2SO_4						

With $BiCl_3$ in 3 M HCl we have both a salt and an acid in the solution. Addition of 6 M NaOH should produce a precipitate (white) of $Bi(OH)_3$, "P". There would also be a substantial exothermic acid-base reaction between H^+ ions in the HCl and OH^- ions in the NaOH, so there would be a very noticeable rise in temperature on mixing, which we denote with "H". The neutralization reaction happens more quickly than does the precipitation (this is usually the case), so the tube will first heat up, and then the precipitate will form as more NaOH is added. We indicate this by "H → P", heat followed by precipitate. There would be no reaction with H_2SO_4.

When 6 M NaOH is mixed with 3 M H_2SO_4 there should be a large heat effect because of the acid-base reaction that occurs (but no precipitate would form), so in the space for NaOH-H_2SO_4 we have "H" (only).

The matrix in Table 40.1 summarizes the reaction information that will be needed for identification of the solutions in the test tubes. Your matrix can be constructed by reasoning like that used in making this one.

B. Identifying Solutions in Isolation, by Simple Observation

When we made the matrix for the six-test-tube system, we noted that the $NiSO_4$ solution would be green, since the Ni^{2+}(aq) ion is green. So, as soon as we see the six text tubes, we can pick out the 0.10 M $NiSO_4$. Let us assume that it is in Test Tube 4. We have now identified one solution out of the six.

C. Selection of Efficient Mixing Tests

Knowing that Test Tube 4 contains $NiSO_4$, we mix the solution in Test Tube 4 with all of the other solutions. For each mixture, we record what we observe. The results might be as follows:

4 + 1	no obvious reaction
4 + 2	no obvious reaction
4 + 3	green precipitate forms
4 + 5	no obvious reaction
4 + 6	white precipitate

Referring to the matrix, we would expect precipitates for mixtures of $NiSO_4$ with NaOH, and for $NiSO_4$ with $BaCl_2$. The former precipitate would be green, the latter white. Clearly the solution in Test Tube 3 is 6 M NaOH. The solution in Test Tube 6 must be 0.10 M $BaCl_2$. At this point three solutions have been identified.

Because 6 M NaOH reacts with several of the solutions, and because we know now that it is in Test Tube 3, we mix the solution in Test Tube 3 with those that remain unidentified, with the following results:

3 + 1	solution becomes hot
3 + 2	solution gets hot, then white precipitate forms with more of 2
3 + 5	apparently no reaction

Again, referring to the matrix, we would expect heat evolution upon mixing NaOH with H_2SO_4, as well as when NaOH reacts with the HCl in the $BiCl_3$. In addition, we would expect to get a white precipitate with $BiCl_3$. It is obvious that Solution 1 is 3 M H_2SO_4. Solution 2 must be 0.10 M $BiCl_3$ in 3 M HCl. By elimination, Solution 5 must be $Al(NO_3)_3$. We repeat the 3 + 5 mixture, since there should have been an initial precipitate that dissolved in excess NaOH. This time, using 1 drop of Solution 3 added to 1.0 mL of Solution 5, we get a precipitate. In an excess of Solution 3 the precipitate dissolves as the aluminum hydroxide complex ion is formed.

WEAR YOUR SAFETY GLASSES WHILE
PERFORMING THIS EXPERIMENT

Experimental Procedure

Using the approach we have outlined in our example, carry out tests as necessary for the identification of the solutions in your set. After you have tentatively assigned each solution to its test tube, carry out some confirmatory mixing reactions, so that you have at least two observations that independently support your conclusions about each tube.

> **DISPOSAL OF REACTION PRODUCTS.** Dispose of all reaction products in the waste container, unless directed otherwise by your instructor.

Experiment 40

Observations and Analysis: The Ten Test Tube Mystery

A. Construction of a Reaction Matrix

Construct your reaction matrix on the back of the ASA page, following the instructions on the front of the ASA.

B. Identifying Solutions in Isolation, by Simple Observation

Solutions Identified by Observation in Isolation

Test tube #	Observations	Conclusions	Solution identity

C. Selection of Efficient Mixing Tests

Mixing Tests: First Series

Test tubes combined	Observations	Conclusions

Mixing Tests: Second Series

Test tubes combined	Observations	Conclusions

(continued on following page)

Confirmatory Tests (combine at least two strategically chosen pairs of tubes not previously combined, to double-check your identifications)

Test tubes combined	Prediction(s)	Observations	Conclusion(s) supported

Final identifications: # 1 _____   # 2 _____   # 3 _____

4 _____   # 5 _____   # 6 _____   # 7 _____

8 _____   # 9 _____   # 10 _____

Which unknown set did you have (if applicable)? _____

Experiment 40

Advance Study Assignment: The Ten Test Tube Mystery

1. Construct the reaction matrix for your set of ten solutions. This should go on the reverse side of this page. This matrix can be made using information in Appendix II or from other references, or, if your instructor allows it, you can go into the lab prior to the scheduled session and work with known solutions to obtain the needed information.

2. Which solutions should you be able to identify by simple observation, without mixing any solutions?

3. Outline the procedure you will follow in identifying the solutions that will require mixing tests. Be as specific as you can about what you will look for and what conclusions you will be able to draw from your observations.

Experiment 41

Preparation of Aspirin

One of the simpler organic reactions that we can carry out is the formation of an ester from an acid and an alcohol:

$$R-\overset{\displaystyle O}{\overset{\|}{C}}-OH + HO-R' \rightarrow R-\overset{\displaystyle O}{\overset{\|}{C}}-O-R' + H_2O \qquad (1)$$

$$\text{an acid} \qquad\quad \text{an alcohol} \qquad\qquad \text{an ester}$$

In the equation, R and R' are H atoms or organic fragments like CH_3, C_2H_5, or more complex aromatic groups. There are many esters, because there are many organic acids and alcohols, but they all can be formed, in principle at least, by Reaction 1. The driving force for the reaction is, in general, not very great, so that we end up with an equilibrium mixture of ester, water, acid, and alcohol.

There are some esters that are solids because of their high molecular weight or other properties. Most of these esters are not soluble in water, so they can be separated from the mixture by crystallization. This experiment deals with an ester of this sort, the substance commonly called aspirin.

Aspirin has been used since the beginning of the twentieth century to treat headaches and fever. It remains one of the most widely used medicines in the world, with a daily production of about one hundred tons, or about thirteen regular-strength pills per year for each person on Earth. Its effectiveness is actually two-fold: as an anti-inflammatory, which explains its pain-killing properties; and as an anticoagulant, which is important for its use in the prevention of heart attacks and strokes. It works by inhibiting an enzyme that catalyzes formation of precursors of prostaglandins, rather large organic molecules that regulate inflammation of tissues, and of thromboxanes, which stimulate platelet aggregation, the first step in blood clotting. Researchers on prostaglandins have received two Nobel Prizes, the first in 1970, to their discoverer, and the second in 1982, for work relating to the effects of aspirin on prostaglandin behavior.

Aspirin can be made by the reaction of the hydroxyl (—OH) group in the salicylic acid molecule with the carboxyl (—COOH) group in acetic acid:

acetic acid   salicylic acid   aspirin (salicylic acid acetate)

A better preparative method, which we will use in this experiment, employs acetic anhydride in the reaction instead of acetic acid. The anhydride can be considered to be the product of a reaction in which two acetic acid molecules combine, with the elimination of a molecule of water. The anhydride will react with the water produced in the esterification reaction and will tend to drive the reaction to the right. A catalyst (normally sulfuric or phosphoric acid) is also used, and speeds up the reaction.

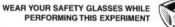

acetic anhydride   salicylic acid   aspirin   acetic acid

The aspirin you will prepare in this experiment is relatively impure and should certainly not be taken internally, even if the experiment gives you a headache!

There are several ways in which the purity of your aspirin can be estimated. Probably the simplest way is to measure its melting point. If the aspirin is pure, it will melt sharply at the literature value of the melting point. If it is impure, the melting point will be lower than the literature value by an amount that is roughly proportional to the amount of impurity present.

A more quantitative measure of the purity of your aspirin sample can be obtained by determining the mass percentage of salicylic acid it contains. Salicylic acid is the most likely impurity in the sample because, unlike acetic acid, it is not very soluble in water. Salicylic acid forms a highly colored magenta complex with Fe(III). By measuring the absorption of light by a solution containing a known amount of aspirin in excess Fe^{3+} ion, you can easily determine the percentage of salicylic acid present in the aspirin.

WEAR YOUR SAFETY GLASSES WHILE PERFORMING THIS EXPERIMENT

Experimental Procedure

Weigh a 50-mL Erlenmeyer flask on a triple-beam or top-loading balance and add 2.0 g of salicylic acid. Measure out 5.0 mL of acetic anhydride in your graduated cylinder, and pour it into the flask in such a way as to wash any crystals of salicylic acid on the walls down to the bottom. Add about five drops of $85\%_{mass}$ phosphoric acid to serve as a catalyst. **CAUTION:** **Both acetic anhydride and phosphoric acid are reactive chemicals that can give you a bad chemical burn, so use due caution in handling them. If you get any of either chemical on your hands or clothes, wash thoroughly with soap and water. Acetic anhydride is a powerful lachrymator, so keep its fumes away from your eyes!**

Clamp the flask in place in a beaker of water on a hotplate, or supported on a wire gauze on a ring stand over a Bunsen burner. Heat the water to $75 \pm 10°C$, stirring the liquid in the small flask occasionally with a stirring rod. Maintain this temperature for about fifteen minutes, by which time the reaction should be complete. *Cautiously,* add 2 mL of water to the flask to decompose any excess acetic anhydride. There will be some hot acetic acid vapor evolved as a result of the decomposition, so keep your eyes and nose away.

When the liquid has stopped giving off vapors, remove the flask from the water bath and add 20 mL of water. Let the flask cool for a few minutes in air, during which time crystals of aspirin should begin to form. Put the flask in an ice bath to hasten more complete crystallization and increase the yield of product. If crystals are slow to appear, it may be helpful to scratch the inside of the flask with a stirring rod. Leave the flask in the ice bath for at least 5 minutes, stirring or swirling the flask periodically to help it cool completely.

Collect the aspirin by filtering the cold liquid through a Büchner funnel using suction. Turn off the suction and pour about five milliliters of ice-cold deionized water over the crystals; after about fifteen seconds turn on the suction again to remove the wash liquid along with most of the impurities. Repeat the washing process with another five-milliliter sample of ice-cold water. Draw air through the funnel for a few minutes to help dry the crystals and then transfer them to a piece of dry, weighed filter paper. Weigh the sample on the paper to ± 0.1 g.

Test the solubility properties of the aspirin by taking samples of the solid the size of a pea on your spatula, putting them in separate 1.0-mL samples of each of the following solvents, and stirring:

1. toluene, $C_6H_5CH_3$ (nonpolar aromatic)
2. heptane, C_7H_{16} (nonpolar aliphatic)

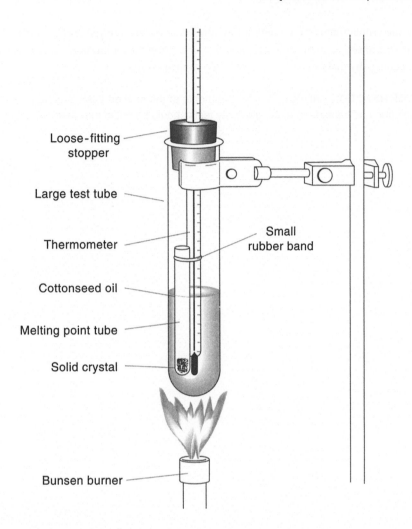

Loose-fitting stopper

Large test tube

Thermometer

Cottonseed oil

Melting point tube

Solid crystal

Small rubber band

Bunsen burner

Figure 41.1 The apparatus shown above may be used to determine the melting point of your aspirin. **It is critical that the stopper not seal the large test tube completely.**

3. ethyl acetate, $C_2H_5OCOCH_3$ (aliphatic ester)
4. ethyl alcohol, C_2H_5OH (polar aliphatic, hydrogen bonding)
5. acetone, CH_3COCH_3 (polar aliphatic, non-hydrogen-bonding)
6. water, H_2O (highly polar, hydrogen bonding)

To determine the melting point of the aspirin, add a small amount of your prepared sample to a melting point tube, as directed by your instructor. (This will be made from 5-mm-diameter glass tubing if the apparatus shown in Figure 41.1 is used; with most commercial melting point devices the tube will be much smaller.) Bring the solid down to the bottom by tapping the tube on the bench top, using enough solid to give you a depth of about five millimeters. If your lab has a commercial melting point apparatus, your instructor will show you how to use it. Otherwise, set up the apparatus shown in Figure 41.1. Fasten the melting point tube to the thermometer with a small rubber band, which should be above the surface of the oil. The thermometer bulb and sample should be about two centimeters above the bottom of the tube. Heat the oil bath *gently,* especially after the temperature gets above 100°C. As the melting point is approached, the crystals will begin to soften. Report the melting point as the temperature at which the last crystals disappear. You want the temperature to be rising *slowly* as you reach this point!

To analyze your aspirin for its salicylic acid impurity, weigh out 0.10 ± 0.01 g of your sample into a weighed 100-mL beaker. Dissolve the solid in 5 mL of $95\%_{mass}$ ethanol. Add 5 mL of 0.025 M $Fe(NO_3)_3$ in 0.5 M HCl, and 40. mL of deionized water. Make all volume measurements with a graduated cylinder. Stir the solution to mix all reagents.

Rinse out a spectrophotometer tube with a few milliliters of the solution and then fill the tube with that solution. Measure the absorbance of the solution at 525 nm. From the calibration curve or equation provided, calculate the percentage by mass of salicylic acid in the aspirin sample.

DISPOSAL OF REACTION PRODUCTS. The contents of the suction flask may be poured down the drain. The toluene and heptane from the solubility tests should be put in the waste container.

Name _____   **Section** _____

Experiment 41

Data and Calculations: Preparation of Aspirin

Mass of salicylic acid used   _____ g

Volume of acetic anhydride used   _____ mL

Mass of acetic anhydride used
(density = 1.08 g/mL)   _____ g

Mass of aspirin obtained   _____ g

Theoretical yield of aspirin   _____ g

Percent yield of aspirin (by mass)   _____ % by mass

Melting point of aspirin   _____ °C

Absorbance of aspirin solution   _____

Mass percent salicylic acid impurity   _____ %$_{mass}$

(continued on following page)

Observed solubility properties of aspirin:

Toluene   _____   Ethyl alcohol _____

Heptane   _____   Acetone   _____

Ethyl acetate _____   Water   _____

S = soluble   I = insoluble   SS = slightly soluble

Circle those characteristics listed below which would be likely to be present in a good solvent for aspirin.

      organic   aliphatic   polar   hydrogen bonding

      inorganic   aromatic   nonpolar   nonhydrogen bonding

Experiment 41

Advance Study Assignment: Preparation of Aspirin

1. Calculate the theoretical yield of aspirin to be obtained in this experiment, starting with 2.0 g of salicylic acid and 5.0 mL of acetic anhydride (density = 1.08 g/mL).

_____ g

2. If 1.9 g of aspirin were obtained in this experiment, what would the percentage yield by mass be?

_____ % (by mass)

3. The name "acetic anhydride" implies that the compound will react with water to form acetic acid. Write the balanced equation for the reaction by which this occurs.

4. Identify R and R′ in Equation 1 when the ester, aspirin, is made from salicylic acid and acetic acid.

5. Write the balanced equation for the reaction by which aspirin decomposes in an aqueous ethanol solution.

Experiment 42

Rate Studies on the Decomposition of Aspirin

Once aspirin has been prepared, it is relatively stable; in tablet form, stored in a tight container, it resists decomposition for years. But if it is exposed to moist air or heat, it will slowly break down by the reverse of the reaction we used to make aspirin in Experiment 41:

$$\text{Aspirin} + \text{Water} \rightarrow \text{Salicylic Acid} + \text{Acetic Acid} \tag{1}$$

This reaction is one of many good arguments against storing medications in a bathroom medicine cabinet! Cool, dry places are far kinder to most medications.

In this experiment we will study the rate at which aspirin in aqueous solution at 80°C will decompose. Like many organic reactions, this one is slow at room temperature. Raising the temperature allows us to carry out the reaction in a few minutes rather than several hours. To analyze the reaction mixture we will use the same method as in Experiment 41, where we find the amount of salicylic acid impurity in the aspirin sample. Iron(III) forms a very intensely colored violet complex with salicylic acid, so we can use the amount of light the reaction mixture absorbs to find out how far the reaction has gone. This will allow us to determine the rate of reaction at different concentrations. Our goal will be to use the observed rates to establish the order of the reaction, the rate constant, and the activation energy. The theory and terminology we will use are found in Experiment 20, which you should review now, even though you may have performed the experiment earlier in this course.

Under the conditions of this experiment, the rate expression for the decomposition of aspirin in hot water can be simplified to just

$$\text{rate} = k \, [\text{aspirin}]^n \tag{2}$$

where n is the order of the reaction in aspirin.

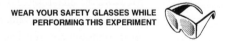

WEAR YOUR SAFETY GLASSES WHILE PERFORMING THIS EXPERIMENT

Experimental Procedure

Obtain a thermometer.

A. Measuring the Rate of Reaction

Step 1. Setting the Stage

Set up a hot-water bath, using a 600-mL beaker about three-quarters full of water. As a heater, use an electric hot plate with magnetic stirrer if such is available. Otherwise, use a Bunsen burner, supporting the beaker on a wire gauze on an iron ring. If you use a burner, you should put a larger iron ring around the beaker, near the top, to prevent it from tipping over. Clamp this larger ring to the ring stand as well. Prepare an ice bath by filling a 400-mL beaker with crushed ice; add water until the level is at the top of the ice. Clean three medium (18 × 150 mm) test tubes. Rinse them with a few milliliters of acetone; pour the acetone into one of the tubes, shake it, then pour this same acetone into the next tube and repeat; pour the acetone into the last test tube and repeat a final time. Dry the tubes with a gentle stream of compressed air. Number the test tubes and put them in a test tube rack. Put a small funnel in Tube 1.

Step 2. Weighing the Sample

While the water is heating, weigh out on an analytical balance about sixty milligrams of aspirin to ± 0.0001 g. This takes a bit of doing, since the sample is very small. It is probably easiest to make the weighing on a piece of weighing paper. Tare the paper, and use a small spatula to add a very small amount of aspirin. Add more aspirin, a very small amount at a time, until you have 60 ± 5 mg (0.060 ± 0.005 g) in total. Record the mass. Then, carefully, remove the paper from the balance. Fold it at the edge and transfer the sample to the funnel. Tap the paper and the funnel to get as much sample as possible into Tube 1.

Step 3. Preparing the Reaction Mixture

With a rinsed 10-mL graduated cylinder, or a pipet, add 10.0 mL of deionized water to the aspirin in Tube 1. The aspirin will not dissolve in the water; some will stay at the bottom of the tube, and some will float on the water. Put your stirring rod in the tube.

Step 4. Carrying Out the Reaction

By now the water in the 600-mL beaker should be getting warm. If it is boiling, slowly add tap water or ice to lower the temperature to 80 ± 1°C, as measured to ± 0.1°C on your thermometer. **CAUTION:** **Hot water can give you a bad burn, so be careful not to tip over the beaker.** Adjust your heat source as necessary to hold the bath steady at 80 ± 1°C for at least five minutes. When you have managed that (and it may take a little time) you are ready to make your first run. Pick up the tube with the reaction mixture in it, and, noting the time to the nearest second or starting a stopwatch, put the tube in the water bath. Stir the mixture continuously, to help dissolve the aspirin and warm the reaction mixture up quickly. When the aspirin has all dissolved, which should happen within a minute, keep the rod in the tube, stirring occasionally. Measure and record the temperature of the bath at 1-minute intervals as the reaction proceeds; it will fall initially, but should come back to the original temperature fairly quickly. Leave the tube in the bath for 6.0 minutes. Then remove the tube and put it in the ice-water bath to stop the reaction. Stir the reaction mixture to hasten the cooling. Within a minute the solution should be cool, and you can put the tube back in the test tube rack.

Step 5. Adding the Indicator

Drain the graduated cylinder or pipet, and use it to measure out 5.0 mL of 0.025 M $Fe(NO_3)_3$ into the reaction mixture. The solution should turn violet as the iron-salicylic acid complex forms. Stir the mixture until it is of uniform color.

Step 6. Finding the Salicylic Acid Concentration

Measure the absorbance of the solution at 525 nm on a spectrophotometer. If this is your first time measuring absorbance, the procedure is described in Appendix IV. After adjusting the spectrophotometer for 0 and 100% transmittance, rinse the spectrophotometer tube twice with a small amount of the solution from Step 5. Then add a few milliliters of that solution to the tube and record the absorbance you obtain. Find the concentration of the salicylic acid in the reaction mixture, using the graph you are given or the equation you are given relating absorbance and concentration. The graph or equation will provide the salicylic acid concentration in the reaction mixture *before* the indicator was added. By Equation 1, that concentration will equal the change that occurs in the aspirin concentration.

Step 7. Correcting for Warm-Up Time

Once the temperature of the reaction mixture is up to 80.°C, the reaction rate will obey Equation 2; but during the first minute, when the aspirin is dissolving and the mixture is warming up, the reaction is much slower. To correct for this, you might run the same size sample at 80.°C, for 1 minute, and subtract the absorbance for that sample from the total absorbance. You could then use 5 minutes as the time the reaction occurred. The authors actually did this, but found that if they took account of the warm-up period by using the total absorbance and 5.5 minutes for the time of reaction, the rates were essentially the same as with the more accurate, but more complex, procedure. So, when calculating the rate, we will adopt this simpler approach: we will use 5.5 minutes as the time, along with the total absorbance observed in Step 6.

B. Determining the Order of the Reaction and the Rate Constant

Reset your water bath to the temperature you used in Part A, and prepare a new ice bath.

Repeat Steps 1 through 7, again at 80 ± 1°C, but using about forty milligrams of sample, weighed to ± 0.0001 g, in Test Tube 2. Be sure to rinse out your graduated cylinder or pipet with deionized water before measuring out the 10.0 mL of water. The results you obtain with this sample, along with those from Part A, will allow you to find the order of the reaction and the rate constant at 80.°C. On the report page carry out the calculations for the order and the rate constant of the reaction. Calculate the mean value of the rate constant and the standard deviation. (See Appendix IX.) Then proceed to Part C.

C. Finding the Activation Energy

The last piece of information to be obtained in this experiment is the activation energy for the reaction, which is the energy that must be present in the key step when the reaction proceeds. It is found by measuring the dependence on temperature of the rate constant for the reaction.

We will use nearly the same procedure as in Part A, Steps 1 through 7, except that we will carry out the reaction in Test Tube 3 at 70.°C instead of 80.°C. The sample size should be about sixty milligrams, weighed accurately. Bring the water bath to 70 ± 1°C and hold it there for 5 minutes. Make a new ice bath, and make sure that you rinse your pipet or graduated cylinder before measuring out the water.

At this lower temperature, the reaction will go much more slowly than in Part A. In order to get easily measured absorbance values, run the reaction for 12 minutes rather than 6 minutes. The aspirin will be slower to dissolve, so the warm-up period will be 2 minutes rather than 1 minute. It will require a considerably longer time to dissolve the aspirin, but, with continuous stirring, it should go into solution in 2 minutes. Record the temperature at 2-minute intervals. Record the absorbance that you obtain. To correct for warm-up time, calculate the reaction rate using the *measured* absorbance but assume the effective time to be *11 minutes*.

With the data you have obtained, you can proceed to evaluate the energy of activation, E_a, for the reaction. This is done using the rate constants at 80.°C and 70.°C and the Arrhenius equation, as in Experiment 20:

$$\ln k = -E_a/RT + \text{a constant} \tag{3}$$

Recall that $\ln k$ is the natural logarithm of the rate constant, R is the gas constant, and T is the temperature in Kelvins.

DISPOSAL OF REACTION PRODUCTS.   All of the reaction products may be poured down the sink, with the water running as you pour.

Experiment 42

Data and Calculations: Rate Studies on the Decomposition of Aspirin

Let Asp = aspirin and Sal = salicylic acid (MM Asp = 180.2 g/mol)

	Part A Tube 1	Part B Tube 2	Part C Tube 3
1. Mass of aspirin in tube [grams]	_____	_____	_____
Moles of aspirin in tube	_____	_____	_____
[Asp] in moles/L (solution volume = 10.0 mL)	_____	_____	_____

(Temperature measurements are on following page)

2. Absorbance, Abs, of solution	_____	_____	_____
[Sal] in reaction mixture	_____M	_____M	_____M
Change in [Asp], $\Delta[\text{Asp}] = -[\text{Sal}]$:	_____M	_____M	_____M

Note that $\Delta[\text{Asp}] \ll [\text{Asp}]$

3. Rate of reaction $= -\Delta[\text{Asp}]\Delta t = k[\text{Asp}]^n$ **(4)**

Rate in moles/(L·min), with $\Delta t = 5.5$ min	_____	_____	
Rate in moles/(L·min), with $\Delta t = 11$ min			_____

4. (Rate in Tube 1)(Rate in Tube 2) _____

5. ([Asp] in Tube 1)([Asp] in Tube 2) _____

The reaction is _____ order (Use Eq. 4, above, for Tube 1 and Tube 2)
(Select nearest integer, and give reasoning.)

6. Value of rate constant k	_____	_____	_____

Mean value of k at 80.°C _____ Standard deviation _____

(continued on following page)

7. $\dfrac{k_{\text{mean}} \text{ at } 80.°C}{k \text{ at } 70.°C}$ _____

8. Temperature measurements during reactions:

time [min]	0	1	2	3	4	5	6	Mean
or	0	2	4	6	8	10	12	temperature

t, °C in Tube 1   _____ _____ _____ _____ _____ _____ _____   _____

in Tube 2   _____ _____ _____ _____ _____ _____ _____   _____

in Tube 3   _____ _____ _____ _____ _____ _____ _____   _____

9. $\ln k = -E_a/RT + \text{a constant}$ **(3)**

Use mean temperatures in K, not °C, in Equation 3

$$\ln (k_{80}/k_{70}) = \underline{\hspace{3cm}} = \underline{\hspace{3cm}}$$

Solve for E_a

The activation energy, $E_a = \underline{\hspace{3cm}}$ J $= \underline{\hspace{3cm}}$ kJ

Experiment 42

Advance Study Assignment: The Rate of Decomposition of Aspirin

1. A student studies the rate at which aspirin decomposes by the reaction

$$\text{Asp} + H_2O \rightarrow \text{Sal} + \text{acetic acid}$$

   in which Asp = aspirin and Sal = salicylic acid. She weighs out 59.4 mg of aspirin and dissolves it, making a 10.0-mL solution in water. She heats the solution for 5.0 minutes at 90.°C (which means 90 ± 1°C), and finds that 8.0% of the aspirin is converted to salicylic acid and acetic acid.

   a. How many moles of aspirin were in the initial solution? What was the initial molarity of the aspirin? The molar mass of aspirin is 180.2 g/mol.

   _____ moles _____ M

   b. What is the rate of reaction? (See Equation 4 on the report page.)

   _____ moles/(L·min)

   c. In a similar experiment with a smaller amount of aspirin, and thus a lower [Asp], she finds that once again 8% of the aspirin decomposes in 5 minutes at 90.°C. What is the order of the reaction? How do you know?

   _____

   d. What is the rate constant for the reaction?

   _____

(continued on following page)

e. In another experiment she determines that the activation energy, E_a, for the decomposition reaction is 95 kJ/mol. How long would it take for 10% of an aspirin sample to decompose at body temperature, 37°C?

_____ minutes

Experiment 43

Analysis for Vitamin C

Vitamin C, known chemically as ascorbic acid, is an important component of a healthy diet. In the mid-eighteenth century the British Royal Navy found that the addition of citrus fruit to sailors' diets prevented the malady called scurvy. Humans are one of the few members of the animal kingdom unable to synthesize vitamin C, resulting in the need for its regular ingestion in order to remain healthy. The National Academy of Sciences has established a threshold of 75 mg/day for (nonpregnant) female adults (90 mg/day for male adults) as the Reference Daily Intake (RDI). Linus Pauling, a chemist whose many contributions to chemical bonding theory should be well known to you, recommended a level of 500 mg/day to help ward off the common cold. He also suggested that large doses of vitamin C are helpful in preventing cancer.

The vitamin C content of foods can be easily determined by oxidizing ascorbic acid, $C_6H_8O_6$, to dehydro-L-ascorbic acid, $C_6H_6O_6$:

$$C_6H_8O_6 \rightarrow C_6H_6O_6 + 2\,H^+ + 2\,e^-$$

$$\text{vitamin C} \qquad \text{dehydro-L-ascorbic acid} \qquad\qquad (1)$$

This reaction is very slow for ascorbic acid in the dry state but occurs readily when in contact with moisture. A reagent that is particularly good for the oxidation is an aqueous solution of iodine, I_2. Since I_2 is not very soluble in water, we dissolve it in a solution of potassium iodide, KI, in which the I_2 exists mainly as I_3^-, a (much more soluble) complex ion. The reaction with ascorbic acid involves I_2, which is reduced to I^- ion:

$$2\,e^- + I_2(aq) \rightarrow 2\,I^-(aq) \qquad\qquad (2)$$
$$\text{yellow} \quad \text{colorless}$$
$$\text{to red}$$

In the overall reaction, one mole of ascorbic acid requires one mole of I_2 for complete oxidation.

When the colored I_2 solution is added to the colorless ascorbic acid solution, the characteristic yellow to red-brown (depending on concentration) iodine color disappears because of the above reaction. Although we could use the first permanent appearance of the yellow color of dilute iodine to mark the endpoint of our titration, better results are obtained when starch is added as an indicator. Starch reacts with I_2 to form an intensely colored blue complex. In the titration I_2 reacts preferentially with ascorbic acid, and so its concentration remains very low until the ascorbic acid is all oxidized. At that point, the I_2 concentration begins to go up, and the reaction with the indicator occurs:

$$I_2(aq) + starch \rightarrow \text{starch-}I_2\text{ complex(aq)} \qquad\qquad (3)$$
$$\text{yellow} \qquad\qquad\qquad \text{intense blue}$$

Because an I_2 solution cannot be prepared accurately by direct weighing, it is necessary to standardize the I_2 against a reference substance of known purity. We will use pure ascorbic acid for this reference; that is, as our primary standard. Once it is standardized, you can use the iodine solution for the direct determination of vitamin C in any kind of sample.

Experimental Procedure

A. Standardization of the Iodine Titrant

Obtain a buret and an unknown vitamin C sample. Weigh out accurately on the analytical balance three pure ascorbic acid samples (not your unknown!) of approximately one-tenth of a gram into clean 250-mL Erlenmeyer flasks. Dissolve each sample in approximately one hundred milliliters of water.

Clean your buret thoroughly. Draw about one-hundred milliliters of the stock I_2 solution into a 400-mL beaker and add approximately one-hundred and fifty milliliters of water. Stir thoroughly and cover with a piece of aluminum foil. Rinse the buret with a few milliliters of the I_2 solution three times. Drain and then fill the buret with the I_2 solution.

After taking an initial reading of the buret (you may find looking toward a light source will make it easier to see the bottom of the meniscus), add 1.0 mL of starch indicator to the first ascorbic acid sample and titrate it with the iodine solution. Note the disappearance of the I_2 color as you swirl the flask gently and continuously during the titration. Continue adding the iodine titrant, using progressively smaller volume increments, until the sample solution turns (and remains) a distinct blue. After reading the buret, titrate the other two sample solutions using the same technique—being sure to add the starch indicator first and to read your buret before and after each titration.

B. Analysis of an Unknown Containing Vitamin C

Given your experience standardizing the titrant, you should be able to devise an analogous procedure to determine the vitamin C content of your unknown sample. You will need to select a sample size, and you may need to carry out an initial treatment of the sample. In particular, if your instructor assigns you a fruit juice sample, it will be desirable to first filter the sample through cheesecloth and then rinse the filter with water to bring through any remaining vitamin C.

In choosing the sample sizes, it may be helpful to calculate an iodine solution parameter called the titer—the number of milligrams of ascorbic acid that reacts with 1.0 mL of iodine solution. This number is easily found from the I_2 concentration and the balanced chemical reaction. It is desirable to have the volume of I_2 for each titration be at least 15 mL. Using a small initial sample will give you an indication of how much to scale up for your final titrations.

If your unknown was a solid, report your results in percent vitamin C in the solid, by mass. For liquid samples, report how many milligrams of vitamin C are present per 100 mL. In each case, calculate the mean and standard deviation of the measured concentrations of your unknown, as well as the sample size required to give the nonpregnant female RDI of vitamin C.

DISPOSAL OF REACTION PRODUCTS. All reaction products from this experiment may be diluted with water and poured down the drain.

Experiment 43

Data and Calculations: Analysis for Vitamin C

A. Standardization of Iodine Solution

Sample	I	II	III
Mass of pure ascorbic acid sample, g	_____	_____	_____
Moles of ascorbic acid (MM = 176.14 g/mol)	_____	_____	_____
Initial buret reading [mL]	_____	_____	_____
Final buret reading [mL]	_____	_____	_____
Volume of I_2 titrant added [mL]	_____	_____	_____
Moles of I_2 consumed	_____	_____	_____
Molarity of I_2 [moles/L]	_____	_____	_____
Titer of I_2 [(mg Asc)/(mL I_2 solution)]	_____	_____	_____

(continued on following page)

B. Analysis of an Unknown Sample Containing Vitamin C

Sample	I	II	III
Mass or volume of unknown [g or mL]	_____	_____	_____
Initial buret reading [mL]	_____	_____	_____
Final buret reading [mL]	_____	_____	_____
Volume of I_2 titrant added [mL]	_____	_____	_____
Moles of iodine added	_____	_____	_____
Moles of vitamin C in sample	_____	_____	_____
Mass of vitamin C in sample [mg]	_____	_____	_____
Concentration of vitamin C: $\%_{mass}$ in solid sample or [mg/100 mL] in liquid sample	_____	_____	_____

Mean concentration _____ $\%_{mass}$ or [mg/100 mL]

Standard deviation _____ $\%_{mass}$ or [mg/100 mL]

Amount of sample that will furnish nonpregnant female RDI, mL or mg   _____

Experiment 43

Advance Study Assignment: Analysis for Vitamin C

1. Write a balanced equation for the reaction between I_2 and ascorbic acid. Identify the oxidizing agent and the reducing agent.

2. A solution of I_2 was standardized with ascorbic acid (Asc). Using a 0.1032-g sample of pure ascorbic acid, 25.15 mL of I_2 titrant was required to reach the starch endpoint.

   a. What is the molarity of the iodine solution?

   _____ M

   b. What is the titer of the iodine solution?

   _____ (mg Asc)/(mL I_2 solution)

(continued on following page)

3. A sample of fresh grapefruit juice was filtered and titrated with the I_2 solution above. A 100.-mL (the decimal point here indicates 100. has three significant figures: that is, it denotes 100 ± 1 mL) sample of the juice took 11.86 mL of the iodine titrant to reach the starch endpoint.

a. What is the concentration of vitamin C in the juice in (mg vitamin C)/(100 mL of juice)?

_____ mg/100 mL

b. What quantity of this juice will provide the male RDI amount of vitamin C?

_____ mL

Experiment 44

Fundamentals of Quantum Mechanics

For a great many years, scientists relied on their first-hand experience with the physical world to guide their scientific interpretation of it: if an idea didn't make physical sense in light of their own experience, they discarded it, and it was often a scientist's interactions with the physical world (like Newton's falling apple) that guided and inspired the creation of their theories. However, in 1900 the physicist Max Planck, as part of an effort to make electric light bulbs more efficient, was attempting to mathematically describe blackbody radiation (the light emitted from something glowing white-hot, like an incandescent light bulb filament), and he just could not get it to work. In desperation, he made a mathematical assumption that at the time he thought was physically ridiculous: he assumed that light energy could not come in just any amount, but rather that it had to be a multiple of a minimum quantity—he assumed light was quantized. Remarkably, this made the mathematics work out, and with this assumption he was able to mathematically describe blackbody radiation perfectly. The idea of quantization eventually became the underpinning of a whole new way of looking at the world, called quantum mechanics. It is a strange view of things, one that at first does not seem to connect with our direct experience of the physical world at all. But it underlies our understanding of a lot of modern technology, from lasers to computer chips, as well as today's understanding of chemical bonding, so it is worth trying to gain a better feel for it.

As you will learn in this experiment, there are some situations where quantization is something we can experience first-hand. You will also get a pretty good workout doing this experiment, so come to lab prepared to sweat!

We now understand that everything is quantized: from the speed at which a bicycle or a baseball can travel to the color of light that can be emitted by the relaxation of an excited atom. However, for macroscopic objects, the spacing between allowed levels is so small that we can never pick up on the quantization. That is why the concept of quantized states seems so foreign to us. (Consider the mass of water in a bucket: in a way, it is quantized, in that you can only add or take away a discrete number of water molecules. You can't add one-third of an H_2O molecule, or take half of one away. But because the mass of a single water molecule is so incredibly small, we don't observe this quantization on the human scale. In our experience, the mass of water in a bucket can be varied continuously and set to any amount we wish. The quantization only becomes noticeable when working with an incredibly small mass of water—that is, on a submicroscopic [or "quantum"] scale.) There are a few everyday things that exhibit quantization we can observe first hand, and almost all of these things have some relationship to musical instruments. In this experiment you will study one of these: the oscillation of a string.

Consider a rubber band stretched between two fixed points. If you flick this rubber band, it will vibrate back and forth for a while before coming to rest again. If you listen carefully, you will even hear it make a tone as it vibrates. What's interesting is that this tone has the same pitch whether you flick the band gently or with a lot of force. (The sound may be louder or softer depending on how hard you flick, but the pitch—the note it plays—is the same.) The pitch remains (essentially) constant throughout the time the rubber band vibrates. This is because the vibration is quantized. So too are the vibrations of a guitar string, a piano string, the reed in a woodwind instrument like a clarinet, and the air in the cavity of a brass instrument like a trombone. That is why these are musical instruments: they produce a reliable, fixed tone, whether played loudly or softly.

As with most musical instruments, we can tune a rubber band. Its pitch depends on how long it is, the mass of the material it is made of, and how much tension it is under. Generally, the vibrations of a rubber band are too rapid for us to count easily (which is why we can hear them), so we are going to use a different, more slowly moving model to learn some of the general rules that apply to quantized systems. Our model system will effectively consist of spinning a series of jump ropes (or something similar), sometimes in ways that wouldn't work so well for the actual game of jump rope.

Experimental Procedure

You will be given a 4 m section of standard "rope" marked with a black line at the 3 m point—this may actually be lab tubing or something similar. You will study the oscillation of this rope—how it moves back and forth, up and down, or around and around, over and over again. (In this experiment the oscillation will be around and around: turning in a circle as in a game of jump rope.) We will investigate the frequency of oscillation under different conditions—how often, per unit of time, the repeated pattern of motion returns to the starting point. We will find that the rope has a series of favored frequencies, called resonant frequencies, under a given set of conditions. Some conditions, such as the length and weight of the rope, matter, while others, like the color of the rope or whether you are left- or right-handed, should have no effect.

A. The Effect of Quantum State on Resonant Frequency

Take one end of your rope and give the other end to your partner. Hold the very end of the rope: this ensures that the length of the oscillating rope is well defined. Also hold the rope consistently at waist level, just in front of you. Have your partner move away from you until the rope is stretched out in a straight line between you, then move together until the rope is suspended just barely above the ground between you. The tension on the rope matters, and we want to keep it constant, so in this part always hold your rope such that when it is stationary, it is just barely touching the ground. Try to stay in that position as you do the experiment.

Now, with your partner holding their hand still, rotate your end of the rope such that the entire rope starts to spin, as it would in a game of jump rope. You may have to make exaggerated loops with your arm to get started, but once you have, try to move the hand spinning the rope only a little. Spin the rope fast enough that the arc it describes is constant all the way around—that is, don't go too slowly. Once you have the rope going in a regular, stable pattern, have your partner time (with his or her free hand so that his or her end of the rope remains stationary) how long it takes for the rope to go around 50 times. Stop and record this on the report page.

Set up again, but this time when you spin the rope, get it going a good deal faster. If you do this right, you will end up having two segments of rope spinning between you and your partner, with a stationary point in the middle. This stationary point is called a node and it indicates that the oscillation of the rope is now in a higher quantum state than it was when you were spinning it more slowly. Specifically, you are now in the first excited state, while before (with no nodes) you were in the ground state. You will find it takes a lot more energy to maintain the oscillation in the excited state—you will get tired doing this! So without delay, your partner should start timing and again determine how long it takes for the rope to go around 50 times. Stop and record this on the report page.

Finally, set up as before but this time spin the rope still faster so you get *two* nodes. (This is called the second excited state.) Again, time how long it takes the rope to go around 50 times and record the result.

Optional   If you have it in you, try to spin the rope so fast you get *even more* nodes. (These are called higher-order excited states, like the third, the fourth, and so on.) This will be hard to do! You don't have to record anything here, but it will give you a clear sense of the relationships among nodes, excited states, and energy. ■

B. The Effect of Length on Resonant Frequency

Repeat the previous steps (time the ground state, first excited state, and, if possible, the second excited state) while holding the rope at the 3 meter mark, again controlling the tension by keeping the rope just taught enough that it barely touches the ground when it is at rest. You may want to coil up the extra rope and hold it so that it doesn't get tangled in the working section of the rope. Record your results in the report page.

C. The Effect of Tension on Resonant Frequency

For this part, take your standard rope and prepare to do the experiment again, holding it at the end as you did in Part A. However, once you have the rope in the position you used previously, put one of your feet immediately behind the other (heel to toe) and step back that far (about ten inches), such that there is more tension in the rope and it is hanging a little above the ground. Repeat the steps in Part A and note whether the frequency of oscillation is higher or lower than it was before.

Optional ## D. The Effect of Mass on Resonant Frequency

For this last part, repeat Part A with an alternate rope (made of a different material) to determine whether a change in mass (at approximately the same length and tension) causes an increase or decrease in resonant frequency. (If it is not obvious, your instructor may tell you whether the standard or the alternate rope has a larger mass per unit of length.) ■

Name _____ **Section** _____

Experiment 44

Data and Calculations: Fundamentals of Quantum Mechanics

A. The Effect of Quantum State on Resonant Frequency

Time required for standard rope to complete 50 cycles in ground (lowest energy, zero node) state: _____ sec

Time required for standard rope to complete 50 cycles in first excited (one node) state: _____ sec

Time required for standard rope to complete 50 cycles in second excited (two node) state: _____ sec

Resonant frequency of standard rope in ground (lowest energy, zero node) state: _____ Hz

Resonant frequency of standard rope in first excited (one node) state: _____ Hz

Resonant frequency of standard rope in second excited (two node) state: _____ Hz

B. The Effect of Length on Resonant Frequency

Time required for standard rope held at 3-m mark to complete 50 cycles in ground (lowest energy, zero node) state: _____ sec

Time required for standard rope held at 3-m mark to complete 50 cycles in first excited (one node) state: _____ sec

Time required for standard rope held at 3-m mark to complete 50 cycles in second excited (two node) state: _____ sec

Resonant frequency of standard rope held at 3-m mark in ground (lowest energy, zero node) state: _____ Hz

Resonant frequency of standard rope held at 3-m mark in first excited (one node) state: _____ Hz

Resonant frequency of standard rope held at 3-m mark in second excited (two node) state: _____ Hz

(continued on following page)

C. The Effect of Tension on Resonant Frequency

Time required for elevated standard rope to complete 50 cycles in ground (lowest energy, zero node) state: _____ sec

Time required for elevated standard rope to complete 50 cycles in first excited (one node) state: _____ sec

Time required for elevated standard rope to complete 50 cycles in second excited (two node) state: _____ sec

Resonant frequency of elevated standard rope in ground (lowest energy, zero node) state: _____ Hz

Resonant frequency of elevated standard rope in first excited (one node) state: _____ Hz

Resonant frequency of elevated standard rope in second excited (two node) state: _____ Hz

Optional   D. The Effect of Mass on Resonant Frequency

Time required for alternate rope to complete 50 cycles in ground (lowest energy, zero node) state: _____ sec

Time required for alternate rope to complete 50 cycles in first excited (one node) state: _____ sec

Time required for alternate rope to complete 50 cycles in second excited (two node) state: _____ sec

Resonant frequency of alternate rope in ground (lowest energy, zero node) state: _____ Hz

Resonant frequency of alternate rope in first excited (one node) state: _____ Hz

Resonant frequency of alternate rope in second excited (two node) state: _____ Hz

Experiment 44

Advance Study Assignment: Fundamentals of Quantum Mechanics

1. This experiment investigates the phenomenon of resonance, specifically in the oscillations of a length of rope, as an example of the quantization that underlies quantum mechanics. What three physical characteristics of the rope are intentionally varied in the course of this experiment (including optional Part D)?

2. When spun quickly enough, a rope will have one or more stationary points along its length, points that do not move or rotate. What is the technical term for these stationary points?

3. The lowest-energy oscillation in a given system is called the ground state. This is the state obtained when the rope is spun just quickly enough to keep it rotating consistently. If the rope is spun quite a bit faster, at the correct speed, it can enter a higher-energy state with one or more of the stationary points from Question 2 along its length. What is the technical term for such a higher-energy state?

4. Frequency is measured in units of Hertz, abbreviated Hz. One Hertz is one oscillation per second. If it takes 76 seconds for a rope to go around 50 times, what is the frequency of its oscillation, in Hertz? Be sure to show your work!

Appendix I

Vapor Pressure and Density of Liquid Water

Temperature [°C]	Vapor pressure [mm Hg]	Liquid density [g/mL]	Temperature [°C]	Vapor pressure [mm Hg]	Liquid density [g/mL]
0.0	4.59	0.99979	25.0	23.8	0.99700
0.5	4.75	0.99982	25.5	24.5	0.99687
1.0	4.93	0.99985	26.0	25.2	0.99674
2.0	5.30	0.99989	26.5	26.0	0.99661
3.0	5.69	0.99992	27.0	26.8	0.99647
4.0	6.10	0.99993	27.5	27.6	0.99633
5.0	6.54	0.99992	28.0	28.4	0.99619
6.0	7.02	0.99989	28.5	29.2	0.99605
7.0	7.52	0.99986	29.0	30.1	0.99590
8.0	8.05	0.99980	29.5	31.0	0.99576
9.0	8.61	0.99974	30.0	31.9	0.99561
10.0	9.21	0.99965	32.0	35.7	0.99499
12.0	10.5	0.99945	34.0	39.9	0.99433
14.0	12.0	0.99920	36.0	44.6	0.99364
15.0	12.8	0.99906	38.0	49.8	0.99292
15.5	13.2	0.99898	40.0	55.4	0.99218
16.0	13.6	0.99890	45.0	72.0	0.99017
16.5	14.1	0.99882	50.0	92.6	0.98800
17.0	14.5	0.99873	55.0	118.2	0.98566
17.5	15.0	0.99864	60.0	149.6	0.98316
18.0	15.5	0.99855	65.0	187.8	0.98052
18.5	16.0	0.99846	70.0	234.0	0.97773
19.0	16.5	0.99836	75.0	289.5	0.97481
19.5	17.0	0.99826	80.0	355.6	0.97177
20.0	17.5	0.99816	85.0	434.0	0.96859
20.5	18.1	0.99806	90.0	526.4	0.96530
21.0	18.7	0.99795	92.0	567.7	0.96394
21.5	19.2	0.99784	94.0	611.6	0.96257
22.0	19.8	0.99773	96.0	658.3	0.96118
22.5	20.5	0.99761	98.0	708.0	0.95978
23.0	21.1	0.99750	99.0	734.0	0.95906
23.5	21.7	0.99738	99.5	747.2	0.95871
24.0	22.4	0.99725	100.0	760.7	0.95835
24.5	23.1	0.99713			

Wagner and Pruss, *J Phys Chem Ref Data*, 31–2: 387–535, 2002, in NIST Chemistry WebBook, NIST Standard Reference Database Number 69, Eds. P. J. Linstrom and W. G. Mallard, National Institute of Standards and Technology, Gaithersburg MD, 20899, http://webbook.nist.gov (retrieved June 28, 2010 and December 13, 2018).

Appendix II

Summary of Solubility Properties of Ions and Solids

Cation (color in aqueous solution)	Cl⁻, Br⁻ I⁻, SCN⁻	SO_4^{2-}	CrO_4^{2-}	PO_4^{3-}	*$C_2O_4^{2-}$	CO_3^{2-}
Na^+, K^+, NH_4^+	S	S	S	S	S	S
Ba^{2+}	S	I	A	A^-	A	A^-
Ca^{2+}	S	S^-	S	A^-	A	A^-
Mg^{2+}	S	S	S	A^-	A^-	A^-
Fe^{3+} (yellow)	*S	S	A^-	A	S	D: A^-
Cr^{3+} (violet‡)	S	S	A^-	A	S	A^-
Al^{3+}	S	S	A^-, C	A, C	A^-, C	D: A^-, C
Ni^{2+} (green)	S	S	S	A^-, C	A, C	A^-, C
Co^{2+} (pink)	S	S	A^-	A^-	A^-, C	A^-
Zn^{2+}	S	S	A^-, C	A^-, C	A^-, C	A^-, C
Mn^{2+} (pale pink)	S	S	S	A^-	A^-	A^-
Cu^{2+} (blue)	**S	S	A^-, C	A^-, C	A, C	A^-, C
Cd^{2+}	S	S	A^-, C	A^-, C	A, C	A^-, C
Bi^{3+}	A	A^-	A	A	A	A^-
Hg^{2+}	†S	S	A	A^-	A	A^-
Sn^{2+}, Sn^{4+}	A, C	A, C	A, C	A, C	A, C	A, C
Sb^{3+}	A, C	A, C	A, C	A, C	A^-, C	A, C
Ag^+	††C	S^-	A, C	A, C	A, C	A^-, C
Pb^{2+}	C, HW	C	C	A, C	A, C	A^-, C
Hg_2^{2+}	O^+	A	A	A	O	A

Key: S, soluble in water; no precipitate on mixing cation, 0.10 M, with anion, 1.0 M

S^-, slightly soluble; tends to precipitate on mixing cation, 0.10 M, with anion, 1.0 M, but precipitation may be slow and/or small in quantity

HW, soluble in hot water

A^-, soluble in 1.0 M $HC_2H_3O_2$ (acetic acid)

A, soluble in acid (6 M HCl or some other nonprecipitating, nonoxidizing acid)

A^+, soluble in 12 M HCl

O, soluble in hot 6 M HNO_3 (a strong, oxidizing acid)

O^+, soluble in aqua regia, a (dangerous!) aqueous oxidizing acid and complexing agent combination

CAUTION: Acidic solutions are corrosive, more so when hot, and even more so if they are oxidizing. Use the above with great care.

C, soluble in solutions containing a good complexing ligand (see rightmost column in table on next page)

D, unstable; decomposes to a product with solubility as indicated

I, insoluble in any common solvent or solution

*FeI_3 is unstable; it decomposes to FeI_2 and I_2.

**CuI_2 is unstable; it decomposes to CuI and I_2.

†HgI_2 is insoluble, but dissolves in excess I^-.

††AgCl and AgSCN dissolve in 6 M NH_3, but AgBr and AgI do not.

‡Other colors are often observed, particularly green, in the presence of chloride, sulphate, and most other anions that can substitute in as ligands; ligand substitution is often slow at room temperature.

Nomenclature: SCN^- = thiocyanate ion; CrO_4^{2-} = chromate ion; $C_2O_4^{2-}$ = oxalate ion; SO_3^{2-} = sulfite ion; S^{2-} = sulfide ion; O^{2-} = oxide ion; ClO_3^- = The chlorate ion is ClO_3^-; CN^- is the cyanide ion. $C_2H_3O_2^-$ = acetate ion; NO_2^- = nitrite ion; $C_2O_4^{2-}$ = oxalate ion; $S_2O_3^{2-}$ = thiosulfate ion; CN^- = cyanide ion

Cation (color in aqueous solution)	SO_3^{2-}	S^{2-}	OH^-, $*O^{2-}$	NO_3^-, ClO_3^- $C_2H_3O_2^-$, NO_2^-	Complexing agents**
Na^+, K^+, NH_4^+	S	S	S	S	—
Ba^{2+}	A	S	S^-	S	—
Ca^{2+}	A^-	D: A^-	S^-	S	—
Mg^{2+}	S	D: A^-	A^-	S	—
Fe^{3+} (yellow)	D: S	D: A^-	A^-	S	—
Cr^{3+} (violet‡)	S	D: A^-	A^-	S	†(OH^-)
Al^{3+}	A^-, C	D: A^-, C	A^-, C	S	OH^-
Ni^{2+} (green)	A^-	O	A^-, C	S	NH_3
Co^{2+} (pink)	A^-	O	A^-	S	$C_2O_4^{2-}$, ††(NH_3)
Zn^{2+}	S	A^-, C	A^-, C	S	OH^-, NH_3
Mn^{2+} (pale pink)	S	A^-	A^-	S	—
Cu^{2+} (blue)	A^-, C	O	A^-, C	S	NH_3
Cd^{2+}	A^-, C	A	A^-, C	S	NH_3
Bi^{3+}	A	A^+, O	A^-	A^-	Cl^-
Hg^{2+}	D: O	O^+	A^-	S	—
Sn^{2+}, Sn^{4+}	A, C	A, C	A, C	A, C	OH^-, Cl^-
Sb^{3+}	A, C	A, C	A, C	A, C	OH^-, Cl^-
Ag^+	A, C	O	A^-, C	S	‡‡See note
Pb^{2+}	A, C	O	A^-, C	S	OH^-
Hg_2^{2+}	D: O	D: O^+	D: O	S	—

Key: S, soluble in water; no precipitate on mixing cation, 0.10 M, with anion, 1.0 M

  S^-, slightly soluble; tends to precipitate on mixing cation, 0.10 M, with anion, 1.0 M, but precipitation may be slow and/or small in quantity

  A^-, soluble in 1.0 M $HC_2H_3O_2$ (acetic acid)

  A, soluble in acid (6 M HCl or some other nonprecipitating, nonoxidizing acid)

  A^+, soluble in 12 M HCl

  O, soluble in hot 6 M HNO_3 (a strong, oxidizing acid)

  O^+, soluble in aqua regia, a (dangerous!) aqueous oxidizing acid and complexing agent combination

  CAUTION: **Acidic solutions are corrosive, more so when hot, and even more so if they are oxidizing. Use the above with great care.**

  C, soluble in solutions containing a good complexing ligand (see rightmost column in the table above)

  D, unstable; decomposes to a product with solubility as indicated

*Oxides behave like hydroxides, but may be slow to dissolve.

**Oxalate forms complexes with many metal ions and may serve as an effective complexing agent as well.

†Cr^{3+} can, under some conditions, form complexes with OH^-.

††Co^{2+} can, under some conditions, form complexes with NH_3.

‡Other colors are often observed, particularly green, in the presence of chloride, sulphate, and most other anions that can substitute in as ligands; ligand substitution is often slow at room temperature.

‡‡Most silver salts can be complexed with (in order of increasing complexing strength): [additional] NH_3, SCN^-, $S_2O_3^{2-}$, or CN^-. **CAUTION:** **USE CN^- IN BASE ONLY!!!**

Nomenclature: SCN^- = thiocyanate ion; CrO_4^{2-} = chromate ion; $C_2O_4^{2-}$ = oxalate ion; SO_3^{2-} = sulfite ion; S^{2-} = sulfide ion; O^{2-} = oxide ion; ClO_3^- = The chlorate ion is ClO_3^-; CN^- is the cyanide ion. $C_2H_3O_2^-$ = acetate ion; NO_2^- = nitrite ion; $C_2O_4^{2-}$ = oxalate ion; $S_2O_3^{2-}$ = thiosulfate ion; CN^- = cyanide ion

Appendix IIA

Some Properties of the Cations in Groups I, II, and III

In the experiments on qualitative analysis, we used the chemical and physical properties of the cations in Groups I, II, and III to separate the ions from each other. Here we have summarized most of the properties that were employed, and have included some others that help distinguish the cations we studied.

Lead Ion, Pb^{2+}

Most lead compounds are insoluble in water; the only exceptions are $Pb(NO_3)_2$ and $Pb(C_2H_3O_2)_2$. Lead chloride is slightly soluble in water at room temperature, while it is much more soluble in hot water and in solutions where the Cl^- concentration is high. The lead(II) ion forms stable complexes with OH^-, $C_2H_3O_2^-$, and, to a lesser degree, with Cl^-. $PbSO_4$, an acid-insoluble sulfate, is white and will dissolve in 6 M NaOH or in a concentrated solution of acetate ions via the formation of the $[Pb(OH)_4]^{2-}$(aq) complex ion or $Pb(C_2H_3O_2)_2$(aq), a neutral but soluble species which does not appreciably dissociate in solution. $PbCrO_4$ is a bright yellow insoluble solid, often used as a confirmatory test for lead. It will dissolve in 6 M NaOH.

Silver Ion, Ag^+

Most silver compounds are insoluble, the nitrate being just about the only exception. Silver acetate is slightly soluble. Silver chloride is white and insoluble in water, but soluble in 6 M NH_3, due to the formation of $Ag(NH_3)_2^+$ complex ion. These properties of AgCl are usually used in the confirmatory test for silver ion. They may also be used to test for the presence of Cl^- ion. In NaOH solution, Ag^+ precipitates as brown Ag_2O; it is soluble in nitric acid and in concentrated ammonia solutions.

Mercurous Ion, Hg_2^{2+}

No salts of the Hg(I) ion are soluble in water. Mercury(I) nitrate solutions always contain fairly high concentrations of HNO_3. Calomel, Hg_2Cl_2 (containing the stable Hg_2^{2+} cation, which is the dimer of the odd-electron unstable Hg^+ ion), precipitates upon addition of HCl to any solution of Hg(I) ion. It is white and essentially insoluble in all common solvents and solutions. If treated with 6 M NH_3, it turns gray-black due to a reaction to form Hg metal (black) and white, insoluble, $HgNH_2Cl$. This is the definitive test for the Hg(I) ion.

Bismuth Ion, Bi^{3+}

There are no water-soluble bismuth salts. The usual solution is highly acidic and contains $Bi(NO_3)_3$ or $BiCl_3$. Bismuth ion forms complexes with Cl^- but not with OH^- or NH_3. $Bi(OH)_3$ is white and insoluble in water but soluble in strong acids. If you add a few drops of acidic $BiCl_3$ solution to water, a cloudy white precipitate of BiOCl forms (containing the BiO^+ cation, the hydrolysis product of Bi^{3+}). This is the usual test for the Bi^{3+} ion. Another confirmatory test is to add 0.10 M $SnCl_2$ solution to $Bi(OH)_3$ in a strongly basic solution. Bi(III) will be reduced to black Bi metal.

Tin Ion, Sn^{2+} or Sn^{4+}

No common tin compounds are water-soluble. The usual 0.10 M tin solution contains $SnCl_2$ in 0.10 M HCl. Tin exists in either the +2, called stannous, or +4, called stannic, state. The Sn(II) species, particularly in basic solution, is a good reducing agent, and can reduce Bi(III) and Hg(II) to the metals. Tin in either oxidation state forms many complex ions and in solution is usually present as a complex ion. In basic solution Sn(II) exists as $Sn(OH)_4^{2-}$. In HCl the complex ion has the formula $SnCl_4^{2-}$. Tin compounds are white, for the most part. SnS is brown, and its formation at pH 0.5 as in Experiment 37 is good evidence for the presence of tin. The classic confirmatory test is to add $SnCl_2$ in HCl to a solution of $HgCl_2$. The tin(II) ion reduces the mercury(II) to form a precipitate of white Hg_2Cl_2 (which may darken to gray if the reduction proceeds all the way to metallic mercury). A less toxic and more environmentally friendly confirmatory test for Sn(II) uses the organic compound Janus Green (JG), which is easily reduced by Sn^{2+}. A color change from blue to either violet-red (JG^-) or colorless (JG^{2-}) confirms the presence of Sn(II). (The color change observed depends upon the relative amounts of Sn^{2+} and JG that are present.)

Antimony Ion, Sb^{3+}

There are no common antimony salts that are soluble in water. Antimony salts are not easily dissolved. The usual solution of Sb(III) contains $SbCl_3$ in 3 M HCl, in which the antimony exists as the $SbCl_6^{3-}$ complex ion. Antimony also forms a stable complex ion with hydroxide ion, $Sb(OH)_6^{3-}$. If a few drops of $SbCl_3$ in acidic solution are added to water, a white precipitate of SbOCl (containing the SbO^+ cation) forms, very similar to that observed with $BiCl_3$. The confirmatory test for antimony is the formation of the characteristic red-orange sulfide on precipitation of Sb_2S_3 by addition of thioacetamide to antimony solution at pH 0.5. If the cation is Bi(III), the precipitate will be black Bi_2S_3.

Copper Ion, Cu^{2+}

Copper(II) is colored in most of its compounds and solutions. The hydrated ion, $Cu(H_2O)_4^{2+}$ is blue, but in the presence of other complexing agents, the Cu(II) ion may be green or dark blue. The sulfate, chloride, nitrate, and acetate are water-soluble. If 6 M NH_3 is added to a solution of Cu(II) ion, light blue $Cu(OH)_2$ initially precipitates, but in excess ammonia the hydroxide readily dissolves because of formation of the very characteristic dark blue copper(II) ammonia complex ion, $Cu(NH_3)_4^{2+}$. This ion serves as an excellent confirmatory test for the presence of copper. Copper ion is readily reduced to the metal by the more active metals, such as zinc.

Nickel Ion, Ni^{2+}

Nickel is another of the colored aqueous cations. The chloride, nitrate, sulfate, and acetate are water-soluble and form green solutions containing the $Ni(H_2O)_6^{2+}$ complex ion. Nickel forms other complex ions, but the one most commonly observed is the blue $Ni(NH_3)_6^{2+}$ ion, formed on addition of 6 M NH_3 in excess to Ni(II) solutions. The color is much less intense than that of the copper ammonia complex ion. The usual test for the presence of nickel is the formation of a rose-red precipitate of nickel dimethylglyoxime on addition of dimethylglyoxime to a slightly basic solution containing Ni^{2+}.

Iron(III) Ion, Fe^{3+}

Iron has two common oxidation states, +2 and +3. Iron(II), called ferrous ion, is less commonly observed, since it is readily oxidized in air to the +3 state. We will limit our work to iron(III), or ferric, salts. Iron(III) salts are often colored, usually yellow in aqueous solution. The color is due to hydrolysis, since the $Fe(H_2O)_6^{3+}$ is itself colorless. The nitrate, chloride, and sulfate are water-soluble but tend to hydrolyze to form basic salts that may require a slightly acidic solution if they are to dissolve completely. Iron(III) hydroxide is very insoluble and is the color of rust. Iron forms several complex ions, but the one most encountered in qualitative analysis is the $FeSCN^{2+}$ ion (you may have determined the equilibrium constant for the formation of this species in Experiment 23), which is dark red and often used in the confirmatory test for Fe(III) ion. Iron does not form complex ions with NH_3 or OH^-.

Chromium Ion, Cr³⁺

Chromium(III), as its name implies, is colored ("chroma" is the Greek word for color); it may be violet, green, or gray-blue in solution. The chromium +3 ion usually exists as a complex which, unlike many complex ions, is sometimes slow to undergo ligand exchange. Chromium exists in two common oxidation states, +3 and +6. In the latter state it is found as the yellow chromate, CrO_4^{2-}, anion, or, if the solution is acidic, as the orange $Cr_2O_7^{2-}$ anion. The chloride, nitrate, sulfate, and acetate of Cr^{3+} are water-soluble, but the dissolution process may be more rapid in acidic solution. If excess 6 M NaOH is added to a solution of Cr(III), the initial gelatinous gray-green hydroxide precipitate may dissolve, and one may obtain a dark green solution due to the formation of the $Cr(OH)_4^-$ complex ion. With 6 M NH_3 the same insoluble gray-green hydroxide is initially formed, which may slowly dissolve to yield a pink to violet complex ion. Boiling a solution of either complex causes insoluble $Cr(OH)_3$ to re-form. In most qualitative analysis schemes chromium is oxidized to the +6 state, where in acidic solution it forms a characteristic, but fleeting, deep blue solution on addition of H_2O_2. Where chromium(III) is the only cation in a sample, the formation of the dark green hydroxide complex is definitive. If this test is inconclusive, oxidize Cr(III) as in Experiment 38 and use the alternate confirmatory test described there.

Aluminum Ion, Al³⁺

Aluminum salts are typically white. The nitrate, chloride, sulfate, and acetate are water-soluble but do show a tendency to form basic salts if no acid is added. Aluminum forms a very stable hydroxide complex ion, $Al(OH)_4^-$, so it is difficult to precipitate the otherwise insoluble $Al(OH)_3$ by addition of NaOH to Al(III) solutions. That hydroxide will come down on addition of 6 M NH_3 to a solution containing Al^{3+} buffered with NH_4^+ ion. The precipitate is characteristically light, translucent, and gelatinous. A good confirmatory test for aluminum involves first precipitating $Al(OH)_3$ from a solution of the sample, then dissolving the solid in very dilute acetic acid. Add a few drops of catechol violet reagent, HCV. If aluminum is present, the solution will turn blue due to the formation of the aluminum complex ion, $AlCV^{2+}$.

Appendix III

Standard Atomic Weights of the Elements (Scaled relative to Carbon-12 ≡ 12 g/mol)

	Symbol	Atomic number	Atomic mass*		Symbol	Atomic number	Atomic mass*
Aluminum	Al	13	26.98	Molybdenum	Mo	42	95.95
Antimony	Sb	51	121.76	Neodymium	Nd	60	144.24
Argon	Ar	18	39.95	Neon	Ne	10	20.18
Arsenic	As	33	74.92	Nickel	Ni	28	58.69
Barium	Ba	56	137.33	Niobium	Nb	41	92.91
Beryllium	Be	4	9.012	Nitrogen	N	7	14.01
Bismuth	Bi	83	208.98	Osmium	Os	76	190.23
Boron	B	5	10.81	Oxygen	O	8	16.00
Bromine	Br	35	79.90	Palladium	Pd	46	106.42
Cadmium	Cd	48	112.41	Phosphorus	P	15	30.97
Calcium	Ca	20	40.08	Platinum	Pt	78	195.08
Carbon	C	6	12.01	Potassium	K	19	39.10
Cerium	Ce	58	140.12	Praseodymium	Pr	59	140.91
Cesium	Cs	55	132.91	Protactinium	Pa	91	231.04
Chlorine	Cl	17	35.45	Rhenium	Re	75	186.21
Chromium	Cr	24	52.00	Rhodium	Rh	45	102.91
Cobalt	Co	27	58.93	Rubidium	Rb	37	85.47
Copper	Cu	29	63.55	Ruthenium	Ru	44	101.07
Dysprosium	Dy	66	162.50	Samarium	Sm	62	150.36
Erbium	Er	68	167.26	Scandium	Sc	21	44.96
Europium	Eu	63	151.96	Selenium	Se	34	78.97
Fluorine	F	9	19.00	Silicon	Si	14	28.09
Gadolinium	Gd	64	157.25	Silver	Ag	47	107.87
Gallium	Ga	31	69.72	Sodium	Na	11	22.99
Germanium	Ge	32	72.63	Strontium	Sr	38	87.62
Gold	Au	79	196.97	Sulfur	S	16	32.06
Hafnium	Hf	72	178.49	Tantalum	Ta	73	180.95
Helium	He	2	4.003	Tellurium	Te	52	127.60
Holmium	Ho	67	164.93	Terbium	Tb	65	158.93
Hydrogen	H	1	1.008	Thallium	Tl	81	204.38
Indium	In	49	114.82	Thorium	Th	90	232.04
Iodine	I	53	126.90	Thulium	Tm	69	168.93
Iridium	Ir	77	192.22	Tin	Sn	50	118.71
Iron	Fe	26	55.85	Titanium	Ti	22	47.87
Krypton	Kr	36	83.80	Tungsten	W	74	183.84
Lanthanum	La	57	138.91	Uranium	U	92	238.03
Lead	Pb	82	207.2	Vanadium	V	23	50.94
Lithium	Li	3	6.94	Xenon	Xe	54	131.29
Lutetium	Lu	71	174.97	Ytterbium	Yb	70	173.05
Magnesium	Mg	12	24.31	Yttrium	Y	39	88.91
Manganese	Mn	25	54.94	Zinc	Zn	30	65.38
Mercury	Hg	80	200.59	Zirconium	Zr	40	91.22

*Standard atomic weights describe the typical average molar masses of elements as found on Earth. Only elements with sufficiently stable isotopes that a characteristic isotopic abundance can be defined have standard atomic weights.

Appendix IV

Making Measurements—Laboratory Techniques

For centuries people who made measurements set up systems of units to describe length, area, volume, and mass. These units often had rather romantic names that were created for a specific purpose. There was a myriad of such names, and among them we might mention:

acre	barrel	bolt	bushel
carat	chain	cord	cubit
dram	ell	em	fathom
furlong	gill	grain	hand
league	noggin	pace	perch
quire	rod	stone	ream

These days many of these units are still used, often in connection with only one application, such as horse racing (furlong), a gemstone's weight (carat), or printing (em).

Scientists realized early on that it would be advantageous to have one unified system of measurement to be used by all: several systems were suggested and used by your authors when they were young students. These were mostly based on the metric approach, with different sizes related to one another by factors of 10. Thus, we have the meter, equal to <u>100</u> centimeters, equal to <u>1000</u> millimeters. In 1960, a group of scientists set up the International System of Units, or SI, a coherent system of seven base units, which can be used to describe most measurements one is likely to make, and which can be used together in many natural laws. The five of these units we will use, and their abbreviations, are:

meter (m)　　kilogram (kg)　　second (s)　　Kelvin (K)　　mole (mol)

Although in principle we might expect to use these units directly in reporting data, some of them turn out to have awkward magnitudes in many situations. So we use SI units when it is convenient, but for the first two, we tend to work more with related metric units, or with unrelated, but conveniently sized, units. We will discuss our approach in the following sections. (Perhaps not surprisingly, none of the units in our very first list are SI units, or metrically related to SI units.)

Mass

One of the most important measurements you will be making involves determining the mass of a sample. The unit we will almost always use is the gram, rather than the kilogram, since a kilogram (about two pounds) is just too large for bench-scale chemistry. The gram, abbreviated g, is not a base unit in SI (despite its non-prefixed name) but is easily converted to the kilogram when necessary.

Measurement of mass is almost always done by determining the force exerted on a sample by gravity. This force is proportional to the mass of the sample. The device that is used is called a balance, because the mass of the sample is established by balancing it against a standard mass or force. Several balances that were used in the past century are shown in Figure IV.1. The balance on the left was used until about 1950. Using it required a set of weights that were put on one pan until the two pans came to the same height; with care one could weigh a sample to ± 0.0001 grams, but it would take several minutes to make one weighing. There were a lot of drawbacks to such a balance, but they were used for over a century in the form shown or more primitive ones. After World War II, instrument design developed rapidly. The balance in the middle is a mechanical one, but the weights are internal and added to or taken from a pan inside the balance. Final readings are taken

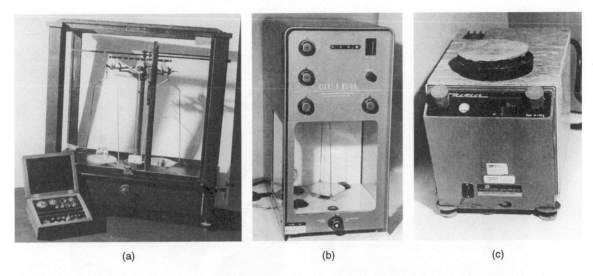

(a) (b) (c)

Figure IV.1 (a) A mechanical analytical balance from circa 1940—good to ± 0.0001 g, maybe. (b) A mechanical analytical balance circa 1960—maximum load 120 g, good to ± 0.0001 g. (c) A top-loading semi-automatic balance circa 1970—good to about ± 1 mg. *(Sherman Schultz)*

from a set of dials that add or take off weights and an optically projected scale that furnishes masses in the milligram range. This type of balance is still used in some schools, and if you have one in your laboratory, your instructor will show you how it operates. The balance on the right is called a top-loading balance; mass is read directly from dials and an optical scale; this kind of balance gives rapid results but is limited in its precision to milligrams at best.

In recent years, electronic balances have been developed that furnish the mass directly with a digital readout (see Fig. IV.2). They do not have any weights but instead balance the sample mass by a magnetic force that can be accurately related to mass. These balances can be accurate to ± 0.0001 g and may arrive at the mass of a sample very quickly. In a real sense, they are the ultimate weighing machines. In using such a balance, you first briefly depress the control bar. This will zero the balance, and it will read 0.0000 g. Place your sample on the balance pan, close the balance doors if it has any, and read the mass when the balance gives a steady reading. If your sample is to be in a container, you can find its mass by weighing the empty container, then briefly depressing the control bar to re-zero, or tare, the balance. Then when you add your sample, its mass will appear directly as the digital readout. There are many brands of these balances, so if your lab has one, your instructor will describe the details of the operation of your balance before you use it.

Here are some guidelines for successful balance operation, which apply to the weighing of a sample on any balance:

1. Be certain the balance has been "zeroed" (reads 0 grams) before you place anything on the balance pan.
2. Never weigh chemicals directly on a balance pan. Always use a suitable container or weighing paper. In several experiments you will weigh a sample tube and measure the mass of sample removed, by difference.
3. Be certain that air currents are not disturbing the balance pan. In the case of an analytical balance, always shut the balance case doors when making measurements you are recording.
4. Never put hot, or even warm, objects on the balance pan. The temperature difference will change the density of the air surrounding the balance, and thus give inaccurate measurements.
5. Record your measurement, to the proper number of significant figures, directly on your report page or in your notebook.
6. After finishing your measurements, be certain that the balance registers zero again and close the balance doors. Brush out any chemicals you may have spilled.
7. Be gentle with your balance. It is a sensitive, rather delicate instrument, and, like a person, responds best when treated properly.

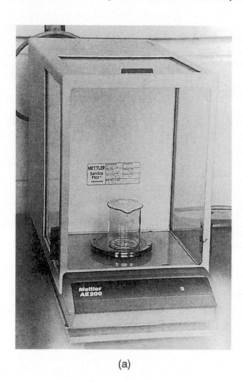

(a)   (b)

Figure IV.2 (a) A modern electronic automatic analytical balance, circa 1985. Its maximum load is 120 g, and it is reliable to ± 0.0001 g. The beaker weighs 27.5056 g. (b) An electronic top-loading balance with a maximum load of 3.6 kg and reliable to ± 0.01 g. Current balances of each of these types differ only in that they automatically re-calibrate themselves to account for changes in the ambient temperature. *(Mary Lou Wolsey)*

Volume

Another very common measurement we make in the laboratory is that of volume. The base SI unit of volume is the cubic meter (m^3). Again, this is a much larger scale than you ordinarily encounter in a chemistry lab. A large bathtub might contain a single cubic meter of water, roughly 250 gallons. The units of volume we ordinarily use are the liter, L, and the milliliter, mL. A liter equals 0.001 m^3, and a mL equals 1×10^{-6} m^3, so both are derived from the SI unit.

Another name for the milliliter, mL, is the cubic centimeter, cm^3, sometimes called the cc, pronounced "see see," a term often used in medicine.

$$1 \text{ mL} = 1 \text{ cc} = 1 \text{ cm}^3$$

If we are asked to add 150 mL of water to a beaker during an experiment in which we are carrying out a chemical reaction, we don't need high accuracy and might use a beaker or flask on which there are some volume markings (as shown in Fig. IV.3, on the next page), which are reliable to about five percent. Somewhat more precise volumes can be obtained with a graduated cylinder, which can yield measurements within about one percent of a needed volume.

A. Pipets

Still more precise volume measurements are often required; these may be obtained with pipets (also spelled pipettes, but in this manual we will use the shorter spelling), which are available in many different volumes, from 1 mL up to 100 mL. In several experiments in this manual, pipets are used. Pipets are calibrated to deliver the specified volume when the meniscus at the liquid level is coincident with the horizontal line etched around the upper pipet stem. The following steps make for proper use of a pipet:

1. The pipet must be clean. When it drains, there should be no drops left on the pipet wall. It does not need to be dry.
2. Always use a pipet bulb, not your mouth, when pulling liquid into a pipet. You shouldn't run the risk of getting the liquid or its vapor into your mouth.

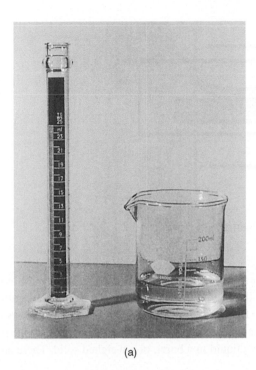
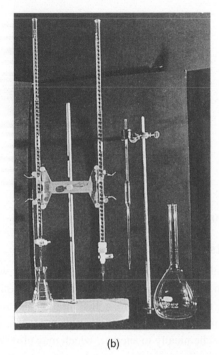

(a)   (b)

Figure IV.3   (a) A 25-mL graduated cylinder, good to ± 0.2 mL, and a 250-mL beaker with graduations, good to ± 20 mL. (b) A pair of 50-mL burets, a 10-mL volumetric transfer pipet, and a 1000-mL volumetric flask. The Class A versions of these are reliable to ± 0.05 mL for the burets, ± 0.02 mL for the pipets, and ± 0.3 mL for the volumetric flasks. The uncertainties are twice as large for Class B volumetric glassware. *(Sherman Schultz)*

3. Place the pipet bulb on the upper end of the pipet and squeeze the air out. Immerse the tip of the pipet into the liquid and draw up enough liquid to get some into the main body of the pipet. Remove the bulb and place your index finger over the top of the pipet stem. Hold the pipet in the horizontal position and swirl the liquid around inside, rinsing the upper stem and body. Drain the liquid into a beaker; repeat the rinsing twice with small volumes of the liquid. This ensures that the contents of the pipet will be the liquid you wish to work with, and not water or a previously-used reagent. Finally, using the bulb, fill the pipet, drawing liquid a centimeter or two above the calibration mark. Put your index finger over the top of the stem and, carefully, with the tip of the pipet against the side, let liquid flow out into a beaker until the bottom of the meniscus is just at the level of the calibration mark. Holding your finger tightly on the top of the pipet stem, transfer the pipet tip to the receiving container, with the tip touching the wall. Lift your index finger to release the liquid and let the liquid flow into the container. Hold the tip against the wall for about ten seconds (longer for a thick liquid) after it appears that the liquid has stopped flowing, to ensure the correct amount of liquid drains out. Almost all glass pipets are designed "to deliver" and will be marked with the code "TD." These will accurately deliver the indicated amount of an *aqueous* liquid when a small amount remains in the tip—that last bit should *not* be blown out with a pipet bulb.

   It is now possible to obtain autopipets and repipets, which more readily deliver precise volumes of reagents. These speed up the pipetting procedure enormously, and you may have them in your laboratory. They too require proper technique and are calibrated to work with aqueous liquids. They are generally made "To Contain" (TC), however, and so every last drop in their tips must be expelled in order to obtain precise volumes. Their plungers have a two-position stroke to make this readily accomplished.

   It takes skill and practice to make good use of <u>any</u> kind of pipet. (Autopipets and even repipets are NOT foolproof exceptions to this statement!) You are *strongly* encouraged to *practice* with any measuring device that is new to you, before relying on it for quantitative measurements. With volumetric apparati, using a balance to confirm the mass of pure water transferred in a practice measurement is an easy way to do so.

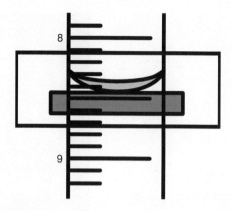

Figure IV.4 Reading a buret. The volume reading is 8.46 mL.

B. Burets

A buret (also spelled burette, but in this manual we will use the shorter spelling) is a long calibrated tube fitted with a stopcock to control the release of liquid reagents contained inside of them (see Figure IV.3b). Burets are used in a procedure called titration and are frequently used in pairs. This procedure allows us to add precisely measured volumes of reagents to reach a so-called endpoint, at which the amount of one reagent is equivalent chemically to another, which may also be a liquid in a buret, or a weighed solid. There are several titration experiments in this manual.

As with a pipet, working with a buret requires skill and practice. The following procedure should be helpful:

1. Check the buret for cleanliness by filling it with deionized water and allowing it to drain out. A clean buret will leave an unbroken film of water on the interior walls, with no beading or droplets. If necessary, clean your buret with soapy water and a long buret brush. Rinse with tap water, and finally with deionized water.
2. Rinse the buret three times with a few milliliters of the reagent to be used. Tip the buret to wash the walls with the reagent, and drain through the tip into a beaker. With the stopcock closed, fill the buret with reagent to a level a little above the top graduation, making sure to fill the tip completely. Open the stopcock carefully and let the liquid level go just below the top zero line. Read the level to the nearest 0.01 mL. To do this, place a white card with a sharp black rectangle on it (a piece of black tape is ideal) behind the buret, so that the reflection of the bottom of the meniscus is just above the upper black line, and at eye level (see Fig. IV.4). Then make the initial volume reading.

   Add the reagent until you get to the endpoint of the titration, which is usually established by a color change of a chemical called an indicator. To hit the endpoint accurately, you need to add reagent in very small quantities when the titration is nearly complete, while you can add it more rapidly and in larger quantities at the beginning of a titration. Quickly twisting a stopcock through the open position can often be a more foolproof way to add small volumes of titrant than is trying to deliver drops slowly by barely opening the stopcock, but both techniques are widely used. Swirl the container to ensure the reactants are well-mixed. Often you get a clue from the indicator that you are near the endpoint. If you go past it, you can use your first titration as a guide for a second or third trial. Read the final liquid level as before, to ± 0.01 mL.

C. Volumetric Flasks

A volumetric flask is one that has been especially calibrated to hold a specified volume: 10.00 mL, 25.00 mL, etc. These flasks are used when accurate dilutions are required in analytical experiments. Place the solute or solution in the previously cleaned, but not necessarily dry, flask. Fill the flask about half full of solvent and swirl the contents until any solid has dissolved. Then add solvent to bring the volume up to the mark on the flask. After the liquid volume reaches the lower neck of the flask, add the solvent with a wash bottle or Pasteur pipet. The last volume increments should be added dropwise until the bottom of the meniscus is coincident with the mark on the flask. Stopper the flask and mix thoroughly by inverting at least 13 times, allowing the bubble of trapped air to move all the way from the top of the neck to the bottom of the flask, and back, each time.

D. Pycnometers

The most precise volume measurements are made with a pycnometer, such as the one used in the first experiment in this manual. A pycnometer is simply a container with a very well-defined volume, such as a small flask fitted with a ground glass stopper. With such a device, and a liquid with an accurately known density, you can determine the volume of the pycnometer to ± 0.001 mL, which is much better than you can do with a pipet or buret. Knowing that volume, you can measure the density of an unknown liquid with an accuracy equal to that of the density of the calibrating liquid, usually within 0.01%.

Temperature

The SI unit of temperature is the Kelvin, which is equal in size to the Centigrade degree, °C. At 0°C, the Kelvin temperature is 273.15 K. The degree sign, °, is not used when reporting Kelvin temperatures.

Measurement of temperature is always done with a thermometer. Your lab drawer probably contains a standard laboratory thermometer suitable for temperature readings between -10°C and $+100$°C, with 1°C graduations. We will use this thermometer in experiments where high accuracy is not needed.

The typical glass thermometer contains a bulb connected to a very fine capillary tube. It contains a liquid, usually alcohol or mercury, which fills the bulb and part of the capillary. As the temperature goes up, the liquid expands and the level rises in the capillary. To make the graduations, one would in principle find the level at 0°C and at 100°C, and divide that interval into 100 equal parts. With mercury, the scale fits the one obtained from ideal gas behavior quite well. With other liquids, you must calibrate the thermometer at several places if you are to obtain reliable temperatures. Your standard thermometer can be read to ± 0.2°C if you are careful, but the actual error is likely to be greater than that.

In some experiments it is desirable to be able to get more precise temperature values than are possible with a standard liquid-filled thermometer. You may be furnished with a digital thermometer that has a wide range and reads directly to ± 0.1°C. The sensing tip on that kind of thermometer contains a thermistor, a temperature-sensitive resistor, which has electrical resistance that varies markedly with temperature. In recent years these thermometers have come into wide usage.

Glass thermometers are fragile and relatively expensive, so be careful when using them. If you slip a split rubber stopper around the thermometer and support it with a clamp you will minimize breakage. Should you ever break a mercury thermometer, contact your lab instructor. Liquid mercury has a low vapor pressure, but that vapor is very toxic, so it is important to clean up mercury spills completely, which is not easy. With liquid-filled thermometers, do not heat them above their maximum readable temperature, since that can ruin the thermometer or even cause it to explode.

When making temperature readings, allow enough time for the level to become steady before noting the final temperature. This will probably take less than a minute, so if you check the reading over a period of time you should be able to get a reliable value.

To obtain temperatures with an accuracy better than ± 0.1°C is not a simple matter. Some mercury thermometers read to ± 0.01°C, but they have large bulbs and very fine capillaries, and are very expensive: over $250 each. Still higher precision can be obtained using thermistors: by making careful resistance measurements, one can get temperatures accurate to ± 0.001°C.

Pressure

In studying the behavior of gases we must be able to measure the gas pressure. One way to accomplish this is by comparing the pressure of interest with that exerted by the atmosphere, using a device called a manometer. A manometer is a glass U-tube partially filled with a liquid, usually water or mercury.

The atmospheric pressure is measured with a mercury barometer (or, increasingly, its electronic equivalent). This consists of a straight glass tube about eighty centimeters long, which is initially completely filled with mercury. The tube is then inverted while the open end is immersed in a pool of mercury. The mercury level in the tube falls. The observed height of the Hg column above the pool, as read from a scale behind the tube, is equal to the pressure exerted by the atmosphere, and is called the atmospheric pressure. We can

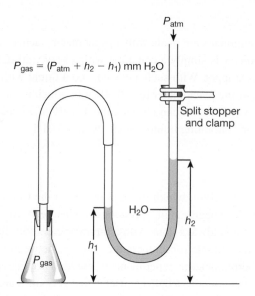

Figure IV.5 Measuring the pressure of a gas with a water manometer.

also measure the atmospheric pressure with an aneroid barometer, which consists of a spiral-wound, flexible-walled flat metal tube containing a fixed amount of air. As the external pressure changes, the tube flexes, which turns a needle that records the pressure.

Atmospheric pressure is often reported in mm Hg, but it may be given in other units, such as atmospheres, by scaling the value in mm Hg against 760 mm Hg, the standard atmospheric pressure. In SI the pressure is given in either Pascals or bars. One standard atmosphere equals 1.01325×10^5 Pascals, or 1.0325 bars. In this manual we will report gas pressures in mm Hg or mm H_2O, since these are most easily measured directly. We can convert pressures in mm Hg to mm H_2O, or vice versa, by using the conversion factor 1 mm Hg = 13.57 mm H_2O.

Having found the atmospheric pressure, it is a simple matter to measure the gas pressure in a flask such as that shown Figure IV.5. In Experiment 8, the pressure inside the flask is equal to the atmospheric pressure, in mm H_2O, plus the pressure exerted by a water column of effective height $h_2 - h_1$, the difference between the water levels in the right and left arms of the U-tube. If you know those heights in mm,

$$P_{gas} = (P_{atm} + h_2 - h_1) \text{ mm } H_2O$$

When reading the water levels in a manometer, take the height measurements at the bottom of the water meniscus in each of the two arms, which you can read to ± 1 mm or better.

Light Absorption

In many chemical experiments the interaction of light with a sample can be used to furnish very useful information. In this laboratory course we will often determine the concentration of a species in a solution by measuring the amount of light at a given wavelength (of a given color) that is absorbed by the sample. In an instrument called a spectrophotometer one can do this readily and precisely. A spectrophotometer contains a device called a monochromator (usually incorporating a diffraction grating), which allows only a specific wavelength of light to pass through a sample. The absorbance, A, of the sample at that wavelength can be read directly from a meter on the instrument. By a famous equation called Beer's Law, the absorbance is proportional to the concentration, c, of the sample, usually expressed as a molarity:

$$A = K \times c$$

where K is a constant that depends on the sample, the container, and the wavelength of the light. Most samples obey Beer's Law. To find the concentration of a species in a sample, one measures the absorbance, at an appropriate wavelength, of a solution containing a known concentration of the species. From that absorbance,

Figure IV.6   A Spectronic 20 spectrophotometer. The absorbance of a solution can be reliably read to within ± 1%. *(Sherman Schultz)*

or several absorbances obtained at known concentrations, we can make a calibrating graph of concentration vs absorbance and then use that graph to analyze unknown solutions containing that species. Your instructor will probably furnish you with such a graph in some of the experiments you perform.

A classic spectrophotometer is a Spectronic 20 (see Fig. IV.6). This instrument has two knobs on the front face. On the left is a zero transmittance adjustment knob and on the right a 100% transmittance adjustment knob. On the top surface at the left is a covered sample chamber, and at the right there is a wavelength selection knob. The absorbance is read from the upper meter or display register.

To use this type of spectrophotometer, turn it on about fifteen minutes before you need to operate it, to give the internal light sources time to warm up. (Most spectrophotometers require a warm up delay, but its length depends on the light source[s] they employ.) Set the wavelength to one where the sample absorbs a moderate amount of the incident light. Since the sample is probably colored, this wavelength will most likely be in the visible region of the spectrum. With the sample compartment closed, turn the zero adjustment knob until the display indicates 0% transmission. Open the sample chamber and insert a cuvet (a special liquid sample holder with reproducibly transparent sides; also spelled cuvette, but in this manual we will use the shorter spelling) that is about three-fourths full of a "blank" reference solution (usually pure water). Shut the sample chamber and turn the 100% adjustment knob until the display reads 100% transmission (this will be zero on the absorbance scale). Repeat the adjustment steps until no further changes are necessary. Rinse the cuvet several times with the solution being studied, then fill the sample tube about three-fourths full, insert it in the sample chamber, and close the chamber. Read the absorbance from the display. Then use a calibration graph to obtain the concentration of the species in the solution.

pH Measurements

The pH of a solution is used to describe the concentration of the H^+ ion in that solution. To determine pH, one can use acid-base indicators, which change color as the pH changes. Each indicator has a characteristic pH range over which the color change occurs, so it can be used to indicate whether the pH is higher, lower, or where it falls in that range. By employing several indicators, we can fix the pH of a solution within about half of a pH unit.

More accurate pH measurements can be made with a pH meter. The pH meter is a very high-resistance voltmeter which measures the difference in voltage between a reference electrode (usually a so-called calomel electrode) and a glass indicating electrode whose potential is a function of the H^+ ion concentration in the solution in which it is immersed. The meter indicates the pH directly. The two electrodes may be separate but more commonly are together in one combination electrode. The electrode must be kept wet, in an appropriate storage solution, when not in use.

When you are asked to find the pH of a solution with a pH meter, you may first need to calibrate the pH meter, although it is likely that will have been done for you prior to the laboratory session. To calibrate the meter you will need a buffer solution with a well-defined pH to serve as a reference. Put about twenty-five milliliters of the buffer in a small beaker. Remove the electrode from the storage solution, rinse it with a stream of deionized water from your wash bottle, and then gently blot the electrode once with a clean tissue to draw off excess rinse water. NEVER WIPE THE ELECTRODE! Place the electrode in the buffer. (The electrode is fragile and expensive, so treat it gently.) Allow the system to equilibrate for a minute or two, or until the pH reading becomes steady. Then use the pH adjust knob to bring the displayed pH to that of the reference buffer solution, or press the button used to indicate to the meter that the value has stabilized and can be used.

Remove the electrode from the buffer, rinse it with deionized water as before, blot with a tissue, and place the electrode in the solution being studied. Allow time for equilibration, and record the pH. Rinse the electrode and return it to the storage solution. Used properly, the average pH meter can reliably indicate pH to within one tenth of a pH unit. Many will indicate to the nearest hundredth of a unit but are not reliable to that digit unless exquisite care is employed and a temperature correction is applied.

Separation of Precipitates from Solution

A. Decantation

Decantation is used to remove a liquid from a precipitate by simply pouring it off. Allow the solution to sit for a period of time until the precipitate has settled to the bottom of its container. (Some precipitates will not settle out without centrifugation—see below.) Position your stirring rod across the top of the container with one end protruding beyond the lip. With the index finger of one hand holding the rod in place, pour the liquid slowly down the stirring rod into a receiving vessel. Try not to disturb the precipitate as the last bits of liquid are poured off.

B. Filtration—The Büchner Funnel

Filtration is used when you wish to recover a precipitate, in pure form, from a liquid. This is first required in Experiment 3. You will use a Büchner funnel, which contains a piece of filter paper of appropriate size covering the holes at the base of the funnel. The Büchner funnel is connected by a rubber stopper or adaptor to the top of a side-arm filter flask. The side arm of the flask is connected via a series of rubber tubes and a safety trap to a water aspirator, a central vacuum system accessible via a jet on your lab bench, or another vacuum source (see Fig. 3.2). When the vacuum source is turned on, the reduced pressure created in the system by the vacuum source will draw air through the funnel.

When you are ready to filter, moisten the filter paper with water, *then* turn on the vacuum source. With the water faucet or valve to the vacuum source in the fully open position, pour the slurry of solution and precipitate down a stirring rod, as described under "Decantation," above. Wash out any remaining precipitate with a slow stream of liquid from your wash bottle. The procedure may call for washing the precipitate on the Büchner funnel with an appropriate volatile liquid before passing air through the filter cake for several minutes to complete drying. Turn off the water or vacuum source, disconnect the vacuum tube from the filter flask, and remove the funnel.

C. Centrifugation

Centrifugation is used to aid in the separation of a precipitate from a solution in a test tube. Spinning the test tube at several hundred revolutions per minute effectively increases the force of gravity experienced by the tube to many times normal earth gravity (see Fig. IV.7).

To use the centrifuge, be certain that the test tube containing the precipitate is not overly full—the liquid should be at least 3 cm below the top of the tube. Check to see that the test tube has no cracks, as these will cause the tube to break during centrifugation. Place the test tube containing your precipitate and solution in one of the centrifuge tubes. Place a similar tube containing the same amount of water as you have in your

Figure IV.7 A lab centrifuge. *(Sherman Schultz)*

sample tube in the centrifuge slot that is opposite your sample, to serve as a counterweight and keep the rotor of the centrifuge balanced. Turn on the centrifuge, allow it to spin for about a minute, and then turn it off. Keep your hands away from the spinning centrifuge top. After the spinning top has come to rest, remove the tube. Decant the supernatant (clarified) liquid from the precipitate in your sample tube, as described above under "Decantation."

Qualitative Analysis—A Few Suggestions

In the course of your laboratory program, you may do several experiments involving qualitative analysis. In those experiments we will use small (13 × 100 mm) test tubes and small beakers as sample containers. Separations of precipitates from solutions will (or can) be accomplished by centrifugation.

In many procedures you will be told to add 1.0 mL, or 0.5 mL, to a mixture. This is best accomplished by first finding out what 1.0 mL looks like in the test tube being used. So, measure out 1.0 mL in a small graduated cylinder, and pour that into a test tube, noting the height reached by the liquid in the tube. From then on, use that height as a guide when you need to add an approximate volume; half that height indicates 0.5 mL.

Many of the steps in qualitative analysis involve heating a mixture in a boiling-water bath. To make the bath, fill a 250-mL beaker about two-thirds full of water. Support the beaker on a wire gauze on an iron ring and heat the water to the boiling point with a Bunsen burner, adjusting the flame so as to keep the water at the simmering point. Or, place the beaker on a hotplate, adjusting the heat setting to accomplish the same result. Use this bath to heat the test tube when that is called for. You can remove the tube with your test tube clamp.

In separating a solid from a solution, we first centrifuge the mixture. The supernatant (clarified) liquid is decanted into a test tube or discarded, depending on the procedure. The solid must then be washed free of any remaining liquid. To do this, add the indicated wash liquid and stir with a glass stirring rod. Centrifuge again and discard the wash. Keep a set of stirring rods in a 400-mL beaker filled with deionized water. Return a rod to the beaker when you are done using it, and it will be ready when you next need it.

Pay attention to what you are doing, and don't just follow the directions as though you were making a cake. Try to keep in mind what happens to the various cations in each step, so that you don't end up pouring the material you want down the drain and keeping a waste product. When you need to put a fraction aside while you are working on another part of the mixture, make sure you know where you put it by labeling the test tube or rack with the number of the step in which you will return to analyze that fraction.

Appendix V

Mathematical Considerations—Significant Figures and Making Graphs

In the laboratory you will be carrying out experiments of various sorts. Some of them will involve almost no mathematics. In others, you will need to make calculations based on your experimental results. The calculations are not difficult, but it is important you make them properly. In particular, you should not report a result that implies an accuracy that is greater, or lesser, than is consistent with the accuracy of your experimental data. In making such calculations we resort to the use of significant figures as a guide to proper reporting of our results. In the first part of this appendix, we will discuss significant figures and how to use them.

In several experiments we will carry out a series of measurements on the same system under different conditions. For example, in Experiment 8 we measure the pressure of a sample of gas at different temperatures. In such experiments it is helpful to draw a graph representing our data, since it may reveal some properties of the sample that are not at all obvious if we just have a table of data. The second part of this appendix will discuss how to construct and interpret a graph.

Significant Figures

In the laboratory we make many kinds of measurements. The precision and accuracy of the measurement depends on the device we use to make it. Using an analytical balance, we can measure mass to ± 0.0001 g, so if we have a sample weighing 2.4965 g, we have *five* meaningful figures in our result. These figures are called, reasonably enough, significant figures. If we measure the volume of a sample using a graduated cylinder, the precision and accuracy of the volume we obtain depends on the cylinder we use. If we have a volume of 6 mL and measure it with a 100-mL cylinder, we would have difficulty distinguishing between a volume of 6 mL and 7 mL, and would be unable to say more than that the volume was 6 mL. That volume reading has only *one* significant figure. With a 10-mL cylinder, we could measure the volume more precisely, report a volume of 6.4 mL, and have confidence that both figures in our result were meaningful, and that it had *two* significant figures. Many measurements, perhaps most, can be interpreted as we have here.

Significant figures indicate the precision with and accuracy to which measurements were or are to be taken, and often implicitly indicate the devices used. For example, suppose there is an instruction to add 1.00 mL of water to a reaction vessel. This could not be accomplished to the indicated precision and accuracy (three significant figures) with a beaker, or even with a graduated cylinder. A pipet is a precisely made device and can deliver 1.00 mL of liquid if used properly. In such a case, we would need to use a pipet, or another mechanism capable of providing the precision and accuracy called for. On the other hand, when instructions call for adding 1.0 mL of water to a reaction vessel, a precision and accuracy of only two significant figures is called for, and a graduated cylinder is not only appropriate, but the wise choice. It will allow the quantity to be measured out more quickly and efficiently than with a pipet, and the instructions indicate that the amount need not be more precise than that. (Counting drops from a dropper bulb or basing the volume on the liquid height in a test tube would also suffice in this example.) If you were asked to add 1 mL of water to a reaction vessel, it would technically indicate that anything between 0 and 2 mL would suffice (which is rarely true, since 0 mL amounts to adding nothing, making the step unnecessary). We will instead use written-out words for such approximate quantities, calling for "one milliliter of water" in the example at hand. In these cases there is no need to measure at all, once you have developed a sense of approximate amounts at the lab bench.

Some numbers are exact and have no inherent error. There will be a whole number, like 12, or 19, students in your lab group. There is no way there will be 14.5 students, unless there is something strange going on. Conversion factors, used to convert one set of units to another, often contain exact numbers: 1 meter = 100 cm.

Both numbers are exact, with no uncertainty at all. With volumes, 1 liter = 1000 mL. Again, both numbers are exact. If we convert from one measurement system to another, then usually only one of the numbers in the conversion factor is exact: 1 mole = 6.022×10^{23} molecules. Here the 1 is exact, but the second number is not, and has four significant figures. One way to indicate that numbers are exact is to underline them.

   If you are in doubt as to the number of significant figures in a given number, there is a method for finding out that usually works. We write the number in scientific notation, expressing the number as the product of a number between 1 and 10, multiplied by a power of ten.

$$26.042 = 2.6042 \times 10^1 \quad 0.0091 = 9.1 \times 10^{-3} \quad 605.20 = 6.0520 \times 10^2$$

The first number in the product has the number of significant figures we seek. So, the first of the numbers above has 5 significant figures, the second has 2, and the third has 5 (a trailing zero is significant). If a number is written by a well-trained scientist, you should be able to count on it being written with attention to significant figures. If you saw 2400, you would understand it to have 2 significant figures; if the scientist had intended three significant figures, she would have written 2.40×10^3. If she wished to indicate four significant figures, she would have written 2.400×10^3 or 2400. (the explicit decimal point makes the place-holding zeroes significant), and she would have written 2.4000×10^3 or 2400.0 to indicate five significant figures.

   Students usually have little trouble deciding on the number of significant figures in a piece of data. The difficulty arises when the piece of data is used in a calculation, and they have to express the result using the proper number of significant figures. Say, for example, you are weighing a sample in a beaker and you find that:

Mass of sample plus beaker =	25.4329 g
Mass of beaker	= 24.6263 g
Mass of sample	= 0.8066 g

Even though the two measured masses contain 6 significant figures, the mass of the sample contains only 4. What matters with addition or subtraction is the left-most decimal place that is not significant; that decimal place must also be insignificant in the sum or difference, because the sum or difference cannot be known with any greater precision (in this case ± 0.0001 g) than any of the values being added or subtracted. In another case, where we mix some components to make a solution, weighing each of the components separately on different balances, we get:

Mass of beaker	= 25.5329 g
Mass of water	= 14.0 g
Mass of salt	= 6.42 g
Total mass	= ?

Here we have several pieces of data, each with a different degree of precision. The first mass is reliable to ± 0.0001 g, the second to ± 0.1 g, and the third to ± 0.01 g. If we take the sum, we get 45.9529 g, but we certainly can't report that value, since it could be off, not by ± 0.0001 g, but by ± 0.1 g, the possible error in the mass of the water. We can't improve the quality of a result by a calculation, so, using good sense, we can only report the mass to ± 0.1 g. Since the measured mass is closer to 46.0 g than to 45.9 g, we round up to 46.0 as the reported mass, and have a mass with 3 significant figures. Generalizing, *in adding or subtracting numbers, round off the result so that it has the same last decimal place as in the least precise measured quantity*. Round off the last significant digit based on the digits that follow it.

   When *multiplying* or *dividing* measured quantities, the rule is simple: *The number of significant figures in the result is equal to the number of significant figures in the quantity with the fewest significant figures.*

   In the first experiment in this manual, we measure the density of a liquid using a pycnometer, which is simply a flask with a well-defined volume. We find that volume by weighing the empty pycnometer and then weighing the pycnometer again when it is full of water. We might obtain the following data:

Mass of pycnometer plus water	60.8867 g
Mass of empty pycnometer	31.9342 g
Mass of water	28.9525 g

We are asked to find the volume of the pycnometer, given that the density of water is 0.9973 g/mL:

$$\text{density} = \frac{\text{mass}}{\text{volume}} \quad \text{so} \quad \text{volume} = \frac{\text{mass}}{\text{density}} = \frac{28.9525 \text{ g}}{0.9973 \text{ g/mL}} = 29.03_{0883} \text{ mL}$$

The mass contains 6 significant figures, the density has 4, and the results from a calculator have 8 or more. Since the density has the fewest significant figures (4), we report the volume to 4 significant figures, as 29.03 mL. To avoid rounding error, sometimes a few insignificant figures are carried along through calculations, subscripted, as shown in the example above.

In most cases, the reasoning required to correctly report a calculated value is no more complicated than that used here. Remember, you can't improve the accuracy in a result by means of a calculation. If you note the quality of your data, you should be able to report your result with the same quality. As with most activities, a little practice helps, and so does a little common sense.

Graphing Techniques

In many chemical experiments we find that one of our measured quantities is dependent on another. If we change one, we change the other. In such an experiment, we find data under several sets of conditions, with each data point associated with two measured quantities. In such cases, we can represent these points on a two-dimensional graph, which shows in a continuous way how one quantity depends on another.

Interpreting a Graph

The first graph in this lab manual appears in Experiment 3 and is a graph of this kind. It is reproduced in Figure V.1. The graph describes how the solubility of two compounds, KNO_3 and $CuSO_4 \cdot 5 H_2O$, each depends on temperature. The temperature, which was the controlled variable when the data for the graph was collected, is shown along the horizontal axis, called the abscissa, over a range from 0 to 100°C. The solubility, in grams per 100 grams water, was the measured quantity in the experiment that produced the data and is on the vertical axis, sometimes called the ordinate. The small ✕ symbols along each of the two curved lines in the graph are data points, showing measured solubility values at specific temperatures.

The graph has some implied features that are not shown. We made the graph on a piece of graph paper on which there was a grid of parallel lines. On the vertical lines, the temperature is constant. If you draw a vertical line upward from the 50 hatch mark on the vertical axis, the temperature on that line is always 50°C. On a vertical

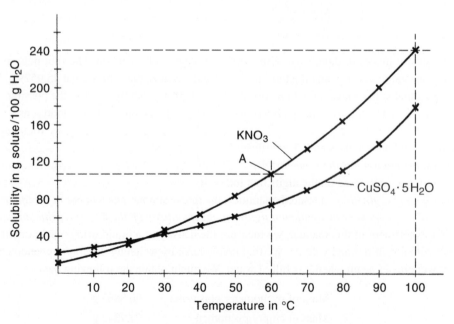

Figure V.1

line at 73°C, three-tenths of the distance between the 70 and 80 hatch marks, it is always 73°C, and so on. Along horizontal lines the solubility is constant. A horizontal line drawn through the 80 hatch mark represents a solubility of 80 grams per 100 grams of water all along its length. When we obtained the data for the solubility of KNO_3, we found that at 60°C we could dissolve 105 g KNO_3 in 100 g of water. We entered the data point we show as A on the graph at 60°C and (105 g KNO_3/100 g H_2O). The rest of the data points for the two compounds were entered in the same way. We then drew a smooth curve through those points and removed the grid from the image.

Given a graph like that we have shown, you can extract the data by essentially working backward. To find the solubility of KNO_3 at 60°C, draw the dashed line up from the 60 hatch mark. The solubility is given by the point (marked "A") at which that line intersects the KNO_3 graph line. Draw a horizontal line from that intersection to the vertical axis and read off the solubility of KNO_3, which you can see is about (105 g / 100 g H_2O). If you wanted to find the amount of water you would need to dissolve 21 g KNO_3 at 100°C, you would draw a vertical line up from the 100 hatch mark up to the KNO_3 line; then draw a horizontal line at the intersection with the KNO_3 curve. You would find that at 100°C, about 240 g KNO_3 will dissolve in 100 g H_2O. Then a simple conversion factor calculation would give you the amount of water needed to dissolve 21 g KNO_3. Clearly, there is a lot of information in Figure V.1 if you interpret it completely. Its meaning is almost, but not quite, obvious.

Making a Graph

Now let us construct a graph from some experimental data. Given the grid in Figure V.2, we measure the pressure of a sample of air as a function of temperature. One of our students obtained the following set of data:

Pressure in mm Hg	674	739	784	821
Temperature in °C	0.0	24.5	42.0	60.1

To make the graph, we need to decide what goes where. Since we are controlling the temperature and measuring the pressure, we put the pressure on the vertical axis, and the temperature on the horizontal axis, labeling those axes. The temperature goes from 0°C to 60°C; we want the data we graph to fill the grid, not just be in

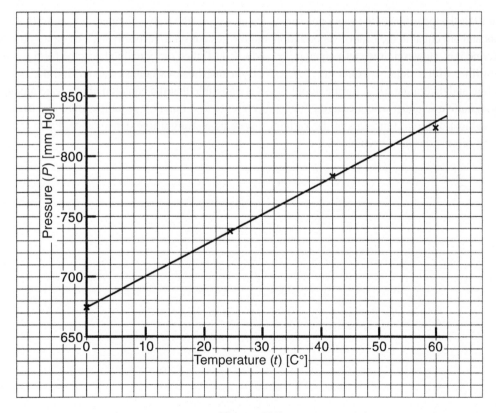

Figure V.2

one corner, so we select a temperature interval between grid lines that will fill the grid. Since we have about thirty vertical grid lines available, and need to cover a 60°C interval, we put the 10° hatch marks at intervals of five grid lines, with 2° between grid lines. Since there are about twenty horizontal grid lines, and we need to cover a pressure change of 150 mm Hg, we make the lowest pressure 650 mm Hg and the interval between grid lines equal to 10 mm Hg. We put in the hatch marks at 50-mm intervals. Having laid out the graph, we insert each of the data points, using ✗'s or dots. The points fall on a nearly straight line, so we draw such a line, minimizing as best we can the sum of the distances from the data points to the line. If we wish, we can find the equation for that line, which, since it is straight, is of the form $y = mx + b$ where m is the slope of the line and b is the value of y when x equals zero. Plugging two (well-separated) points on the line,

$(x_1, y_1) = (0, 674)$ and $(x_2, y_2) = (60, 830)$, into the point-point formula for a line, $y - y_1 = \left(\dfrac{y_2 - y_1}{x_2 - x_1}\right)(x - x_1)$,

and simplifying, our student found that the equation was $y = 2.6x + 674$. This equation can be used to find another useful piece of information about the properties of gases, but we need not go into that here.

A key to successful construction of graphs is to properly select the variables and assign them to the two axes. (The variable you control should be on the horizontal axis, and the quantity of interest that responds to changes in the variable you control should be on the vertical axis.) Label those axes. Then, given the range of the variables, select grid intervals that will cause your data to (nearly) fill the area of the grid. Put in the hatch marks for each variable at appropriate intervals and label them with the values of the variable. Take the data points and carefully plot them on the grid. If the line is theoretically straight, draw a straight line through the data points in such a way as to minimize the sum of the distances from the points to the line.

Many natural laws are not as simple as the one relating gas pressure to temperature. However, by a few rather easy tricks, involving mathematical manipulation of variables, one can obtain data points that will lie on a straight line. For example, in Experiment 15, we find that the vapor pressure of a liquid depends on temperature according to the following equation:

$$\ln VP = \frac{-\Delta H_{vap}}{RT} + C$$

where $\ln VP$ is the logarithm of the vapor pressure to the base e, T is the Kelvin temperature, and ΔH_{vap} and R are constants and are equal to the heat of vaporization and the gas constant, respectively. The equation looks formidable, and it is. However, if we let x equal $\ln VP$ and $y = 1/T$, then the equation takes the form $y = mx + b$, a straight line, with a slope m, equal to $-\Delta H_{vap}/R$. By using those variables, we can find the slope of the straight line and determine ΔH_{vap} for the liquid. It is very possible that the equation above was initially found by an imaginative scientist who measured the vapor pressure as a function of temperature and tried plotting the data against different functions of the variables, seeking a straight line dependence. From such efforts many great discoveries have been made.

Appendix VI

Suggested Locker Equipment

2 beakers, 30 or 50 mL
2 beakers, 100 mL
2 beakers, 250 mL
2 beakers, 400 mL
1 beaker, 600 mL
2 Erlenmeyer flasks, 25 or 50 mL
2 Erlenmeyer flasks, 125 mL
2 Erlenmeyer flasks, 250 mL
1 graduated cylinder, 10 mL
1 graduated cylinder, 25 or 50 mL
1 funnel, long or short stem
1 thermometer, $-10°C$ to $150°C$
2 watch glasses, 75, 90, or 100 mm
1 crucible and cover, size #0
1 evaporating dish, small

2 medicine droppers
2 medium test tubes, 18×150 mm
8 small test tubes, 13×100 mm
4 micro test tubes, 10×75 mm
1 test tube brush
1 file
1 spatula
1 test tube clamp, Stoddard or equivalent
1 test tube rack
1 pair of lab tongs
1 sponge
1 towel
1 plastic wash bottle
1 chemical casserole, small

Appendix VII

Introduction to Excel

Over the years a type of software called a spreadsheet has come into general use for processing scientific data obtained in the laboratory, or, indeed, data involved in business activities. Essentially, a spreadsheet contains many cells in which data can be entered and manipulated. It is particularly useful when there are several, or many, experimental values for a property, such as the volume of a gas obtained at different temperatures under constant pressure, and the same calculation needs to be conducted on those values over and over. Carrying out such calculations by hand, with a calculator, is time consuming and prone to error. Setting up a spreadsheet to conduct the same calculations takes more time, but a spreadsheet will not make calculation mistakes and will instantly recalculate everything in the event you find a data entry error and need to change a number. This often makes it very worthwhile to spend the time required to set up a spreadsheet, and doing so is also a skill that will serve you well in the future in nearly any line of work. In this introduction we will offer a short primer on how to take advantage of the most widely used spreadsheet package, called Excel, in some of its simpler applications. To get started, you will need to have a computer on which the software has been installed. There are many versions of this program, but we try here to provide instructions for a wide range of versions. These instructions are specific to Excel but should be readily adapted to other spreadsheet packages, with only minor changes.

Basic Layout

Click on the Excel icon on your computer. If there is none, your instructor will tell you how to call it up. You should see a screen on which there is a grid containing many rectangular boxes, called cells. Each cell has an associated label, with a letter indicating its column, and a number indicating its row; the letter + number combination referring to a given cell is called a cell reference. If you click on cell B5, it will be highlighted. If you put the cursor in that cell, type a number or a word, and click the enter or return key on the keyboard, that data will appear in the cell. You can change the width of the cell by clicking and holding on the grid line in question, up where it intersects with the rectangle containing a letter (called the column heading). Double-clicking on it will automatically adjust the width of the column to accommodate the widest entry in that column.

Setting Up Columns of Data

Let's proceed by entering the heading "Temperature [C]" in cell A1. Change the cell width, if necessary, to make the title fit. In cell A3, type in the number 20 and press enter or return. The highlighted cell will automatically become A4. Enter the formula "=A3+20" in that cell, press return or enter, and 40 should appear in A4. Click cell A4 *once* with the mouse, and then move the cursor to the lower right-hand corner of that cell. You will find that the cursor will change from a white plus symbol to a solid black plus symbol or an open square. While the cursor is a black plus or a square, click and drag the cursor down to select all the cells in which you want to copy that formula; in this case, let's go down to cell A12. When you release the mouse button, the numbers 60, 80, 100, and so on, up to 200, will appear in column A. This method is very useful if you have a column of data that are related by a formula.

Now enter the heading "Temperature [K]" in cell B1, adjusting the width of that column as necessary to make it fit. In cell B3, enter the formula "=A3+273", press enter, and 293 will appear in cell B3. Using the same approach used for column A, copy this formula down into cells B4 through B12 to produce the Kelvin temperatures in column B for each of the Celsius temperatures in column A.

In cell C1, type the heading "Volume [L]". We will use the Ideal Gas Law, $PV = nRT$, to find the volume (in liters) of $\underline{1}$ mole of gas at $\underline{1}$ atmosphere pressure and the Celsius temperatures in column A. By the Ideal Gas Law, $V = nRT/P$; in our example, $n = \underline{1}$ mol, $P = \underline{1}$ atm, $R = 0.0821$ L·atm/(mole·K), and T is the Kelvin temperature. So in column C, V equals $\underline{1}*0.0821*T/\underline{1}$, or just $0.0821*T$. (The mathematical symbols used for operations in Excel do not correspond exactly to those we use elsewhere: $+$ and $-$ do mean add and subtract, but $*$ means multiply, $/$ means divide, and $\wedge$ means raise to a power. We write 2^3 as $2\wedge3$ in Excel. A number like 3.45×10^5 should be entered and will be displayed as 3.45E5 in Excel.)

To calculate the volumes in column C, first enter the gas constant into a cell and label it: type "$R =$" into cell A14, and "0.0821" in cell B14. We should specify the units for R, so in cell C14, enter "L atm/(mol K)". In cell C3, type the formula "=$B$14*B3" and press enter. (Here's a handy shortcut: Rather than typing in B3 from the keyboard, you can click on that cell in the spreadsheet when you get to that point typing in the formula. This can be very helpful when you are constructing your own formulas and know which cell you want to use; you can just click on it to get the right cell reference!) In cell C3 the number 24.055 should appear (possibly with more or fewer significant figures shown, as Excel knows nothing of significant figures and will simply display what it has sufficient room to display). Highlight cell C3 by clicking on it once, position the cursor over the lower right-hand corner of the cell, and when the black plus symbol or square appears, click and drag down to cell C12 and release. The volumes, in liters, of $\underline{1}$ mole of an ideal gas at the various temperatures should appear in column C.

You may wonder why we used $B$14 instead of just B14 in the formula in cell C3. Try replacing $B$14 with B14 in the formula in cell C3, and see what happens! At first, everything will look fine, but if you drag down the cursor to copy this cell, the program will increment not only B3 to B4, but also B14 to B15, and zeroes will appear in column C, since B15 is zero. In many formulas, you will find that you will wish to have all or part of a cell reference remain unchanged when a formula is copied. You can accomplish this by preceding the letter, or number, or both, in a cell reference with a dollar sign ($).

Some Comments on Excel

In Excel, the program will assign the contents of a cell to one of three categories: a number, text, or a formula. Numbers and text only cause trouble when one is mistaken for the other. To avoid this, when entering numerical values you intend to use in subsequent calculations or a plot, be certain to enter *only* the number—do not, for example, append it with units within the same cell. (Specifying units is an excellent habit, but to do so in a spreadsheet use a column heading or an adjacent cell, as demonstrated above.) Formulas most often cause trouble when Excel does not realize you are entering one. You must get into the habit of beginning every formula you enter into Excel with $=$ or $+$, lest Excel interpret your entry as text. (Some spreadsheets will assume a formula, and you may need to indicate text entries by beginning them with a double or single quote mark [" or ']). The $=$ (equals sign) used to denote a formula must be the first character in a cell, with no space to its left. The equation for the formula may be complex, but it must include the Excel symbol(s) for any operation to be carried out, and should not have any spaces in it. All references to particular cells must be to cells with known numerical contents. Parentheses should be carefully used to make clear to Excel the desired order of the mathematical operations you wish to carry out.

If you mis-type or mis-click (and this is inevitable!) you may find nothing happens when you hit enter, or you may get one of several error messages, which may or may not be helpful. If you get into trouble, the delete (Del) or escape (Esc) buttons may return the screen to its previous state, that before the error occurred. If not (or if these buttons cause something else undesired to happen), using the "Undo" option (indicated by a toolbar icon with an arrow looping up and then back down and to the left, or from the Edit menu in most versions of Excel, both in the upper left corner of the screen) will often put you back on track. Do not be reluctant to seek help, either from your lab instructor or a fellow student more familiar with Excel than you are. Working with Excel can be frustrating, but sooner or later you will able to use it effectively, and learning to use it will pay huge dividends.

Using Excel to Make Graphs

In some cases it is helpful to create graphs that show the relationships between experimentally measured variables, and a spreadsheet package can make this much easier to do than it is by hand.

Given two columns of data for two related variables, like temperature and volume, we can use Excel to construct a graph (Excel calls it a "chart") that shows how the variables are related.

Before you proceed, click on View on the top line of the screen, and select Normal from the options presented.

Next, choose Chart from the Insert menu on the top edge of the screen. If you are using a newer version of Excel, you will need to select the type of chart you want at this point—in science, the default selection you should make is Scatter, and specifically the flavor that displays only the data points. (Note: Hovering over an icon in Excel, that is, holding the mouse stationary over it for a few seconds, will bring up an explanation of the meaning of that icon.) If you are using a version of Excel dating from before 2007 (your authors find these superior, frankly), choosing Chart from the Insert menu will bring up a screen showing several possible graph types, from which you will want to select XY Scatter and then click on Next >.

Now you need to specify the data you wish to use in the chart. In an older version of Excel, click on the Series tab; in a newer version, choose Select Data from the top of the screen. Click on Add. In the data entry regions that appear (in the newest versions of Excel, you will first have to click on "Series1" under "Legend Entries" and click the "Edit" button), labeled Name, X Values, and Y Values, delete the contents of any boxes that are not already empty by selecting the contents and hitting backspace or delete.

Now, to specify the y values for the chart, put the cursor in the Y Values box and click and drag the mouse down over the column of data containing the dependent variable, in our case the volume, in column C (specifically, cells C3 through C12). When you release the mouse button, that data range will appear in the Y Values box (as "=Sheet1!$C$3:$C$12").

To specify the x values, put the cursor in the X Values box and click and drag the mouse down over the independent variable data, in our case the Kelvin temperatures in column B (specifically, cells B3 to B12). When you release the mouse button, that cell range should appear in the X Values box (as "=Sheet1!$B$3:$B$12"). A small graph showing the relationship between the variables will appear on the screen. The scale will probably be incorrect, but we will fix that soon. For now, hit enter or return on the keyboard to go on to the next step.

If you are using a newer version of Excel (from 2007 or later), click on OK, as you are done specifying the data you wish to use, and then click on the first of the "Chart Layouts" that appear at the top of your screen (in the newest versions, you will need to click on "Add Chart Element"). Now, no matter what version of Excel you are using, click on "Chart Title " (in the newest versions, specify "Above Chart" and then double-click the heading that appears inside the chart box). Type "One mole of an ideal gas at one atmosphere pressure" as the title of the graph [do not hit enter or return yet]. Now click once on the X Axis title box and type in the name of the independent variable, "Temperature [K]". Click once on the Y Axis title box and type the name of the dependent variable, "Volume [L]". (In the newest versions of Excel you will have to select each of these things in turn from the "Add Chart Element" pulldown, where they are under "Axis Titles" as "Primary Horizontal" and "Primary Vertical", respectively.)

In newer versions of Excel, click on the Chart Layout tab at the top of the screen (in the newest versions, the "Add Chart Element" button), then click on Gridlines. With older versions of Excel, simply click on the Gridlines tab of the Chart Options box you are currently in. Add or remove the gridlines as seems reasonable.

If you are using an older version of Excel, you will now want to click on Next >, to get to Step 4, and decide where to place the graph, either on a separate page or as an object on the spreadsheet (usually, you'll want the latter), before clicking on "Finish" to bring down the final graph, labeled and titled, but probably not scaled as you would like.

To fix the scale, *right*-click (press down on the side of the mouse *opposite* that you normally click) on the numbers of the axis you wish to change and select Format Axis. Click on "Scale" or "Axis Options", whichever is offered to you. Proceed by disabling automatic axis limits and entering appropriate fixed maximum and minimum axis limit values (250 to 500 for x, 20 to 40 for y, in our example). This is easier to do if you can see the chart and the data as you do so—you can move the data entry window out of the way if necessary by clicking and holding on the window name (Format Axis) and dragging the window to a more convenient location. The graph will update as you make changes, which makes it easy to fine-tune your adjustments.

In a chart displaying a single series, such as our example, the legend serves no useful function and can be removed by clicking on it once and pressing the backspace or delete key on the keyboard. (The newest versions of Excel may not automatically put it in, as the older ones did.) This frees up more room for the plot. Were the legend useful, it could be moved onto the plot region by clicking and dragging it there.

To change the physical dimensions and position of the graph you have created, click once on the edge of its frame, to select it. Then position the mouse over any of the "handles" that appear on the frame (a single black square in older versions of Excel, a collection of three tiny dots in middle-aged versions, and a hollow circle in the newest ones) and click and drag to change the dimension to which that handle corresponds. To move the entire plot, click on an empty region of the chart, just inside one of the upper handles, and click and drag. Especially with these operations, it is easy to have things go awry. . . if they do, press the Esc key and then, if needed, use the Undo option.

If you wish to have Excel calculate the optimal line through your data, select your chart (click once on the edge of its frame) and then get to the Trendline menu. (Chart>Add Trendline. . . with most versions, Chart Layout>Trendline with some middle-aged ones, and under the "Add Chart Element" pulldown on the newest versions.) You can specify that Excel should display the equation of this line (as well as customize several other aspects of it) from the Options tab on the Trendline menu.

Appendix VIII

An Introduction to Google Sheets

Google Sheets is a software package like Excel (see Appendix VII), but with several important differences. First, it is a browser-based application, meaning it can be run seamlessly from almost any computing device with a browser built into it (and a large enough screen to be practical). Second, it uses cloud-based storage, which means the files it generates are available on any Internet-connected system (this is possible with the newest versions of Excel as well). Finally, but perhaps most important for education users, it is free. Google Sheets works much like Excel, but the details differ enough that we are including detailed instructions on how to use it to analyze and plot data, as we did for Excel in Appendix VII.

Start by going to Google Sheets in a Web browser: sheets.google.com is the hyperlink. If you are not already logged into a Google account, you will need to do so, or create one, because Sheets stores your data in the cloud, and it needs to know who you are in order to store it, as well as let you access it from other places. Once you are logged in, you will be able to start a new spreadsheet, and in this case you will want it to be a blank one. When you choose that, an empty spreadsheet will open up, and you are ready to begin.

Basic Layout

You are now looking at a screen on which there is a grid containing many rectangular boxes, called cells. Each cell has an associated label, with a letter indicating its column, and a number indicating its row; the letter + number combination referring to a given cell is called a cell reference. If you click on cell B5, it will be highlighted. If you put the cursor in that cell, type a number or a word, and click the enter or return key on the keyboard, that data will appear in the cell. You can change the width of the cell by clicking and holding on the grid line in question, up where it intersects with the rectangle containing a letter (called the column heading).

Setting Up Columns of Data

Let's proceed by entering the heading "Temperature [C]" in cell A1. Change the cell width, if necessary, to make the title fit. In cell A3, type in the number 20 and press enter or return. The highlighted cell will automatically become A4. Enter the formula "=A3+20" in that cell, press enter or return, and 40 should appear in A4. Click cell A4 *once* with the mouse, and then move the cursor to the lower right-hand corner of that cell. You will find that the cursor will change from a white arrow to a thin black plus symbol (if you see a hand icon, you are not in the right spot!). While the cursor is a black plus symbol, click and drag the cursor down to select all the cells in which you want to copy that formula; in this case, let's go down to cell A12. When you release the mouse button, the numbers 60, 80, 100, and so on, up to 200, will appear in column A. This method is very useful if you have a column of data that are related by a formula.

Now enter the heading "Temperature [K]" in cell B1, adjusting the width of that column as necessary to make it fit. In cell B3, enter the formula "=A3+273", press enter, and 293 will appear in cell B3. Using the same approach used for column A, copy this formula down into cells B4 through B12 to produce the Kelvin temperatures in column B for each of the Celsius temperatures in column A.

In cell C1, type the heading "Volume [L]". We will use the Ideal Gas Law, $PV = nRT$, to find the volume (in liters) of $\underline{1}$ mole of gas at $\underline{1}$ atmosphere pressure and the Celsius temperatures in column A. By the Ideal Gas Law, $V = nRT/P$; in our example, $n = \underline{1}$ mol, $P = \underline{1}$ atm, $R = 0.0821$ L·atm/(mole·K), and T is the Kelvin temperature. So in column C, V equals $\underline{1}*0.0821*T/\underline{1}$, or just $0.0821*T$. (The mathematical symbols used for operations in a spreadsheet do not correspond exactly to those we use elsewhere: + and − do mean add and subtract, but * means multiply, / means divide, and ^ means raise to a power. We write 2^3 as 2^3 in a spreadsheet. A number like 3.45×10^5 should be entered and will be displayed as 3.45E5 in a spreadsheet.)

To calculate the volumes in column C, first enter the gas constant into a cell and label it: type "$R =$" into cell A14, and "0.0821" in cell B14. We should specify the units for R, so in cell C14, enter "L atm/(mol K)". In cell C3, write the formula "=\$B\$14*B3" and press enter. (Here's a handy shortcut: Rather than typing in B3 from the keyboard, you can click on that cell in the spreadsheet when you get to that point typing in the formula. This can be very helpful when you are constructing your own formulas and know which cell you want to use: you can just click on it to get the right cell reference!) In cell C3 the number 24.0553 should appear (possibly with more or fewer significant figures shown, as by default Sheets knows nothing of significant figures and will simply display what it has sufficient room to display). Highlight cell C3 by clicking on it once, position the cursor over the lower right-hand corner of the cell, and when the black plus symbol appears, click and drag down to cell C12 and release. The volumes, in liters, of 1 mole of an ideal gas at the various temperatures should appear in column C.

You may wonder why we used \$B\$14 instead of just B14 in the formula in cell C3. Try replacing \$B\$14 with B14 in the formula in cell C3, and see what happens. At first, everything will look fine, but if you drag down to copy this cell, the program will increment not only B3 to B4, but also B14 to B15, and zeroes will appear in column C, since B15 is zero. In many formulas, you will find that you wish to have all or part of a cell reference remain unchanged when a formula is copied. You can accomplish this by preceding the letter, or number, or both, in a cell reference with a dollar sign (\$).

Some Comments on Google Sheets

Spreadsheets will assign the contents of a cell to one of three categories: a number, text, or a formula. Numbers and text only cause trouble when one is mistaken for the other. To avoid this, when entering numerical values you intend to use in subsequent calculations or a plot, be certain to enter *only* the number—do not, for example, append it with units within the same cell. (Specifying units is an excellent habit, but to do so in a spreadsheet use a column heading or an adjacent cell, as demonstrated above.) Formulas most often cause trouble when Google Sheets does not realize you are entering one. You must get into the habit of beginning every formula you enter into Google Sheets with = or +, lest Google Sheets interpret your entry as text. (Some spreadsheets will assume a formula, and you may need to indicate text entries by beginning them with a double or single quote mark [" or ']). The = (equals sign) used to denote a formula must be the first character in a cell, with no space to its left. The equation for the formula may be complex, but it must include the symbol(s) for any operation to be carried out, and should not have any spaces in it. All cell references in equations must be to cells with known numerical contents. Parentheses should be carefully used to make clear to Google Sheets the desired order of the mathematical operations you wish to carry out.

If you mis-type or mis-click (and this is inevitable!), you may find nothing happens when you hit enter, or you may get one of several error messages, which may or may not be helpful. If you get into trouble, the delete (Del) or escape (Esc) buttons may return the screen to its previous state, that before the error occurred. If not (or if these buttons cause something else undesired to happen), using the "Undo" option (indicated by a toolbar icon with an arrow looping to the left, up and then back down, or from the Edit menu, both in the upper left corner of the screen) will often put you back on track. Do not be reluctant to seek help, either from your lab instructor or a fellow student more familiar with computers than you are. Working with a spreadsheet can be frustrating at first, but sooner or later you will able to use it effectively, and learning to use it will pay huge dividends.

Using Google Sheets to Make Graphs

In some cases it is helpful to create graphs that show the relationships between experimentally measured variables, and a spreadsheet package can make this much easier to do than it is by hand.

Given two columns of data for two related variables, like temperature and volume, we can use Sheets to construct a graph (Sheets calls it a "chart") that shows how the variables are related.

Start by clicking on a single empty cell somewhere in the spreadsheet, far from your data (such as E1), so that Sheets does not assume anything about what you want a new chart to contain. Now choose Chart from the Insert menu on the top edge of the screen. Change the Chart type to "Scatter" by clicking on the default type, "Column chart," and picking the "Scatter chart" option instead. (You may have to scroll down through the options, as Scatter is not near the top. Note that in science, you will almost always want your Chart type to be a Scatter chart.) The next steps will be a little easier if we move the chart out of the way, so click once on the chart, then click and hold while dragging it to the right, so that it is no longer over any of your data.

Now you need to specify the data you wish to use in the chart. To specify the y values for the chart, Under Data range, click on the small black 3×3 grid of squares to the right of the SERIES category. When the data entry box appears, click and drag the mouse down over the data for the dependent variable: in our case the volume, in column C (specifically, cells C3 through C12). When you release the mouse button, that cell range should appear in the box (as "Sheet1!C3:C12"). Click OK to confirm this is what you want.

To specify the x values, click on the small black 3×3 grid of squares to the right of the X-AXIS category. When the data entry box appears, click and drag the mouse down over the column of data containing the dependent variable data, in our case the Kelvin temperatures in column B (specifically, cells B3 to B12), or manually enter B3:B12 in the box: both accomplish the same end result.

Now that the chart contains the data of interest, it is time to fine-tune its appearance. In the Chart editor box, click the CUSTOMIZE tab, directly under the box name and to the right of DATA. Click on the "Chart and axis titles" entry to open it up. Now, click on the line under Title text and type "One mole of an ideal gas at one atmosphere pressure" as the title of the graph. Next, click on Chart title and change it to Horizontal axis title. Click under Title text and type in the name of the independent variable, "Temperature [K]". Change Horizontal axis title to Vertical axis title and type the name of the dependent variable, "Volume [L]". Next, click on the Gridlines subheading and add or remove the gridlines as seems reasonable. (We suggest adding minor gridlines on both the horizontal and vertical axes, in this case.)

To fine-tune the scale of your plot, click on the Horizontal axis and/or Vertical axis subheading, as appropriate. Sheets usually does a pretty good job of automatic scaling, but try entering appropriate fixed maximum and minimum axis limit values (250 to 500 for x, 20 to 40 for y, in our example). The graph will update as you make changes, which makes it easy to fine-tune your adjustments.

In a chart displaying a single series, such as our example, a legend serves no useful function and the one automatically placed on the chart can be removed by clicking on the Legend subheading and setting the Position to "None".

To change the physical dimensions and position of the graph you have created, click once just inside its frame, to select it. Then position the mouse over any of the "handles" that appear on the frame (little black squares at the corners and in the middle of each side) and click and drag to change the dimension to which that handle corresponds. To move the entire plot, click on an empty region of the chart, just inside one of the upper handles, and click and drag. Especially with these operations, it is easy to have things go awry; if they do, press the Esc key and then, if needed, use the Undo option.

If you wish to have Google Sheets calculate the optimal line through your data, double-click on one of the data points of the series you want a trendline for, and Sheets will open the Chart editor box and select the Series submenu there. Check the Trendline box, and the best-fit line will appear. You can specify that Sheets should display the equation of this line (as well as customize several other aspects of it) from further down in the Series submenu.

Appendix IX

Statistical Treatment of Laboratory Data

Whenever there are multiple determinations of a measurement (called "replicate measurements"), as is the case in several experiments in this laboratory manual, it is customary for scientists to express a central value and some measure of the deviation of the replicates around that central value—thus indicating something about the reliability of the experimental results. In chemistry, the kinds of experiments that are treated statistically in this fashion generally involve the analysis of a substance to determine the proportions of its chemical constituents, often by either titration or the weighing of precipitates. This branch of chemical science is referred to as quantitative analysis.

It is necessary to have some means of expressing the central value—the best estimate of the real value that one can obtain from the replicate data. You may have heard terms which are frequently used to express the distribution of exam scores—the (arithmetic) mean, median, and mode, along with the range. Let's consider a set of laboratory data, such as four independent measurements of the percent iron by mass in the same ore sample:

Trial number	$\%_{mass}$ Fe
1	15.66
2	15.79
3	15.22
4	15.66

An inspection of this data set shows that the median value (that result about which all others are equally distributed) is 15.66%, the mode (the value having the largest number of replications) is 15.66%, and the mean (the arithmetic mean, or "average") is 15.58%. The mean is calculated by adding up all of the percentage values (62.33%) and dividing by the number of trials (4). The range (difference between high and low values) is 0.57%.

You may have used the terms precision and accuracy in a variety of settings. In science, precision is a measure of the reproducibility of a measurement—the agreement among several determinations of the same value. On the other hand, accuracy properly refers to the closeness of a measurement to the accepted (or true) value. It appears that the above data set has decent precision, but nothing can be said about its accuracy in the absence of additional information. Your instructor will have access to an accepted value. If we assume that the known value for this particular iron ore is 15.88%, it is apparent that the accuracy is poor—all of the measured values fall below the true value, suggesting a *systematic* discrepancy. Your instructor may use a combination of your accuracy and your precision in grading some experiments.

One widely used expression of the precision of a data set is the standard deviation, which, in a manner of speaking, averages the deviations of each individual measurement from the mean value. The standard deviation is calculated by the following formula:

$$\sigma = \sqrt{\frac{\Sigma (x_i - \bar{x})^2}{N - 1}}$$

in which σ is the standard deviation, N is the number of trials, x_i is the value of an individual measurement, $\bar{x}$ is the mean value, and Σ is a mathematical symbol denoting a summation: the addition of the members of a series of terms.

The table below illustrates this calculation.

Trial	$x_i - \bar{x}$	$(x_i - \bar{x})^2$
1	0.08	0.0064
2	0.21	0.0441
3	−0.36	0.1296
4	0.08	0.0064
		$\Sigma = 0.1865$

Dividing 0.1865 by $(4 - 1 =)3$, and then taking the square root, yields 0.25 as the standard deviation, σ, for this data set. A low value of σ is an indication of good precision. While the standard deviation can always be calculated as we just did, you will most likely find it more convenient to do this on your personal calculator (which may have an internal program which will evaluate this quantity) or with a spreadsheet, such as Excel—your instructor or another student may be able to provide some assistance.

In considering the original data set, you might wonder if the low value of 15.22% should have been included. If you were the experimenter who gathered the data, this low value could be thrown out if a known mistake was made, such as spilling some of a solution, or if you know a weighing error was definitely made. However, if such a personal error was not known to have been made, then the value probably should be included. There is a statistical test which can be used as an unbiased guide for rejection of a potential outlier value. This test, called the Q test, is based on the assumption that a sufficiently large data set would approach a normal distribution. If the assumption is made that our data set is large enough that it resembles a normal distribution, then a Q_{rej} (rejection quotient) can be calculated from the following equation

$$Q_{rej} = \frac{\left|x_q - x_n\right|}{r}$$

where x_q is the questionable value, x_n is the nearest neighbor value, and r is the range, inclusive of the questionable value. Note that the absolute value of the difference is used. The calculated rejection quotient, Q_{rej} is then compared with Q_c, the critical quotient, listed in the table below.

N	2	3	4	5	6	7	8	9	10
Q_c	—	0.94	0.76	0.64	0.56	0.51	0.47	0.44	0.41

The value of N corresponds to the number of trials and Q_c, the critical quotient, represents a threshold at the 90% confidence level that the value of n_q, the questionable value, can be discarded. Thus, if $Q_{rej} > Q_c$, the questionable value can be discarded with a 90% degree of confidence, provided the data resembles a normal distribution. Note that no value of Q_c is given for 2 trials. Can you figure out why? HINT: Compare the difference between n_q and n_n with the range.

Let's go back to the data set of iron percentages. We have identified trial 3, with 15.22% Fe, as the questionable one. As we calculate $Q_{rej} = |15.22 - 15.66|/0.57 = 0.44/0.57 = 0.77$, we find that the Q_c value for four trials is 0.76. Therefore $Q_{rej} > Q_c$ (although barely), and we can discard the 15.22% Fe value with a 90% degree of confidence, subject to the validity of the assumption that our dataset follows a normal distribution.

With the 15.22% value excluded, we can calculate a new mean value of 15.70% Fe with a new standard deviation of 0.075%. Note the significantly lower standard deviation and the improved (though still far from stellar) accuracy. You should do these calculations yourself to verify these data and to gain some practice.

The operational introduction to the statistical analysis of experimental data provided here is intentionally brief. You may wish to do additional reading to gain further insight. Go to your library and consult textbooks used for Analytical Chemistry courses—these books usually have a section on statistics in the first few chapters or in an appendix at the end. Other sources include any introductory text on statistics. Alternatively, you can find a number of websites which give an overview of statistics. You may find that your physics and/or biology courses incorporate a treatment of statistics in the laboratory portion of the course.